TensorFlow 智能移动项目(影印版)
Intelligent Mobile Projects with TensorFlow

Jeff Tang 著

南京　东南大学出版社

图书在版编目(CIP)数据

TensorFlow 智能移动项目:英文/(美)杰夫·唐(Jeff Tang)著. —影印本. —南京:东南大学出版社,2019.3
书名原文:Intelligent Mobile Projects with TensorFlow
ISBN 978-7-5641-8290-8

Ⅰ.①T… Ⅱ.①杰… Ⅲ.①人工智能-算法-英文 Ⅳ.①TP18

中国版本图书馆 CIP 数据核字(2019)第 025341 号
图字:10-2018-490 号

© 2018 by PACKT Publishing Ltd.

Reprint of the English Edition, jointly published by PACKT Publishing Ltd and Southeast University Press, 2019. Authorized reprint of the original English edition, 2018 PACKT Publishing Ltd, the owner of all rights to publish and sell the same.

All rights reserved including the rights of reproduction in whole or in part in any form.

英文原版由 PACKT Publishing Ltd 出版 2018。

英文影印版由东南大学出版社出版 2019。此影印版的出版和销售得到出版权和销售权的所有者 —— PACKT Publishing Ltd 的许可。

版权所有,未得书面许可,本书的任何部分和全部不得以任何形式重制。

TensorFlow 智能移动项目(影印版)

出版发行:东南大学出版社
地　　址:南京四牌楼 2 号　　邮编:210096
出 版 人:江建中
网　　址:http://www.seupress.com
电子邮件:press@seupress.com
印　　刷:常州市武进第三印刷有限公司
开　　本:787 毫米×980 毫米　　16 开本
印　　张:25.25
字　　数:494 千字
版　　次:2019 年 3 月第 1 版
印　　次:2019 年 3 月第 1 次印刷
书　　号:ISBN 978-7-5641-8290-8
定　　价:98.00 元

本社图书若有印装质量问题,请直接与营销部联系。电话(传真):025-83791830

To Lisa and Wozi, who showed me that unconditional love and support can live in harmony with occasional need for attention.

mapt.io

Mapt is an online digital library that gives you full access to over 5,000 books and videos, as well as industry leading tools to help you plan your personal development and advance your career. For more information, please visit our website.

Why subscribe?

- Spend less time learning and more time coding with practical eBooks and Videos from over 4,000 industry professionals

- Improve your learning with Skill Plans built especially for you

- Get a free eBook or video every month

- Mapt is fully searchable

- Copy and paste, print, and bookmark content

PacktPub.com

Did you know that Packt offers eBook versions of every book published, with PDF and ePub files available? You can upgrade to the eBook version at www.PacktPub.com and as a print book customer, you are entitled to a discount on the eBook copy. Get in touch with us at service@packtpub.com for more details.

At www.PacktPub.com, you can also read a collection of free technical articles, sign up for a range of free newsletters, and receive exclusive discounts and offers on Packt books and eBooks.

Foreword

The past decade has seen the explosion of both machine learning and smartphones; today, these technologies are finally merging, and the result is an incredible variety of applications that you would have dismissed as far future Science Fiction just a few years ago. Think about it: you have already become accustomed to talking to your phone, asking it for directions, or telling it to schedule an appointment in your agenda. Your phone's camera tracks faces and recognizes objects. Games are becoming more interesting and challenging as the bots gets smarter and smarter. And countless apps use some form of artificial intelligence under the hood, in less obvious ways, such as recommending content that you will enjoy, anticipating your next trips to tell you when to leave, suggesting what to type next, and so on.

Until recently, all the intelligence happened on the server side, which meant that the user had to be connected to the internet, ideally with a fast and stable connection. The latency and service disruptions that this implied were show-stoppers for many applications. But today the intelligence is right there in the palm of your hand, thanks to tremendous hardware improvements and better Machine Learning libraries.

Most importantly, these technologies are now completely democratized: virtually any software engineer can learn to code an intelligent mobile application based on deep neural networks, using TensorFlow, Google's powerful and open source deep learning library. Jeff Tang's great and unique book will show you how to develop on-device TensorFlow-powered iOS, Android, and Raspberry Pi apps by guiding you through many concrete examples with step-by-step tutorials and hard-earned troubleshooting tips: from image classification, object detection, image captioning, and drawing recognition to speech recognition, forecasting time series, generative adversarial networks, reinforcement learning, and even building intelligent games using AlphaZero — the improved technology built on top of AlphaGo that beat Lee Sedol and Ke Jie, the world champions of the game of Go.

This is going to be a super popular book. It's such an important topic, and it's hard to get good reliable information. So roll up your sleeves, you have an exciting journey ahead of you! What intelligent mobile application will you build?

Aurélien Géron
Former lead of YouTube's video classification team and author of the book Hands-On Machine Learning with Scikit-Learn and TensorFlow (O'Reilly, 2017)
Paris, May 11th, 2018

Contributors

About the author

Jeff Tang fell in love with classical AI more than two decades ago. After his MS in CS, he worked on Machine Translation for 2 years and then, to survive the long AI winter, he worked on enterprise apps, voice apps, web apps, and mobile apps at startups, AOL, Baidu, and Qualcomm. He developed a top-selling iOS app with millions of downloads and was recognized by Google as a Top Android Market Developer. He reconnected with modern AI in 2015 and knew that AI will be his passion and commitment for the next two decades. One of his favorite topics is to make AI available anytime anywhere and hence the book.

> *I'd like to thank Larissa Pinto for reaching out on the book idea, Flavian Vaz and Akhil Nair for all the feedback during content editing. Many thanks to Pete Warden, the TensorFlow mobile lead at Google, for his help before and after becoming a technical reviewer of the book, and to Amita Kapoor, another technical reviewer of the book, who also provided valuable feedback. Special thanks to Aurélien Géron, the best-seller author of the book Hands-On ML, for kindly responding to all my emails, sharing his insights, and writing the Foreword for the book from his packed agenda - Merci beaucoup, Aurélien.*
>
> *I also truly appreciate the understanding and support by my family, other than Lisa and Wozi, during the holiday season and all the months while I had to work like crazy day and night on the book - thanks Amy, Anna, Jenny, Sophia, Mark, Sandy, and Ben.*

About the reviewers

Pete Warden is the technical lead of the mobile and embedded TensorFlow group in Google's Brain team.

Amita Kapoor is an associate professor in the department of electronics, SRCASW, University of Delhi since 1996. She is the recipient of the prestigious DAAD fellowship to pursue a part of her research work in Karlsruhe Institute of Technology, Germany. She had been awarded the Best Presentation Award at International Conference Photonics 2008. She is a member of different professional bodies and has more than 40 publications in international journals and conferences. Her present research areas include ML, AI, Neural Networks, Robotics, Buddhism, and Ethics in AI.

Packt is searching for authors like you

If you're interested in becoming an author for Packt, please visit `authors.packtpub.com` and apply today. We have worked with thousands of developers and tech professionals, just like you, to help them share their insight with the global tech community. You can make a general application, apply for a specific hot topic that we are recruiting an author for, or submit your own idea.

Table of Contents

Preface — 1

Chapter 1: Getting Started with Mobile TensorFlow — 11
- Setting up TensorFlow — 12
  - Setting up TensorFlow on MacOS — 13
  - Setting up TensorFlow on GPU-powered Ubuntu — 15
- Setting up Xcode — 20
- Setting up Android Studio — 20
- TensorFlow Mobile vs TensorFlow Lite — 23
- Running sample TensorFlow iOS apps — 23
- Running sample TensorFlow Android apps — 24
- Summary — 25

Chapter 2: Classifying Images with Transfer Learning — 27
- Transfer learning – what and why — 28
- Retraining using the Inception v3 model — 29
- Retraining using MobileNet models — 38
- Using the retrained models in the sample iOS app — 40
- Using the retrained models in the sample Android app — 43
- Adding TensorFlow to your own iOS app — 44
  - Adding TensorFlow to your Objective-C iOS app — 44
  - Adding TensorFlow to your Swift iOS app — 52
- Adding TensorFlow to your own Android app — 58
- Summary — 63

Chapter 3: Detecting Objects and Their Locations — 65
- Object detection–a quick overview — 66
- Setting up the TensorFlow Object Detection API — 68
  - Quick installation and example — 68
  - Using pre-trained models — 69
- Retraining SSD-MobileNet and Faster RCNN models — 75
- Using object detection models in iOS — 81
  - Building TensorFlow iOS libraries manually — 81
  - Using TensorFlow iOS libraries in an app — 82
  - Adding an object detection feature to an iOS app — 86
- Using YOLO2–another object-detection model — 92
- Summary — 96

Chapter 4: Transforming Pictures with Amazing Art Styles — 97

Table of Contents

Neural Style Transfer – a quick overview	98
Training fast neural-style transfer models	99
Using fast neural-style transfer models in iOS	102
Adding and testing with fast neural transfer models	102
Looking back at the iOS code using fast neural transfer models	106
Using fast neural-style transfer models in Android	107
Using the TensorFlow Magenta multi-style model in iOS	114
Using the TensorFlow Magenta multi-style model in Android	122
Summary	126
Chapter 5: Understanding Simple Speech Commands	127
Speech recognition – a quick overview	128
Training a simple commands recognition model	130
Using a simple speech recognition model in Android	134
Building a new app using the model	135
Showing model-powered recognition results	140
Using a simple speech recognition model in iOS with Objective-C	143
Building a new app using the model	143
Fixing model-loading errors with tf_op_files.txt	150
Using a simple speech recognition model in iOS with Swift	151
Summary	155
Chapter 6: Describing Images in Natural Language	157
Image captioning – how it works	158
Training and freezing an image captioning model	160
Training and testing caption generation	160
Freezing the image captioning model	163
Transforming and optimizing the image captioning model	169
Fixing errors with transformed models	170
Optimizing the transformed model	173
Using the image captioning model in iOS	175
Using the image captioning model in Android	185
Summary	191
Chapter 7: Recognizing Drawing with CNN and LSTM	193
Drawing classification – how it works	194
Training, predicting, and preparing the drawing classification model	196
Training the drawing classification model	196
Predicting with the drawing classification model	197
Preparing the drawing classification model	200
Using the drawing classification model in iOS	206
Building custom TensorFlow library for iOS	206
Developing an iOS app to use the model	207
Using the drawing classification model in Android	215

Building custom TensorFlow library for Android	216
Developing an Android app to use the model	218
Summary	227

Chapter 8: Predicting Stock Price with RNN — 229
RNN and stock price prediction – what and how — 230
Using the TensorFlow RNN API for stock price prediction — 232
- Training an RNN model in TensorFlow — 233
- Testing the TensorFlow RNN model — 237

Using the Keras RNN LSTM API for stock price prediction — 239
- Training an RNN model in Keras — 240
- Testing the Keras RNN model — 244

Running the TensorFlow and Keras models on iOS — 246
Running the TensorFlow and Keras models on Android — 254
Summary — 258

Chapter 9: Generating and Enhancing Images with GAN — 259
GAN – what and why — 260
Building and training GAN models with TensorFlow — 262
- Basic GAN model of generating handwritten digits — 262
- Advanced GAN model of enhancing image resolution — 265

Using the GAN models in iOS — 269
- Using the basic GAN model — 272
- Using the advanced GAN model — 274

Using the GAN models in Android — 277
- Using the basic GAN model — 280
- Using the advanced GAN model — 282

Summary — 284

Chapter 10: Building an AlphaZero-like Mobile Game App — 285
AlphaZero – how does it work? — 286
Training and testing an AlphaZero-like model for Connect 4 — 288
- Training the model — 288
- Testing the model — 292
- Looking into the model-building code — 295
- Freezing the model — 296

Using the model in iOS to play Connect 4 — 298
Using the model in Android to play Connect 4 — 311
Summary — 322

Chapter 11: Using TensorFlow Lite and Core ML on Mobile — 323
TensorFlow Lite – an overview — 323
Using TensorFlow Lite in iOS — 325
- Running the example TensorFlow Lite iOS apps — 325
- Using a prebuilt TensorFlow Lite model in iOS — 327

[iii]

Using a retrained TensorFlow model for TensorFlow Lite in iOS	332
Using a custom TensorFlow Lite model in iOS	333
Using TensorFlow Lite in Android	335
Core ML for iOS – an overview	339
Using Core ML with Scikit-Learn machine learning	340
Building and converting the Scikit Learn models	340
Using the converted Core ML models in iOS	343
Using Core ML with Keras and TensorFlow	344
Summary	349
Chapter 12: Developing TensorFlow Apps on Raspberry Pi	351
Setting up Raspberry Pi and making it move	352
Setting up Raspberry Pi	353
Making Raspberry Pi move	356
Setting up TensorFlow on Raspberry Pi	359
Image recognition and text to speech	361
Audio recognition and robot movement	364
Reinforcement learning on Raspberry Pi	367
Understanding the CartPole simulated environment	368
Starting with basic intuitive policy	372
Using neural networks to build a better policy	374
Summary	381
Final words	381
Other Books You May Enjoy	383
Index	387

Preface

Artificial Intelligence (**AI**), the simulation of human intelligence in computers, has a long history. Since its official birth in 1956, AI has experienced several booms and busts. The ongoing AI resurgence, or the new AI revolution, started in 2012 with the breakthrough in deep learning, a branch of machine learning that is now the hottest branch of AI because of deep learning, when a **deep convolutional neural network** (**DCNN**) won the ImageNet Large-Scale Visual Recognition Challenge with an error rate of only 16.4%, compared to the second best non-DCNN entry with an error rate of 26.2%. Since 2012, improved DCNN-based entries have won the ImageNet challenge every year, and deep learning technology has been applied to many hard AI problems beyond computer vision, such as speech recognition, machine translation, and the game of Go, resulting in one breakthrough after another. In March 2016, Google DeepMind's AlphaGo, built with deep reinforcement learning, beat 18-time human world Go champion Lee Sedol 4:1. At Google I/O 2017, Google announced that they're shifting from mobile-first to AI-first world. Other leading companies such as Amazon, Apple, Facebook, and Microsoft have all invested heavily in AI and launched many AI-powered products.

TensorFlow is Google's open source framework for building machine learning AI apps. Since its initial release in November 2015, when there were already several popular open source deep learning frameworks, TensorFlow has quickly become the most popular open source deep learning framework in less than 2 years. New TensorFlow models to tackle all kinds of tasks that'd require human or even superhuman intelligence have been built on a weekly basis. Dozens of books on TensorFlow have been published. More online blogs, tutorials, courses, and videos on TensorFlow have been made available. It's obvious that AI and TensorFlow are hot, but why another book with the word "TensorFlow" in its title?

This is a unique book, and the first one that combines TensorFlow-powered AI with mobile, connecting the world of the brightest future with the world of most prosperous present. We have all witnessed and experienced the iOS and Android smart phone revolution in the past decade, and we're just starting the AI revolution that'll likely have an even deeper impact on the world around us. What can be better than a theme that integrates the best of the two worlds, a book that shows how to build TensorFlow AI apps on mobile, anytime, anywhere?

Preface

It's true that you can build AI apps using many cloud AI APIs that exist out there, and sometimes it makes sense. However, running AI apps completely on mobile devices have the benefits that you can run the apps even when no network connectivity is available, when you can't afford the round trip to a cloud server, or when users don't want to send the data on their phones to anyone else.

It's also true that there're already a few example TensorFlow iOS and Android apps in the TensorFlow open source project that can get you started with mobile TensorFlow. However, if you have ever tried to run a cool TensorFlow model that amazes you on your iOS or Android device, you'll most likely stumble upon many hiccups before you can see the model successfully running on your phone.

This book can save you a lot of time and effort by showing you how to solve all the common problems you may encounter when running TensorFlow models on mobile. You'll get to see more than 10 complete TensorFlow iOS and Android apps built from scratch in the book, running all kinds of cool TensorFlow models, including the latest and coolest **Generative Adversarial Network (GAN)** and AlphaZero-like models.

Who this book is for

If you're an iOS and/or Android developer interested in building and retraining others' cool TensorFlow models and running them in your mobile apps, or if you're a TensorFlow developer and want to run your new amazing TensorFlow models on mobile devices, the book is for you. If you're interested in TensorFlow Lite, Core ML, or TensorFlow on Raspberry Pi, you'll also benefit from the book.

What this book covers

Chapter 1, *Getting Started with Mobile TensorFlow*, discusses how to set up TensorFlow on Mac and Ubuntu and NVIDIA GPU on Ubuntu and how to set up Xcode and Android Studio. We'll also discuss the difference between TensorFlow Mobile and TensorFlow Lite and when you should use them. Finally, we'll show you how to run the sample TensorFlow iOS and Android apps.

Chapter 2, *Classifying Images with Transfer Learning*, covers what is transfer learning and why you should use it, how to retrain the Inception v3 and MobileNet models for more accurate and faster dog breed recognition, and how to use the retrained models in sample iOS and Android apps. Then, we'll show you how to add TensorFlow to your own iOS app, both in Objective-C and Swift, and your own Android app for dog breed recognition.

Chapter 3, *Detecting Objects and Their Locations*, gives a quick overview of Object Detection, and then shows you how to set up the TensorFlow Object Detection API and use it to retrain SSD-MobileNet and Faster RCNN models. We'll also show you how to use the models used in the example TensorFlow Android app in your iOS app by manually building the TensorFlow iOS library to support non-default TensorFlow operations. Finally, we'll show you how to train YOLO2, another popular object detection model, which is also used in the example TensorFlow Android app, and how to use it in your iOS app.

Chapter 4, *Transforming Pictures with Amazing Art Styles*, first gives an overview of neural style transfer with their rapid progress in the last few years. Then, it shows you how to train fast neural style transfer models and use them in iOS and Android apps. After that, we'll cover how to use the TensorFlow Magenta multi-style model in your own iOS and Android apps to easily create amazing art styles.

Chapter 5, *Understanding Simple Speech Commands*, outlines speech recognition and shows you how to train a simple speech commands recognition model. We'll then show you how to use the model in Android as well as in iOS using both Objective-C and Swift. We'll also cover more tips on how to fix possible model loading and running errors on mobile.

Chapter 6, *Describing Images in Natural Language*, describes how image captioning works, and then it covers how to train and freeze an image captioning model in TensorFlow. We'll further discuss how to transform and optimize the complicated model to get it ready for running on mobile. Finally, we'll offer complete iOS and Android apps using the model to generate natural language description of images.

Chapter 7, *Recognizing Drawing with CNN and LSTM*, explains how drawing classification works, and discusses how to train, predict, and prepare the model. Then, we'll show you how to build another custom TensorFlow iOS library to use the model in a fun iOS doodling app. Finally, we'll show you how to build a custom TensorFlow Android library to fix a new model loading error and then use the model in your own Android app.

Chapter 8, *Predicting Stock Price with RNN*, takes you through RNN and how to use it to predict stock prices. Then, we'll inform you of how to build an RNN model with the TensorFlow API to predict stock prices, and how to build a RNN LSTM model with the easier-to-use Keras API to achieve the same goal. We'll test and see whether such models can beat a random buy or sell strategy. Finally, we'll show you how to run the TensorFlow and Keras models in both iOS and Android apps.

Chapter 9, *Generating and Enhancing Images with GAN*, gives an overview of what GAN is and why it has such great potential. Then, it oultines how to build and train a basic GAN model that can be used to generate human-like handwritten digits and a more advanced model that can enhance low resolution images to high resolution ones. Finally, we'll cover how to use the two GAN models in your iOS and Android apps.

Chapter 10, *Building AlphaZero-like Mobile Game App*, begins with how the latest and coolest AlphaZero works, and how to train and test a AlphaZero-like model to play a simple but fun game called Connect 4 in Keras with TensorFlow as backend. We'll then show you the complete iOS and Android apps to use the model and play the game Connect 4 on your mobile devices.

Chapter 11, *Using TensorFlow Lite and Core ML on Mobile*, demonstrates TensorFlow Lite and then shows you how to use a prebuilt TensorFlow model, a retrained TensorFlow model for TensorFlow Lite, and a custom TensorFlow Lite model in iOS. We'll also show you how to use TensorFlow Lite in Android. After that, we'll give an overview of Apple's Core ML and show you how to use Core ML with standard machine learning using Scikit-Learn. Finally, we'll cover how to use Core ML with TensorFlow and Keras.

Chapter 12, *Developing TensorFlow Apps on Raspberry Pi*, first looks at how to set up Raspberry Pi and make it move, and how to set up TensorFlow on Raspberry Pi. Then, we'll cover how to use the TensorFlow image recognition and audio recognition models, along with text to speech and robot movement APIs, to build a Raspberry Pi robot that can move, see, listen, and speak. Finally, we'll discuss in detail how to use OpenAI Gym and TensorFlow to build and train a powerful neural network-based reinforcement learning policy model from scratch in a simulated environment to make the robot learn to keep its balance.

To get the most out of this book

We recommend that you start with reading the first four chapters in order, along with running the accompanying iOS and Android apps available from the book's source code repository at http://github.com/jeffxtang/mobiletfbook. That'll help you ensure that you have the development environments all set up for TensorFlow mobile app development and that you know how to integrate TensorFlow into your own iOS and/or Android apps. If you're an iOS developer, you'll also learn how to use Objective-C or Swift with TensorFlow, and when and how to use the TensorFlow pod or the manual TensorFlow iOS library.

Then, if you need to build a custom TensorFlow Android library, go to Chapter 7, *Recognizing Drawing with CNN and LSTM,* and if you want to learn how to use a Keras model in your mobile app, check out Chapter 8, *Predicting Stock Price with RNN,* and Chapter 10, *Building an AlphaZero-like Mobile Game App.*

If you're more interested in TensorFlow Lite or Core ML, read Chapter 11, *Using TensorFlow Lite and Core ML on Mobile,* and if you're most interested in TensorFlow on Raspberry Pi, or reinforcement learning in TensorFlow, jump to Chapter 12, *Developing TensorFlow Apps on Raspberry Pi.*

Other than that, you can go through chapters 5 to 10 in order to see how to train different kinds of CNN, RNN, LSTM, GAN, and AlphaZero models and how to use them on mobile, maybe running the iOS and/or Android apps for each chapter before looking into the detailed implementation. Alternatively, you can jump directly to any chapter with the model you're most interested in; just be aware that a later chapter may refer to an earlier chapter for some duplicated details, such as steps of adding a TensorFlow custom iOS library to your iOS app, or fixing some model loading or running errors by building a TensorFlow custom library. However, rest assured that you won't be lost, or at least we've done our best to provide user-friendly and step-by-step tutorials, with occasional references to some steps of previous tutorials, to help you avoid all possible pitfalls you may encounter when building mobile TensorFlow apps, while also avoiding repeating ourselves.

When to read the book

AI, or its hottest branch machine learning, or its hottest subbranch deep learning, has enjoyed rapid progress in recent years. New releases of TensorFlow, backed by Google and with the most popular developer community of all open source machine learning frameworks, has also been launched at a faster speed. When we started writing the book in December 2017, the latest TensorFlow release was 1.4.0, released on November 2, 2017, and after that 1.5.0 was released on January 26, 2018, 1.6.0 on February 28, 2018, 1.7.0 on March 29, 2018, and 1.8.0 on April 27, 2018. All the iOS, Android, and Python code in the book has been tested with all those TensorFlow versions. Still, by the time you read the book, the latest TensorFlow version likely will be later than 1.8.0.

It turns out you don't need to worry too much about new releases of TensorFlow; the code in the book will most likely run seamlessly on the latest TensorFlow releases. During our testing run the apps on TensorFlow 1.4, 1.5, 1.6, 1.7, and 1.8, we've made no code changes at all. It's likely that in a later version, more TensorFlow operations will be supported by default, so you won't need to build a custom TensorFlow library, or you'll be able to build the custom TensorFlow library in a simpler way.

Of course, there's no guarantee that all the code will run without any changes in all future TensorFlow versions, but with all the detailed tutorials and troubleshooting tips covered in the book, no matter when you read the book, now or months later, you should have a smooth time reading the book and running the apps in the book with TensorFlow 1.4-1.8 or later.

As we have to stop with a specific TensorFlow version at certain points to get the book published, we'll continue to test run all the code in the book with every new major TensorFlow release, and update the code and test results accordingly on the book's source code repository at http://github.com/jeffxtang/mobiletfbook. If you have any questions about the code or the book, you may also post an issue directly on the repository.

Another concern is about the choice between TensorFlow Mobile and TensorFlow Lite. The book covers TensorFlow Mobile in most chapters (1 to 10). TensorFlow Lite may be the future of running TensorFlow on mobile, it's still in developer preview as of Google I/O 2018—that's why Google expects you to "use TensorFlow Mobile to cover production cases." Even after TensorFlow Lite is officially released, according to Google, "TensorFlow Mobile isn't going away anytime soon"—in fact, with the latest TensorFlow 1.8.0 version we tested before the book's publication, we found that using TensorFlow Mobile gets even simpler.

If the day that TensorFlow Lite fully replaces TensorFlow Mobile in all use cases, with Lite's better performance and smaller size, does finally come, the skills you'll learn from the book will only better prepare you for that day. In the meantime, before that unforeseeable future arrives, you can read the book and get to know how to use the big brother such as TensorFlow Mobile to run all those amazing and powerful TensorFlow models in your mobile apps.

Download the example code files

You can download the example code files for this book from your account at www.packtpub.com. If you purchased this book elsewhere, you can visit www.packtpub.com/support and register to have the files emailed directly to you.

You can download the code files by following these steps:

1. Log in or register at www.packtpub.com.
2. Select the **SUPPORT** tab.

3. Click on **Code Downloads & Errata**.
4. Enter the name of the book in the **Search** box and follow the onscreen instructions.

Once the file is downloaded, please make sure that you unzip or extract the folder using the latest version of:

- WinRAR/7-Zip for Windows
- Zipeg/iZip/UnRarX for Mac
- 7-Zip/PeaZip for Linux

The code bundle for the book is also hosted on the Packt GitHub account at `https://github.com/PacktPublishing/Intelligent-Mobile-Projects-with-TensorFlow`. In case there's an update to the code, it will be updated on the existing GitHub repository.

We also have other code bundles from our rich catalog of books and videos available at `https://github.com/PacktPublishing/`. Check them out!

Conventions used

There are a number of text conventions used throughout this book.

`CodeInText`: Indicates code words in text, database table names, folder names, filenames, file extensions, pathnames, dummy URLs, user input, and Twitter handles. Here is an example: "Install the `matplotlib`, `pillow`, `lxml`, and `jupyter` libraries. On Ubuntu or Mac, you can run."

A block of code is set as follows:

```
syntax = "proto2";
package object_detection.protos;
message StringIntLabelMapItem {
  optional string name = 1;
  optional int32 id = 2;
  optional string display_name = 3;
};

message StringIntLabelMap {
  repeated StringIntLabelMapItem item = 1;
};
```

Any command-line input or output is written as follows:

```
sudo pip install pillow
sudo pip install lxml
sudo pip install jupyter
sudo pip install matplotlib
```

Bold: Indicates a new term, an important word, or words that you see onscreen. For example, words in menus or dialog boxes appear in the text like this. Here is an example: "Now, let's select the **Enhance Image** option, you'll see the result"

Warnings or important notes appear like this.

Tips and tricks appear like this.

Get in touch

Feedback from our readers is always welcome.

General feedback: Email `feedback@packtpub.com` and mention the book title in the subject of your message. If you have questions about any aspect of this book, please email us at `questions@packtpub.com`.

Errata: Although we have taken every care to ensure the accuracy of our content, mistakes do happen. If you have found a mistake in this book, we would be grateful if you would report this to us. Please visit `www.packtpub.com/submit-errata`, selecting your book, clicking on the Errata Submission Form link, and entering the details.

Piracy: If you come across any illegal copies of our works in any form on the Internet, we would be grateful if you would provide us with the location address or website name. Please contact us at `copyright@packtpub.com` with a link to the material.

If you are interested in becoming an author: If there is a topic that you have expertise in and you are interested in either writing or contributing to a book, please visit `authors.packtpub.com`.

Reviews

Please leave a review. Once you have read and used this book, why not leave a review on the site that you purchased it from? Potential readers can then see and use your unbiased opinion to make purchase decisions, we at Packt can understand what you think about our products, and our authors can see your feedback on their book. Thank you!

For more information about Packt, please visit `packtpub.com`.

Getting Started with Mobile TensorFlow

This chapter covers how to get your development environments set up for building all the iOS or Android apps with TensorFlow that are discussed in the rest of the book. We won't discuss in detail all the supported TensorFlow versions, OS versions, Xcode, and Android Studio versions that can be used for development, as that kind of information can easily be found on the TensorFlow website (http://www.tensorflow.org) or via Google. Instead, we'll just talk briefly about sample working environments in this chapter so that we can dive in quickly to look at all the amazing apps you can build with the environments.

If you already have TensorFlow, Xcode, and Android Studio installed, and can run and test the sample TensorFlow iOS and Android apps, and if you already have an NVIDIA GPU installed for faster deep learning model training, you can skip this chapter. Or you can jump directly to the section that you're unfamiliar with.

We're going to cover the following topics in this chapter (how to set up the Raspberry Pi development environment will be discussed in Chapter 12, *Developing TensorFlow Apps on Raspberry Pi*):

- Setting up TensorFlow
- Setting up Xcode
- Setting up Android Studio
- TensorFlow Mobile vs TensorFlow Lite
- Running sample TensorFlow iOS apps
- Running sample TensorFlow Android apps

Setting up TensorFlow

TensorFlow is the leading open source framework for machine intelligence. When Google released TensorFlow as an open source project in November 2015, there were already several other similar open source frameworks for deep learning: Caffe, Torch, and Theano. By Google I/O 2018 on May 10, TensorFlow on GitHub has reached 99k stars, an increase of 14k stars in 4 months, while Caffe has increased only 2k to 24k stars. Two years later, it's already *the* most popular open source framework for training and deploying deep learning models (it also has good support for traditional machine learning). As of January 2018, TensorFlow has close to 85k stars (https://github.com/tensorflow/tensorflow) on GitHub, while the other three leading open source deep learning frameworks, Caffe (https://github.com/BVLC/caffe), CNTK (https://github.com/Microsoft/CNTK), and Mxnet (https://github.com/apache/incubator-mxnet) have over 22k, 13k, and 12k stars, respectively.

If you're a little confused about the buzz words machine learning, deep learning, machine intelligence, and artificial intelligence (AI), here's a quick summary: machine intelligence and AI are really just the same thing; machine learning is a field, also the most popular one, of AI; deep learning is one special type of machine learning, and is also the modern and most effective approach to solving complicated problems such as computer vision, speech recognition and synthesis, and natural language processing. So in this book, when we say AI, we primarily mean deep learning, the savior that took AI from the long winter to the summer. For more information about the AI winter and deep learning, you can check out https://en.wikipedia.org/wiki/AI_winter and http://www.deeplearningbook.org.

We assume you already have a basic understanding of TensorFlow, but if you don't, you should check out the Get Started (https://www.tensorflow.org/get_started) and Tutorials (https://www.tensorflow.org/tutorials) parts of the TensorFlow website or the Awesome TensorFlow tutorials (https://github.com/jtoy/awesome-tensorflow). Two good books on the topic are *Python Machine Learning: Machine Learning and Deep Learning with Python, scikit-learn, and TensorFlow, 2nd Edition* by Sebastian Raschka and Vahid Mirjalili, and *Hands-On Machine Learning with Scikit-Learn and TensorFlow: Concepts, Tools, and Techniques to Build Intelligent Systems* by Aurélien Géron.

TensorFlow can be installed on MacOS, Ubuntu or Windows. We'll cover the steps to install TensorFlow 1.4 from its source on MacOS X El Capitan (10.11.6), macOS Sierra (10.12.6), and Ubuntu 16.04. If you have a different OS or version, you can refer to the TensorFlow Install (https://www.tensorflow.org/install) documentation for more information. By the time you read this book, it's likely a newer TensorFlow version will come out. Although you should still be able to run the code in the book with the newer version, it's not a guarantee, which is why we use the TensorFlow 1.4 release source code to set up TensorFlow on Mac and Ubuntu; that way, you can easily test run and play with the apps in the book.

> Since we wrote the paragraph above in December 2017, there have been four new official releases of TensorFlow (1.5, 1.6, 1.7, and 1.8), which you can download at https://github.com/tensorflow/tensorflow/releases or from the TensorFlow source code repo (https://github.com/tensorflow/tensorflow), and a new version of Xcode (9.3) as of May 2018. Newer versions of TensorFlow, such as 1.8, by default support newer versions of NVIDIA CUDA and cuDNN (see the section *Setting up TensorFlow on GPU-powered Ubuntu* for detail), and you'd better follow the official TensorFlow documentation to install the latest TensorFlow version with GPU support. In this and the following chapters, we may refer to a specific TensorFlow version as an example, but will keep all iOS, Android, and Python code tested and, if needed, updated for the latest TensorFlow, Xcode, and Android Studio versions in the book's source code repo at https://github.com/jeffxtang/mobiletfbook.

Overall, we'll use TensorFlow on Mac to develop iOS and Android TensorFlow apps, and TensorFlow on Ubuntu to train deep learning models used in the apps.

Setting up TensorFlow on MacOS

Generally, you should use a virtualenv, Docker, or Anaconda installation to install TensorFlow in an isolated environment. But as we have to build iOS and Android TensorFlow apps using the TensorFlow source code, we might as well build TensorFlow itself from source, in which case, using the native pip installation choice could be easier than other options. If you want to experiment with different TensorFlow versions, we recommend you install other TensorFlow versions using one of the virtualenv, Docker, and Anaconda options. Here, we'll have TensorFlow 1.4 installed directly on your MacOS system using the native pip and Python 2.7.10.

Follow these steps to download and install TensorFlow 1.4 on MacOS:

1. Download the TensorFlow 1.4.0 source code (`zip` or `tar.gz`) from the TensorFlow releases page on GitHub: https://github.com/tensorflow/tensorflow/releases
2. Uncompress the downloaded file and drag the `tensorflow-1.4.0` folder to your home directory
3. Make sure you have Xcode 8.2.1 or above installed (if not, read the *Setting up Xcode* section first)
4. Open a new Terminal window, then `cd tensorflow-1.4.0`
5. Run `xcode-select --install` to install command-line tools
6. Run the following commands to install other tools and packages needed to build TensorFlow:

   ```
   sudo pip install six numpy wheel
   brew install automake
   brew install libtool
   ./configure
   brew upgrade bazel
   ```

7. Build from the TensorFlow source with CPU-only support (we'll cover the GPU support in the next section) and generate a pip package file with the `.whl` file extension:

   ```
   bazel build --config=opt
   //tensorflow/tools/pip_package:build_pip_package

   bazel-bin/tensorflow/tools/pip_package/build_pip_package
   /tmp/tensorflow_pkg
   ```

8. Install the TensorFlow 1.4.0 CPU package:

   ```
   sudo pip install --upgrade /tmp/tensorflow_pkg/tensorflow-1.4.0-cp27-cp27m-macosx_10_12_intel.whl
   ```

If you encounter any error during the process, googling the error message, to be honest, should be the best way to fix it, as we intend in this book to focus on the tips and knowledge, not easily available elsewhere, gained from our long hours of building and debugging practical mobile TensorFlow apps. One particular error you may see is the `Operation not permitted` error when running the `sudo pip install` commands. To fix it, you can disable your Mac's **System Integrity Protection** (**SIP**) by restarting the Mac and hitting the *Cmd + R* keys to enter the recovery mode, then under **Utilities-Terminal**, running csrutil disable before restarting Mac. If you're uncomfortable with disabling SIP, you can follow the TensorFlow documentation to try one of the more complicated installation methods such as virtualenv.

If everything goes well, you should be able to run on your Terminal Window, Python or preferably IPython, then run `import tensorflow as tf` and `tf.__version__` to see 1.4.0 as output.

Setting up TensorFlow on GPU-powered Ubuntu

One of the benefits of using a good deep learning framework, such as TensorFlow, is its seamless support for using a **Graphical Processing Unit** (**GPU**) in model training. It'd be a lot faster training a non-trivial TensorFlow-based model on a GPU than on a CPU, and currently NVIDIA offers the best and most cost-effective GPUs supported by TensorFlow. And Ubuntu is the best OS for running NVIDIA GPUs with TensorFlow. You can easily buy one GPU for a few hundred bucks and install it on an inexpensive desktop with an Ubuntu system. You can also install NVIDIA GPU on Windows but TensorFlow's support for Windows is not as good as that for Ubuntu.

To train the models deployed in the apps in this book, we use NVIDIA GTX 1070, which you can purchase on Amazon or eBay for about $400. There's a good blog by Tim Dettmers on which GPUs to use for deep learning (http://timdettmers.com/2017/04/09/which-gpu-for-deep-learning/). After you get such a GPU and install it on your Ubuntu system, and before you install the GPU-enabled TensorFlow, you need to install NVIDIA CUDA 8.0 (or 9.0) and the cuDNN (CUDA-Deep Neural Network library) 6.0 (or 7.0), both are supported by TensorFlow 1.4.

Getting Started with Mobile TensorFlow

An alternative to setting up your own GPU-powered Ubuntu with TensorFlow is to use TensorFlow in a GPU-enabled cloud service such as Google Cloud Platform's Cloud ML Engine (https://cloud.google.com/ml-engine/docs/using-gpus). There are pros and cons of each option. Cloud services are generally time-based billing. If your goal is to train or retrain models to be deployed on mobile devices, meaning the models are not super complicated, and if you plan to do machine learning training for a long time, it'd be more cost effective and satisfying to have your own GPU.

Follow these steps to install CUDA 8.0 and cuDNN 6.0 on Ubuntu 16.04 (you should be able to download and install CUDA 9.0 and cuDNN 7.0 in a similar way):

1. Find the NVIDIA CUDA 8.0 GA2 release at https://developer.nvidia.com/cuda-80-ga2-download-archive, and make the selections shown in the following screenshot:

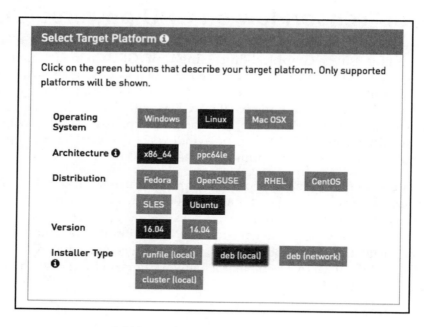

Fig 1.1 Getting ready to download CUDA 8.0 on Ubuntu 16.04

[16]

Chapter 1

2. Download the base installer as shown in the following screenshot:

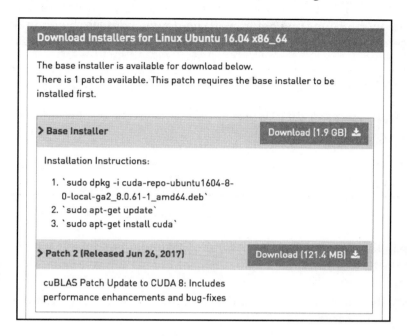

Fig 1.2 Choosing the CUDA 8.0 installer file for Ubuntu 16.04

3. Open a new Terminal and run the following commands (you'll also need to add the last two commands to your .bashrc file for the two environment variables to take effect next time you launch a new Terminal):

```
sudo dpkg -i /home/jeff/Downloads/cuda-repo-ubuntu1604-8-0-local-ga2_8.0.61-1_amd64.deb
sudo apt-get update
sudo apt-get install cuda-8-0
export CUDA_HOME=/usr/local/cuda
export LD_LIBRARY_PATH=/usr/local/cuda/lib64:$LD_LIBRARY_PATH
```

Getting Started with Mobile TensorFlow

4. Download the NVIDIA cuDNN 6.0 for CUDA 8.0 at `https://developer.nvidia.com/rdp/cudnn-download` – you'll be asked to sign up (for free) with an NVIDIA Developer account first before you can download it, as shown in the next screenshot (choose the highlighted **cuDNN v6.0 Library for Linux**):

> Download cuDNN v6.0 [April 27, 2017], for CUDA 8.0
>
> Download packages updated April 27, 2017 to resolve issues related to dilated convolution on Kepler Architecture GPUs.
>
> cuDNN User Guide
>
> cuDNN Install Guide
>
> cuDNN v6.0 Library for Linux

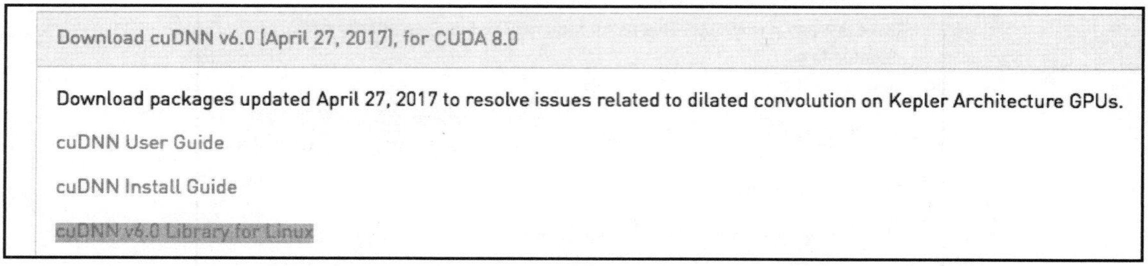

Fig 1.3 Selecting cuDNN 6.0 for CUDA 8.0 on Linux

5. Unzip the downloaded file, assuming it's under the default `~/Downloads` directory, and you'll see a folder named `cuda`, with two subfolders named `include` and `lib64`
6. Copy the cuDNN `include` and `lib64` files to the CUDA_HOME's `lib64` and `include` folders:

```
sudo cp ~/Downloads/cuda/lib64/* /usr/local/cuda/lib64
sudo cp ~/Downloads/cuda/include/cudnn.h /usr/local/cuda/include
```

Now we're ready to install the GPU-enabled TensorFlow 1.4 on Ubuntu (the first two steps given here are the same as those described in the section *Setting up TensorFlow on MacOS*):

1. Download the TensorFlow 1.4.0 source code (`zip` or `tar.gz`) from the TensorFlow releases page on GitHub: `https://github.com/tensorflow/tensorflow/releases`
2. Uncompress the downloaded file and drag the folder to your home directory
3. Download the bazel installer, `bazel-0.5.4-installer-linux-x86_64.sh` at `https://github.com/bazelbuild/bazel/releases`
4. Open a new Terminal Window, then run the following commands to install the tools and packages needed to build TensorFlow:

```
sudo pip install six numpy wheel
cd ~/Downloads
chmod +x bazel-0.5.4-installer-linux-x86_64.sh
./bazel-0.5.4-installer-linux-x86_64.sh --user
```

[18]

5. Build from the TensorFlow source with GPU-enabled support and generate a pip package file with the .whl file extension:

```
cd ~/tensorflow-1.4.0
./configure
bazel build --config=opt --config=cuda
//tensorflow/tools/pip_package:build_pip_package
bazel-bin/tensorflow/tools/pip_package/build_pip_package
/tmp/tensorflow_pkg
```

6. Install the TensorFlow 1.4.0 GPU package:

```
sudo pip install --upgrade /tmp/tensorflow_pkg/tensorflow-1.4.0-cp27-cp27mu-linux_x86_64.whl
```

Now, if all goes well, you can launch IPython and enter the following scripts to see the GPU information TensorFlow is using:

```
In [1]: import tensorflow as tf

In [2]: tf.__version__
Out[2]: '1.4.0'

In [3]: sess=tf.Session()
2017-12-28 23:45:37.599904: I
tensorflow/stream_executor/cuda/cuda_gpu_executor.cc:892] successful NUMA
node read from SysFS had negative value (-1), but there must be at least
one NUMA node, so returning NUMA node zero
2017-12-28 23:45:37.600173: I
tensorflow/core/common_runtime/gpu/gpu_device.cc:1030] Found device 0 with
properties:
name: GeForce GTX 1070 major: 6 minor: 1 memoryClockRate(GHz): 1.7845
pciBusID: 0000:01:00.0
totalMemory: 7.92GiB freeMemory: 7.60GiB
2017-12-28 23:45:37.600186: I
tensorflow/core/common_runtime/gpu/gpu_device.cc:1120] Creating TensorFlow
device (/device:GPU:0) -> (device: 0, name: GeForce GTX 1070, pci bus id:
0000:01:00.0, compute capability: 6.1)
```

Congratulations! You're now ready to train the deep learning models used in the apps in this book. Before we start having fun with our new toy and use it to train our cool models and then deploy and run them on mobile devices, let's first see what it takes to be ready for developing mobile apps.

Setting up Xcode

Xcode is used to develop iOS apps and you need a Mac computer and a free Apple ID to download and install it. If your Mac is relatively older and with OS X El Capitan (version 10.11.6), you can download Xcode 8.2.1 at `https://developer.apple.com/download/more`. Or if you have macOS Sierra (version 10.12.6) or later, installed, you can download Xcode 9.2 or 9.3, the latest version as of May 2018, from the preceding link. All the iOS apps in the book have been tested in both Xcode 8.2.1, 9.2, and 9.3.

To install Xcode, simply double-click the downloaded file and follow the steps on the screen. It's pretty straightforward. You can now run apps on the iOS simulator that comes with Xcode or your own iOS device. Starting Xcode 7, you can run and debug your iOS apps on an iOS device for free, but if you want to distribute or publish your apps, you need to enroll in the Apple Developer Program for $99 a year as an individual: `https://developer.apple.com/programs/enroll`.

Although you can test run many apps in the book with the Xcode simulator, some apps in the book require the camera on your actual iOS device to take a picture before processing it with a deep learning model trained with TensorFlow. In addition, it's generally better to test a model on an actual device for accurate performance and memory usage: a model that runs fine in the simulator may crash or run too slow in an actual device. So it's strongly recommended or required that you test and run the iOS apps in the book on your actual iOS device at least once, if not always.

This book assumes you're comfortable with iOS programming, but if you're new to iOS development, you can learn from many excellent online tutorials such as the iOS tutorials by Ray Wenderlich (`https://www.raywenderlich.com`). We won't cover complicated iOS programming; we'll mainly show you how to use the TensorFlow C++ API in our iOS apps to run the TensorFlow trained models to perform all kinds of intelligent tasks. Both Objective-C and Swift code, the two official iOS programming languages from Apple, will be used to interact with the C++ code in our mobile AI apps.

Setting up Android Studio

Android Studio is the best tool for Android app development, and TensorFlow has great support for using it. Unlike Xcode, you can install and run Android Studio on Mac, Windows, or Linux. For detailed system requirements, see the Android Studio website (`https://developer.android.com/studio/index.html`). Here, we'll just cover how to set up Android Studio 3.0 or 3.0.1 on Mac – all the apps in the book have been tested on both versions.

First, download the Android Studio 3.0.1, or the latest version if it's newer than 3.0.1 and if you don't mind fixing possible minor issues, from the preceding link. You can also download 3.0.1 or 3.0 from its archives at `https://developer.android.com/studio/archive.html`.

Then, double-click the downloaded file and drag and drop the `Android Studio.app` icon to `Applications`. If you have a previously installed Android Studio, you'll get a prompt asking you if you want to replace it with the newer one. You can just select **Replace**.

Next, open Android Studio and you need to provide the path to the Android SDK, which by default is in `~/Library/Android/sdk` if you have a previous version of Android Studio installed, or you can select **Open an existing Android Studio project**, then go to the TensorFlow 1.4 source directory created in the section *Setting up TensorFlow on MacOS*, and open the `tensorflow/examples/android` folder. After that, you can download the Android SDK by either clicking the link to an **Install Build Tools** message or going to **Android Studio's Tools | Android | SDK Manager**, as shown in the following screenshot. From the **SDK Tools** tab there, you can check the box next to a specific version of Android SDK Tools and click the **OK** button to install that version:

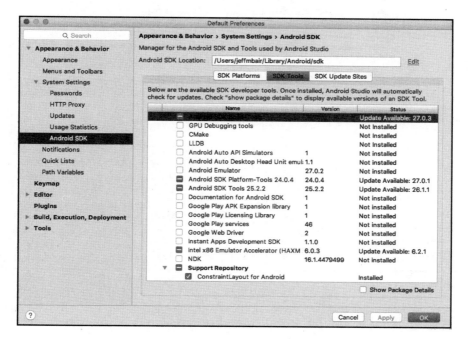

Fig 1.4 Android SDK Manager to install SDK tools and NDK

Finally, as TensorFlow Android apps use the native TensorFlow library in C++ to load and run TensorFlow models, you need to install the Android **Native Development Kit** (**NDK**), which you can do either from the Android SDK Manager shown in the preceding screenshot, or by downloading NDK directly from `https://developer.android.com/ndk/downloads/index.html`. Both the NDK version r16b and r15c have been tested to run the Android apps in the book. If you download the NDK directly, you may also need to set the Android NDK location after opening your project and selecting Android Studio's **File | Project Structure**, as shown in the following screenshot:

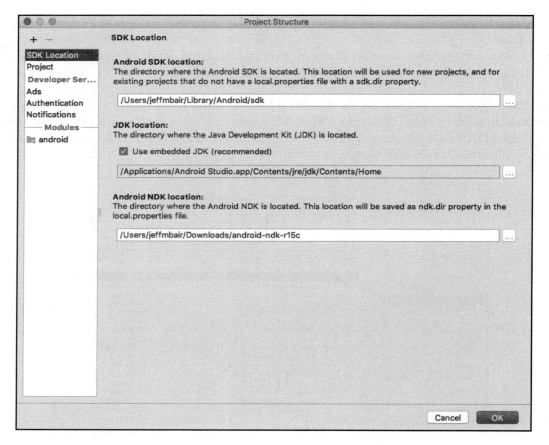

Fig 1.5 Setting project-level Android NDK location

With both Android SDK and NDK installed and set up, you're ready to test run sample TensorFlow Android apps.

TensorFlow Mobile vs TensorFlow Lite

Before we start running sample TensorFlow iOS and Android apps, let's clarify one big picture. TensorFlow currently has two approaches to developing and deploying deep learning apps on mobile devices: TensorFlow Mobile and TensorFlow Lite. TensorFlow Mobile was part of TensorFlow from the beginning, and TensorFlow Lite is a newer way to develop and deploy TensorFlow apps, as it offers better performance and smaller app size. But there's one key factor that will let us focus on TensorFlow Mobile in this book, while still covering TensorFlow Lite in one chapter: TensorFlow Lite is still in developer preview as of TensorFlow 1.8 and Google I/O 2018 in May 2018. So to develop production-ready mobile TensorFlow apps now, you have to use TensorFlow Mobile, as recommended by Google.

Another reason we decided to focus on TensorFlow Mobile now is while TensorFlow Lite only offers a limited support for model operators, TensorFlow Mobile supports customization to add new operators not supported by TensorFlow Mobile by default, which you'll see happens pretty often in our various models of different AI apps.

But in the future, when TensorFlow Lite is out of developer preview, it's likely to replace TensorFlow Mobile, or at least overcome its current limitations. To get yourself ready for that, we'll cover TensorFlow Lite in detail in a later chapter.

Running sample TensorFlow iOS apps

In the last two sections of this chapter, we'll test run three sample iOS apps and four sample Android apps that come with TensorFlow 1.4 to make sure you have your mobile TensorFlow development environments set up correctly and give you a quick preview at what some TensorFlow mobile apps can do.

The source code of the three sample TensorFlow iOS apps is located at `tensorflow/examples/ios:` `simple`, `camera`, and `benchmark`. To successfully run these samples, you need to first download one pretrained deep learning model by Google, called Inception (https://github.com/tensorflow/models/tree/master/research/inception), for image recognition. There are several versions of Inception: v1 to v4, with better accuracy in each newer version. Here we'll use Inception v1 as the samples were developed for it. After downloading the model file, copy the model-related files to each of the samples' `data` folder:

```
curl -o ~/graphs/inception5h.zip
https://storage.googleapis.com/download.tensorflow.org/models/inception5h.z
ip
```

```
unzip ~/graphs/inception5h.zip -d ~/graphs/inception5h
cd tensorflow/examples/ios
cp ~/graphs/inception5h/* simple/data/
cp ~/graphs/inception5h/* camera/data/
cp ~/graphs/inception5h/* benchmark/data/
```

Now, go to each `app` folder and run the following commands to download the required pod for each app before opening and running the apps:

```
cd simple
pod install
open tf_simple_example.xcworkspace
cd ../camera
pod install
open tf_camera_example.xcworkspace
cd ../benchmark
pod install
open tf_benchmark_example.xcworkspace
```

You can then run the three apps on an iOS device, or the simple and benchmark apps on an iOS simulator. If you tap the **Run Model** button after running the simple app, you'll see a text message saying that the TensorFlow Inception model is loaded, followed by several top recognition results along with confidence values.

If you tap the **Benchmark Model** button after running the benchmark app, you'll see the average time it takes to run the model for over 20 times. For example, it takes an average of about 0.2089 seconds on my iPhone 6, and 0.0359 seconds on the iPhone 6 simulator.

Finally, running the camera app on an iOS device and pointing the device camera around shows you the objects the app sees and recognizes in real time.

Running sample TensorFlow Android apps

There are four sample TensorFlow Android apps named TF Classify, TF Detect, TF Speech, and TF Stylize, located in `tensorflow/examples/android`. The easiest way to run these samples is to just open the project in the preceding folder using Android Studio, as shown in the *Setting up Android Studio* section, then make a single change by editing the project's `build.gradle` file and changing `def nativeBuildSystem = 'bazel'` to `def nativeBuildSystem = 'none'`.

Now connect an Android device to your computer and build, install and run the app by selecting Android Studio's **Run | Run 'android'**. This will install four Android apps with the names *TF Classify*, *TF Detect*, *TF Speech*, and *TF Stylize* on your device. TF Classify is just like the iOS camera app, using the TensorFlow Inception v1 model to do real-time object classification with the device camera. TF Detect uses a different model, called **Single Shot Multibox Detector** (**SSD**) with MobileNet, a new set of deep learning models Google released that are targeted in particular to mobile and embedded devices, to perform object detection, drawing rectangles on detected objects. TF Speech uses another different deep learning (speech recognition) model to listen and recognize a small set of words such as Yes, No, Left, Right, Stop and Go. TF Stylize uses yet another model to change the style of the images the camera sees. For more detailed information on these apps, you can check out the TensorFlow Android example documentation at `https://github.com/tensorflow/tensorflow/tree/master/tensorflow/examples/android`.

Summary

In this chapter, we covered how to install TensorFlow 1.4 on Mac and Ubuntu, how to set up a cost effective NVIDIA GPU on Ubuntu for faster model training, and how to set up Xcode and Android Studio for mobile AI app development. We also showed you how to run some cool sample TensorFlow iOS and Android apps. We'll discuss in detail in the rest of the book how to build and train, or retrain each of those models used in the apps, and many others, on our GPU-powered Ubuntu system, and show you how to deploy the models in iOS and Android apps and write the code to use the models in your mobile AI apps. Now that we're all set and ready, we can't wait to hit the road. It'll be an exciting journey, a journey we'd certainly be happy to share with our friends. So why not start with our best friends, and let's see what it takes to build a dog breed recognition app?

2
Classifying Images with Transfer Learning

The sample TensorFlow iOS apps, simple and camera, and the Android app TF Classify, described in the previous chapter all use the Inception v1 model, a pretrained image classification deep neural network model made publicly available by Google. The model is trained for ImageNet (http://image-net.org), one of the largest and best-known image databases with over 10 million images annotated for object classes. The Inception model can be used to classify an image into one of the 1,000 classes, listed at http://image-net.org/challenges/LSVRC/2014/browse-synsets. Those 1,000 object classes include quite a few dog breeds, among many kinds of objects. But the accuracy for recognizing dog breeds is not that high, around 70%, because the model is trained for recognizing a large number of objects, instead of a specific set of objects such as dog breeds.

What if we want to improve the accuracy and build a mobile app on our smart phone that uses the improved model so when we walk around and see an interesting dog, we can use the app to tell us what kind of dog it is.

In this chapter, we'll first discuss why transfer learning, or retraining pretrained deep learning models for such an image classification task, is the most cost-effective way to accomplish the task. Then we'll show you what it takes to retrain some of the best image classification models with a good dog dataset, and how to deploy and run the retrained models in the sample iOS and Android apps we covered in Chapter 1, *Getting Started with Mobile TensorFlow*. Also, we'll cover in step-by-step instructions how to add TensorFlow to your Objective-C or Swift-based iOS and Android apps.

In summary, we're going to cover the following topics in this chapter:

- Transfer learning – what and why
- Retraining using the Inception v3 model
- Retraining using MobileNet models
- Using the retrained models in the sample iOS app
- Using the retrained models in the sample Android app
- Adding TensorFlow to your own iOS app
- Adding TensorFlow to your own Android app

Transfer learning – what and why

We human beings don't learn new things from scratch. Instead, we take advantage of what we have learned as much as possible, consciously or not. Transfer learning in AI attempts to do the same thing—it's a technique that takes a normally small piece of a big trained model and reuses it in a new model for a related task, without the need to access the large training data and computing resources to train the original model. Overall, transfer learning is still an open problem in AI, since in many situations, what takes human beings only a few examples of trial-and-errors before learning to grasp something new would take AI a lot more time to train and learn. But in the field of image recognition, transfer learning has proven to be very effective.

Modern deep learning models for image recognition are typically deep neural networks, or more specifically, deep **Convolutional Neural Networks (CNNs)**, with many layers. Lower layers of such a CNN are responsible for learning and recognizing lower-level features such as an image's edges, contours, and parts, while the final layer determines the image's category. For different types of objects, such as dog breeds or flower types, we don't need to relearn the parameters, or weights, of lower layers of a network. In fact, it'd take many weeks of training from scratch to learn all the weights, typically millions or even more, of a modern CNN for image recognition. Transfer learning in the case of image classification allows us to just retrain the last layer of such a CNN with our specific set of images, usually in less than an hour, leaving the weights of all the other layers unchanged and reaching about the same accuracy as if we'd trained the whole network from scratch for weeks.

The second main benefit of transfer learning is that we only need a small amount of training data to retrain the last layer of a CNN. We'd need a large amount of training data if we have to train millions of parameters of a deep CNN from scratch. For example, for our dog breed retraining, we only need 100+ images for each dog breed to build a model with better dog breed classification than the original image classification model.

If you're unfamiliar with CNN, check out the videos and notes of one of the best resources on it, the Stanford CS231n course *CNN for Visual Recognition* (http://cs231n.stanford.edu). Another good resource on CNN is Chapter 6 of *Michael Nielsen's* online book, *Neural Networks and Deep Learning*: http://neuralnetworksanddeeplearning.com/chap6.html#introducing_convolutional_networks.

In the following two sections, we'll use two of the best pretrained CNN models for TensorFlow and a dog breed dataset to retrain the models and generate better dog breed recognition models. The first model is Inception v3, a more accurate model than Inception v1, optimized for accuracy but with a larger size. The other model is MobileNet, optimized for size and efficiency on mobile devices. A detailed list of pretrained models supported in TensorFlow is at https://github.com/tensorflow/models/tree/master/research/slim#pre-trained-models.

Retraining using the Inception v3 model

In the TensorFlow source that we set up in the previous chapter, there's a Python script, tensorflow/examples/image_retraining/retrain.py, that you can use to retrain the Inception v3 or MobileNet models. Before we run the script to retrain the Inception v3 model for our dog breed recognition, we need to first download the Stanford Dogs Dataset (http://vision.stanford.edu/aditya86/ImageNetDogs), which contains images of 120 dog breeds (you only need to download the Images in the link, not the Annotations).

Untar the downloaded dog `images.tar` file in `~/Downloads`, and you should see a list of folders in `~/Downloads/Images`, as shown in the following screenshot. Each folder corresponds to one dog breed and contains about 150 images (you don't need to supply explicit labels for images as the folder names are used to label the images contained within the folders):

Figure 2.1 Dogset images separated by folders, or labels of dog breed

You may download the dataset and then run the `retrain.py` script on Mac, as it doesn't take too long (less than an hour) for the script to run on the relatively small dataset (about 20,000 images in total), but if you do that on a GPU-powered Ubuntu, as set up in the last chapter, the script can complete in just a few minutes. In addition, when retraining with a large image dataset, running on Mac may take hours or days so it makes sense to run it on a GPU-powered machine.

The command to retrain the model, assuming you have created a `/tf_file` directory and also a `/tf_file/dogs_bottleneck` directory, is as follows:

```
python tensorflow/examples/image_retraining/retrain.py
--model_dir=/tf_files/inception-v3
--output_graph=/tf_files/dog_retrained.pb
--output_labels=/tf_files/dog_retrained_labels.txt
--image_dir ~/Downloads/Images
--bottleneck_dir=/tf_files/dogs_bottleneck
```

The five parameters need a little explanation here:

- `--model_dir` specifies the directory path where the Inception v3 model should be downloaded, automatically by the `retrain.py`, unless it's already in the directory.
- `--output_graph` indicates the name and path of the retrained model.
- `--output_labels` is a file consisting of the folder (label) names of the image dataset, that is used later with the retrained model to classify new images.
- `--image_dir` is the path to the image dataset used to retrain the Inception v3 model.
- `--bottleneck_dir` is used to cache the results generated on bottleneck, the layer before the final layer; the final layer performs classification using those results. During retraining, each image is used several times but the bottleneck values for the image remains the same, even for future reruns of the retrain script. So the first run takes much longer, as it needs to create the bottleneck results.

During the retraining, you'll see 3 values every 10 steps, for a default total of 4,000 steps. The first 20 and last 20 steps and final accuracy look like the following:

```
INFO:tensorflow:2018-01-03 10:42:53.127219: Step 0: Train accuracy = 21.0%
INFO:tensorflow:2018-01-03 10:42:53.127414: Step 0: Cross entropy = 4.767182
INFO:tensorflow:2018-01-03 10:42:55.384347: Step 0: Validation accuracy = 3.0% (N=100)
INFO:tensorflow:2018-01-03 10:43:11.591877: Step 10: Train accuracy = 34.0%
INFO:tensorflow:2018-01-03 10:43:11.592048: Step 10: Cross entropy = 4.704726
INFO:tensorflow:2018-01-03 10:43:12.915417: Step 10: Validation accuracy = 22.0% (N=100)
...
...
INFO:tensorflow:2018-01-03 10:56:16.579971: Step 3990: Train accuracy = 93.0%
INFO:tensorflow:2018-01-03 10:56:16.580140: Step 3990: Cross entropy = 0.326892
INFO:tensorflow:2018-01-03 10:56:16.692935: Step 3990: Validation accuracy = 89.0% (N=100)
INFO:tensorflow:2018-01-03 10:56:17.735986: Step 3999: Train accuracy = 93.0%
INFO:tensorflow:2018-01-03 10:56:17.736167: Step 3999: Cross entropy = 0.379192
INFO:tensorflow:2018-01-03 10:56:17.846976: Step 3999: Validation accuracy = 90.0% (N=100)
INFO:tensorflow:Final test accuracy = 91.0% (N=2109)
```

The train accuracy is the classification accuracy on the images the neural network has used for training, and the validation accuracy is on the images the neural network has not used for training. So the validation accuracy is a more reliable measurement on how accurate the model is, and it normally should be a little less than the train accuracy, but not too much, if the training converges and goes well, that is, if the trained model is neither overfitting nor underfitting.

If the train accuracy gets high but the validation accuracy remains low, it means the model is overfitting. If the train accuracy remains low, it suggests the model is underfitting. Also, the cross entropy is the loss function value and if the retraining goes well, it should overall get smaller and smaller. Finally, the test accuracy is on the images that haven't been used for either training or validation. It's generally the most accurate value we can tell about a retrained model.

As the preceding outputs show, by the end of the retraining, we see the validation accuracy is similar to the train accuracy (90% and 93%, compared to 3% and 21% in the beginning) and the final test accuracy is 91%. The cross entropy also drops from 4.767 in the beginning to 0.379 in the end. So we have a pretty good retrained dog breed recognition model now.

To further improve the accuracy, you can play with the `retrain.py`'s other parameters such as training steps (`--how_many_training_steps`), learning rate (`--learning_rate`), and data augmentation (`--flip_left_right`, `--random_crop`, `--random_scale`, `--random_brightness`). Generally, this is a tedious process that involves a lot of "dirty work" as called by Andrew Ng, one of the best-known deep learning experts, in his *Nuts and Bolts of Applying Deep Learning* speech (video is available at: `https://www.youtube.com/watch?v=F1ka6a13S9I`).

Another Python script you can use to give the retrained model a quick test on your own image (for example, a Labrador Retriever image in `/tmp/lab1.jpg`) is `label_image`, which you can run after first building it as follows:

```
bazel build tensorflow/examples/image_retraining:label_image

bazel-bin/tensorflow/examples/label_image/label_image
--graph=/tf_files/dog_retrained.pb
--image=/tmp/lab1.jpg
--input_layer=Mul
--output_layer=final_result
--labels=/tf_files/dog_retrained_labels.txt
```

You'll see the top five classification results similar to (but, since networks vary randomly, likely not exactly the same as) the following:

```
n02099712 labrador retriever (41): 0.75551
n02099601 golden retriever (64): 0.137506
n02104029 kuvasz (76): 0.0228538
n02090379 redbone (32): 0.00943663
n02088364 beagle (20): 0.00672507
```

The values of `--input_layer` (Mul) and `--output_layer` (final_result) are very important – they have to be the same as what are defined in the model for the classification to work at all. If you wonder how you can get them (from the graph, aka model, file `dog_retrained.pb`), there are two TensorFlow tools that can be helpful. The first one is the appropriately named `summarize_graph`. Here's how you can build and run it:

```
bazel build tensorflow/tools/graph_transforms:summarize_graph

bazel-bin/tensorflow/tools/graph_transforms/summarize_graph --
in_graph=/tf_files/dog_retrained.pb
```

You'll see the summary results similar to these:

```
No inputs spotted.
No variables spotted.
Found 1 possible outputs: (name=final_result, op=Softmax)
Found 22067948 (22.07M) const parameters, 0 (0) variable parameters, and 99
control_edges
Op types used: 489 Const, 101 Identity, 99 CheckNumerics, 94 Relu, 94
BatchNormWithGlobalNormalization, 94 Conv2D, 11 Concat, 9 AvgPool, 5
MaxPool, 1 DecodeJpeg, 1 ExpandDims, 1 Cast, 1 MatMul, 1 Mul, 1
PlaceholderWithDefault, 1 Add, 1 Reshape, 1 ResizeBilinear, 1 Softmax, 1
Sub
```

There's one possible output with the name `final_result`. Unfortunately, sometimes the `summarize_graph` tool doesn't tell us the input name, as it seems to be confused about the nodes used for training. After the nodes used only for training are stripped out, which we'll discuss soon, the `summarize_graph` tool will return the correct input name. Another tool called **TensorBoard** gives us a more complete picture of the model graph. If you have TensorFlow installed directly from binary, you should be able to just run TensorBoard, as by default, it's installed in `/usr/local/bin`. But if you install TensorFlow from source as we did in earlier, you can run the following commands to build TensorBoard:

```
git clone https://github.com/tensorflow/tensorboard
cd tensorboard/
bazel build //tensorboard
```

Now, make sure you have /tmp/retrained_logs, created automatically when you run retrain.py, and run:

```
bazel-bin/tensorboard/tensorboard --logdir /tmp/retrain_logs
```

Then launch the URL http://localhost:6006 on a browser. You'll first see the accuracy graph as shown in the following screenshot:

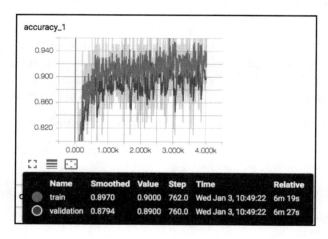

Figure 2.2 Train and validation accuracy of the Inception v3 retrained model

The cross_entropy graph in the following screenshot is just as we described earlier regarding the outputs of running retrain.py:

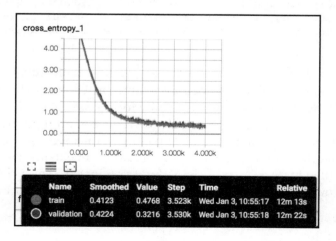

Figure 2.3 Train and validation cross entropy of the Inception v3 retrained model

Chapter 2

Now click the **GRAPHS** tab, and you'll see an operation with the name **Mul** and another with the name **final_result**, as shown here:

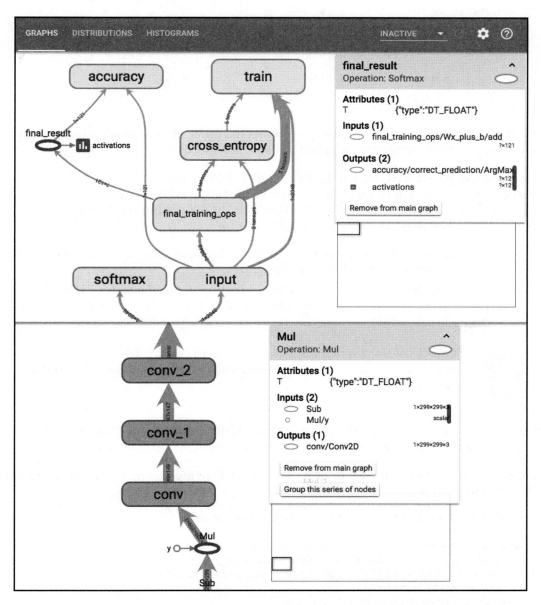

Figure 2.4 The Mul and final_result nodes in the retrained model

Actually, if you prefer a small interaction with TensorFlow, you can try several lines of Python code to find out the names of the output layer and input layer, as shown here in an iPython interaction:

```
In [1]: import tensorflow as tf
In [2]: g=tf.GraphDef()
In [3]: g.ParseFromString(open("/tf_files/dog_retrained.pb", "rb").read())
In [4]: x=[n.name for n in g.node]
In [5]: x[-1:]
Out[5]: [u'final_result']
```

Note that this code snippet won't always work as the order of nodes isn't guaranteed, but it often gives you the information or validation you may need.

Now we're ready to discuss how to further modify the retrained model so it can be deployed and run on mobile devices. The size of the retrained model file `dog_retrained.pb` is too big, about 80 MB, which should go through two steps for optimization before being deployed to mobile devices:

1. **Strip unused nodes**: remove the nodes in the model that are only used during training but not needed during inference.
2. **Quantize the model**: convert all the 32-bit floating numbers for the model parameters to 8-bit values. This would reduce the model size to about 25% of its original size, while keeping the inference accuracy about the same.

TensorFlow's documentation (https://www.tensorflow.org/performance/quantization) offers more details on quantization and why it works.

There are two ways to perform the preceding two tasks: the older way that uses the `strip_unused` tool and the new way that uses the `transform_graph` tool.

Let's see how the older method works: first run the following commands to create a model with all the unused nodes stripped:

```
bazel build tensorflow/python/tools:strip_unused

bazel-bin/tensorflow/python/tools/strip_unused
  --input_graph=/tf_files/dog_retrained.pb
  --output_graph=/tf_files/stripped_dog_retrained.pb
  --input_node_names=Mul
  --output_node_names=final_result
    --input_binary=true
```

If you run the preceding Python code with the output graph, you can find the right input layer name:

```
In [1]: import tensorflow as tf
In [2]: g=tf.GraphDef()
In [3]: g.ParseFromString(open("/tf_files/ stripped_dog_retrained.pb",
"rb").read())
In [4]: x=[n.name for n in g.node]
In [5]: x[0]
Out[5]: [u'Mul']
```

Now run the following command to quantize the model:

```
python tensorflow/tools/quantization/quantize_graph.py
    --input=/tf_files/stripped_dog_retrained.pb
--output_node_names=final_result
--output=/tf_files/quantized_stripped_dogs_retrained.pb
    --mode=weights
```

After this, the model `quantized_stripped_dogs_retrained.pb` is ready to be deployed and used in iOS and Android apps, which we'll see in the following sections of this chapter.

The other way of stripping unused nodes and quantizing the model is to use a tool called `transform_graph`. This is the recommended new way in TensorFlow 1.4 and works fine with the Python `label_image` script but still causes incorrect recognition results when deployed to iOS and Android apps.

```
bazel build tensorflow/tools/graph_transforms:transform_graph
bazel-bin/tensorflow/tools/graph_transforms/transform_graph
   --in_graph=/tf_files/dog_retrained.pb
   --out_graph=/tf_files/transform_dog_retrained.pb
   --inputs='Mul'
   --outputs='final_result'
   --transforms='
     strip_unused_nodes(type=float, shape="1,299,299,3")
     fold_constants(ignore_errors=true)
     fold_batch_norms
     fold_old_batch_norms
     quantize_weights'
```

Using the `label_image` script on test both `quantized_stripped_dogs_retrained.pb` and `transform_dog_retrained.pb` works correctly. But only the first one works correctly in iOS and Android apps.

For detailed documentation on the graph transform tool, see its GitHub README at https://github.com/tensorflow/tensorflow/blob/master/tensorflow/tools/graph_transforms/README.md.

Retraining using MobileNet models

The stripped and quantized model generated in the previous section is still over 20 MB in size. This is because the pre-built Inception v3 model used for retraining is a large-scale deep learning model, with over 25 million parameters, and Inception v3 was not created with a mobile-first goal.

In June 2017, Google released MobileNets v1, a total of 16 mobile-first deep learning models for TensorFlow. These models are only a few MB in size, with 0.47 million to 4.24 million parameters, still achieving decent accuracy (just a bit lower than Inception v3). See its README for more information: https://github.com/tensorflow/models/blob/master/research/slim/nets/mobilenet_v1.md.

The `retrain.py` script discussed in the previous section also supports retraining based on MobileNet models. Simply run a command like the following:

```
python tensorflow/examples/image_retraining/retrain.py
  --output_graph=/tf_files/dog_retrained_mobilenet10_224.pb
  --output_labels=/tf_files/dog_retrained_labels_mobilenet.txt
  --image_dir ~/Downloads/Images
  --bottleneck_dir=/tf_files/dogs_bottleneck_mobilenet
  --architecture mobilenet_1.0_224
```

The generated label file `dog_retrained_labels_mobilenet.txt` actually is the same as the label file generated during retraining using the Inception v3 model. The --architecture parameter specifies one of the 16 MobileNet models, and the value `mobilenet_1.0_224` means using the model with 1.0 as the parameter size (the other three possible values are 0.75, 0.50, and 0.25 – 1.0 for most parameters and accurate but largest size, 0.25 the opposite) and 224 as the image input size (the other three values are 192, 160, and 128). If you add _quantized to the end of the—architecture value, that is --architecture mobilenet_1.0_224_quantized, the model will also be quantized, resulting in the retrained model size of about 5.1 MB. The non-quantized model has a size of about 17 MB.

You can test the model generated previously with `label_image` as follows:

```
bazel-bin/tensorflow/examples/label_image/label_image
--graph=/tf_files/dog_retrained_mobilenet10_224.pb
--image=/tmp/lab1.jpg
--input_layer=input
--output_layer=final_result
--labels=/tf_files/dog_retrained_labels_mobilenet.txt
--input_height=224
--input_width=224
--input_mean=128
--input_std=128

n02099712 labrador retriever (41): 0.824675
n02099601 golden retriever (64): 0.144245
n02104029 kuvasz (76): 0.0103533
n02087394 rhodesian ridgeback (105): 0.00528782
n02090379 redbone (32): 0.0035457
```

Notice that when running `label_image`, the input_layer is named as `input`. We can find this name using the interactive iPython code or the summarize graph tool seen earlier:

```
bazel-bin/tensorflow/tools/graph_transforms/summarize_graph
--in_graph=/tf_files/dog_retrained_mobilenet10_224.pb
Found 1 possible inputs: (name=input, type=float(1), shape=[1,224,224,3])
No variables spotted.
Found 1 possible outputs: (name=final_result, op=Softmax)
Found 4348281 (4.35M) const parameters, 0 (0) variable parameters, and 0 control_edges
Op types used: 92 Const, 28 Add, 27 Relu6, 15 Conv2D, 13 Mul, 13 DepthwiseConv2dNative, 10 Dequantize, 3 Identity, 1 MatMul, 1 BiasAdd, 1 Placeholder, 1 PlaceholderWithDefault, 1 AvgPool, 1 Reshape, 1 Softmax, 1 Squeeze
```

So, when should we use an Inception v3 or MobileNet retrained model on mobile devices? In cases where you want to achieve the highest possible accuracy, you should and can use the retrained model based on Inception v3. If speed is your top consideration, you should consider using a MobileNet retrained model with the smallest parameter size and image input size, in exchange for some accuracy loss.

One tool to give you an accurate benchmark of a model is `benchmark_model`. First, build it as follows:

```
bazel build -c opt tensorflow/tools/benchmark:benchmark_model
```

Then, run it against the Inception v3 or MobileNet v1-based retrain model:

```
bazel-bin/tensorflow/tools/benchmark/benchmark_model
--graph=/tf_files/quantized_stripped_dogs_retrained.pb
--input_layer="Mul"
--input_layer_shape="1,299,299,3"
--input_layer_type="float"
--output_layer="final_result"
--show_run_order=false
--show_time=false
--show_memory=false
--show_summary=true
```

You'll get a pretty long output and at the end there's a line like FLOPs estimate: 11.42 B, meaning it'd take the Inception v3-based retrained model about 11 B FLOPS (floating point operations) to make an inference. An iPhone 6 runs about 2 B FLOPS so it'd take about 5–6 seconds on an iPhone 6 to run the model. Other modern smartphones can run 10 B FLOPS.

By replacing the graph file with the MobileNet model-based retrained model `dog_retrained_mobilenet10_224.pb` and rerunning the benchmark tool, you'll see the FLOPS estimate becomes about 1.14 B, which is about 10 times faster.

Using the retrained models in the sample iOS app

The iOS simple example we see in `Chapter 1`, *Getting Started with Mobile TensorFlow*, uses the Inception v1 model. To make the app use our retrained Inception v3 model and MobileNet model to do better dog breed recognition, we need to make a few changes to the app. Let's first see what it takes to use the retrained `quantized_stripped_dogs_retrained.pb` in the iOS simple app:

1. Double-click the `tf_simple_example.xcworkspace` file in `tensorflow/examples/ios/simple` to open the app in Xcode
2. Drag the `quantized_stripped_dogs_retrained.pb` model file, the `dog_retrained_labels.txt` label file, and the `lab1.jpg` image file we used to test the `label_image` script, and drop to the project's data folder, making sure both **Copy items if needed** and **Add to targets** are checked, as shown in the following screenshot:

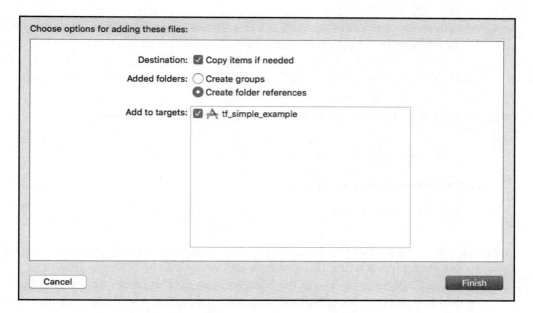

Figure 2.5 Adding the retrained model file and the label file to app

3. Click the `RunModelViewController.mm` file in Xcode, which uses the TensorFlow C++ API to process an input image, run it through the Inception v1 model, and get the image classification result, and change the lines:

```
NSString* network_path = FilePathForResourceName(@"tensorflow_inception_graph", @"pb");
NSString* labels_path = FilePathForResourceName(@"imagenet_comp_graph_label_strings", @"txt");
NSString* image_path = FilePathForResourceName(@"grace_hopper", @"jpg");
```

To the following with the correct model filename, label filename, and test image name:

```
NSString* network_path = FilePathForResourceName(@"quantized_stripped_dogs_retrained", @"pb");
NSString* labels_path = FilePathForResourceName(@"dog_retrained_labels", @"txt");
NSString* image_path = FilePathForResourceName(@"lab1", @"jpg");
```

4. Also in `RunModelViewController.mm`, to match the required input image size for our Inception v3 (from v1) retrained model, change the value 224 in `const int wanted_width = 224;` and `const int wanted_height = 224;` to 299, and the value in both `const float input_mean = 117.0f;` and `const float input_std = 1.0f;` to 128.0f

5. Change the values of the input and output node names from:

   ```
   std::string input_layer = "input";
   std::string output_layer = "output";
   ```

   To the following correct values:

   ```
   std::string input_layer = "Mul";
   std::string output_layer = "final_result";
   ```

6. Finally, you can edit the `dog_retrained_labels.txt` file to remove the leading nxxxx string in each line (for example, remove n02099712 in n02099712 labrador retriever) – on the Mac you can do this by holding down the option key, then making block selection and deletion – so the recognition results will be more readable

Run the app now and click the **Run Model** button, in the Xcode's console window or the app's edit box, you'll see the following recognition results, pretty consistent with the results of running the `label_image` script:

```
Predictions: 41 0.645   labrador retriever
64 0.195   golden retriever
76 0.0261  kuvasz
32 0.0133  redbone
20 0.0127  beagle
```

To use the MobileNet (mobilenet_1.0_224_quantized) retrained model `dog_retrained_mobilenet10_224.pb`, we follow steps similar to the previous ones, and in Steps 2 and 3, we use `dog_retrained_mobilenet10_224.pb`, but in Step 4, we need to keep `const int wanted_width = 224;` and `const int wanted_height = 224;`, and only change `const float input_mean` and `const float input_std` to 128. Finally, in Step 5, we must use `std::string input_layer = "input";` and `std::string output_layer = "final_result";`. These parameters are the same as those used with the `label_image` script for `dog_retrained_mobilenet10_224.pb`.

Run the app again and you'll see similar top recognition results.

Using the retrained models in the sample Android app

To use our retrained Inception v3 model and MobileNet model in Android's TF Classify app is also pretty straightforward. Follow the steps here to test both retrained models:

1. Open the sample TensorFlow Android app, located in `tensorflow/examples/android`, using Android Studio.
2. Drag and drop two retrained models, `quantized_stripped_dogs_retrained.pb` and `dog_retrained_mobilenet10_224.pb` as well as the label file, `dog_retrained_labels.txt` to the `assets` folder of the android app.
3. Open the file `ClassifierActivity.java`, to use the Inception v3 retrained model, and replace the following code:

   ```
   private static final int INPUT_SIZE = 224;
   private static final int IMAGE_MEAN = 117;
   private static final float IMAGE_STD = 1;
   private static final String INPUT_NAME = "input";
   private static final String OUTPUT_NAME = "output";
   ```

   With these lines:

   ```
   private static final int INPUT_SIZE = 299;
   private static final int IMAGE_MEAN = 128;
   private static final float IMAGE_STD = 128;
   private static final String INPUT_NAME = "Mul";
   private static final String OUTPUT_NAME = "final_result";
   private static final String MODEL_FILE =
       "file:///android_asset/quantized_stripped_dogs_retrained.pb";
   private static final String LABEL_FILE =
       "file:///android_asset/dog_retrained_labels.txt";
   ```

4. Or, to use the MobileNet retrained model, replace the code with these lines:

   ```
   private static final int INPUT_SIZE = 224;
   private static final int IMAGE_MEAN = 128;
   private static final float IMAGE_STD = 128;
   private static final String INPUT_NAME = "input";
   private static final String OUTPUT_NAME = "final_result";
   private static final String MODEL_FILE =
       "file:///android_asset/dog_retrained_mobilenet10_224.pb";
   private static final String LABEL_FILE =
       "file:///android_asset/dog_retrained_labels.txt";
   ```

5. Connect an Android device to your computer and run the app on it. Then tap the TF Classify app and point the camera to some dog pictures, and you'll see top results on your screen.

That's all it takes to use the two retrained models in the sample TensorFlow iOS and Android apps. Now that you've seen how to use our retrained models in the sample apps, the next thing you may want to know is how to add TensorFlow to a new or an existing iOS or Android app of your own, so you can start adding the power of AI to your own mobile apps. That's what we'll discuss in detail in the rest of the chapter.

Adding TensorFlow to your own iOS app

In the earlier version of TensorFlow, adding TensorFlow to your own app was very tedious, requiring the use of the manual build process of TensorFlow and other manual settings. In TensorFlow 1.4, the process is pretty straightforward, but still, detailed steps are not well documented in the TensorFlow website. One other thing that's missing is the lack of documentation on how to use TensorFlow in your Swift-based iOS app; the sample TensorFlow iOS apps are all in Objective-C, calling TensorFlow's C++ APIs. Let's see how we can do better.

Adding TensorFlow to your Objective-C iOS app

First, follow these steps to add TensorFlow with the image classification feature to your Objective-C iOS app (we'll start with a new app, but you can just skip the first step if you need to add TensorFlow to an existing app):

1. In your Xcode, click **File | New | Project...**, select **Single View App**, then **Next**, enter `HelloTensorFlow` as **Product Name**, select **Objective-C** as **Language**, then click **Next** and choose a location for the project before hitting **Create**. Close the project window in Xcode (since we'll open the project's workspace file due to its use of pod later).
2. Open a Terminal window, `cd` to where your project is located at, then create a new file named `Podfile`, with the following content:
   ```
   target 'HelloTensorFlow'
   pod 'TensorFlow-experimental'
   ```

Chapter 2

3. Run the command `pod install` to download and install the TensorFlow pod.
4. Open the `HelloTensorFlow.xcworkspace` file in Xcode, then drag and drop the two files (`ios_image_load.mm` and `ios_image_load.h`) that handle image loading from the TensorFlow iOS sample directory `tensorflow/examples/ios/simple` to the `HelloTensorFlow` project folder.
5. Drag and drop the two models, `quantized_stripped_dogs_retrained.pb` and `dog_retrained_mobilenet10_224.pb`, the `label file dog_retrained_labels.txt`, and a couple of test image files to the project folder—after that, you should see something like this:

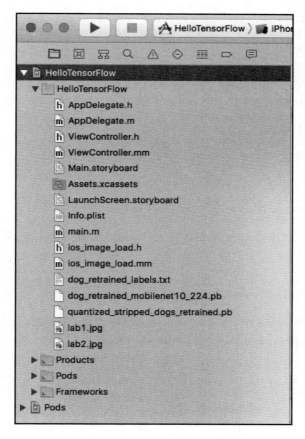

Figure 2.6 Adding utility files, model files, label file and image files

[45]

6. Rename `ViewController.m` to `ViewController.mm`, as we'll mix C++ code and Objective-C code in this file to call TensorFlow C++ API and process the image input and inference results. Then, before `@interface ViewController`, add the following `#include` and function prototype:

```
#include <fstream>
#include <queue>
#include "tensorflow/core/framework/op_kernel.h"
#include "tensorflow/core/public/session.h"
#include "ios_image_load.h"

NSString* RunInferenceOnImage(int wanted_width, int wanted_height, std::string input_layer, NSString *model);
```

7. At the end of `ViewController.mm`, add the following code copied from the `tensorflow/example/ios/simple/RunModelViewController.mm`, with a slight change in the function `RunInferenceOnImage` to accept different retrained models with different input sizes and input layer name:

```
namespace {
    class IfstreamInputStream : public ::google::protobuf::io::CopyingInputStream {
...
static void GetTopN(
...
bool PortableReadFileToProto(const std::string& file_name,
...
NSString* FilePathForResourceName(NSString* name, NSString* extension) {
...
NSString* RunInferenceOnImage(int wanted_width, int wanted_height, std::string input_layer, NSString *model) {
```

8. Still in the `ViewController.mm`, at the end of `viewDidLoad` method, first add the code that adds a label to let users know what they can do with the app:

```
UILabel *lbl = [[UILabel alloc] init];
[lbl setTranslatesAutoresizingMaskIntoConstraints:NO];
lbl.text = @"Tap Anywhere";
[self.view addSubview:lbl];
```

Then the constraints to center the label in the screen:

```
NSLayoutConstraint *horizontal = [NSLayoutConstraint
constraintWithItem:lbl attribute:NSLayoutAttributeCenterX
relatedBy:NSLayoutRelationEqual toItem:self.view
attribute:NSLayoutAttributeCenterX multiplier:1 constant:0];

NSLayoutConstraint *vertical = [NSLayoutConstraint
constraintWithItem:lbl attribute:NSLayoutAttributeCenterY
relatedBy:NSLayoutRelationEqual toItem:self.view
attribute:NSLayoutAttributeCenterY multiplier:1 constant:0];

[self.view addConstraint:horizontal];
[self.view addConstraint:vertical];
```

Finally, add a tap gesture recognizer there:

```
UITapGestureRecognizer *recognizer = [[UITapGestureRecognizer
alloc] initWithTarget:self action:@selector(tapped:)];
[self.view addGestureRecognizer:recognizer];
```

9. In the tap handler, we first create two `alert` actions to allow the user to select a retrained model:

```
UIAlertAction* inceptionV3 = [UIAlertAction
actionWithTitle:@"Inception v3 Retrained Model"
style:UIAlertActionStyleDefault handler:^(UIAlertAction * action) {
        NSString *result = RunInferenceOnImage(299, 299, "Mul",
@"quantized_stripped_dogs_retrained");
        [self showResult:result];
}];
UIAlertAction* mobileNet = [UIAlertAction
actionWithTitle:@"MobileNet 1.0 Retrained Model"
style:UIAlertActionStyleDefault handler:^(UIAlertAction * action) {
        NSString *result = RunInferenceOnImage(224, 224, "input",
@"dog_retrained_mobilenet10_224");
        [self showResult:result];
}];
```

Classifying Images with Transfer Learning

Then create a `none` action and add all three `alert` actions to an alert controller and present it:

```
UIAlertAction* none = [UIAlertAction actionWithTitle:@"None"
style:UIAlertActionStyleDefault
        handler:^(UIAlertAction * action) {}];

UIAlertController* alert = [UIAlertController
alertControllerWithTitle:@"Pick a Model" message:nil
preferredStyle:UIAlertControllerStyleAlert];
[alert addAction:inceptionV3];
[alert addAction:mobileNet];
[alert addAction:none];
[self presentViewController:alert animated:YES completion:nil];
```

10. The result of the inference is shown as another alert controller in the method `showResult`:

```
-(void) showResult:(NSString *)result {
    UIAlertController* alert = [UIAlertController
alertControllerWithTitle:@"Inference Result" message:result
preferredStyle:UIAlertControllerStyleAlert];
    UIAlertAction* action = [UIAlertAction actionWithTitle:@"OK"
style:UIAlertActionStyleDefault handler:nil];
    [alert addAction:action];
    [self presentViewController:alert animated:YES completion:nil];
}
```

The core code related to calling TensorFlow is in the `RunInferenceOnImage` method, modified slightly based on the TensorFlow iOS simple app, consisting of first creating a TensorFlow session and a graph:

```
tensorflow::Session* session_pointer = nullptr;
tensorflow::Status session_status = tensorflow::NewSession(options,
&session_pointer);
...
std::unique_ptr<tensorflow::Session> session(session_pointer);
tensorflow::GraphDef tensorflow_graph;
NSString* network_path = FilePathForResourceName(model, @"pb");
PortableReadFileToProto([network_path UTF8String], &tensorflow_graph);
tensorflow::Status s = session->Create(tensorflow_graph);
```

Then loading the label file and the image file, and converting the image data to appropriate Tensor data:

```
NSString* labels_path = FilePathForResourceName(@"dog_retrained_labels", @"txt");
...
NSString* image_path = FilePathForResourceName(@"lab1", @"jpg");
std::vector<tensorflow::uint8> image_data = LoadImageFromFile([image_path UTF8String], &image_width, &image_height, &image_channels);
tensorflow::Tensor image_tensor(tensorflow::DT_FLOAT,
tensorflow::TensorShape({1, wanted_height, wanted_width, wanted_channels}));
auto image_tensor_mapped = image_tensor.tensor<float, 4>();
tensorflow::uint8* in = image_data.data();
float* out = image_tensor_mapped.data();
for (int y = 0; y < wanted_height; ++y) {
    const int in_y = (y * image_height) / wanted_height;
...
}
```

And finally, calling the TensorFlow session's run method with the image tensor data and the input layer name, getting the returned output results and processing it to get the top five results with confidence values greater than the threshold:

```
std::vector<tensorflow::Tensor> outputs;
tensorflow::Status run_status = session->Run({{input_layer, image_tensor}}, {output_layer}, {}, &outputs);
...
tensorflow::Tensor* output = &outputs[0];
const int kNumResults = 5;
const float kThreshold = 0.01f;
std::vector<std::pair<float, int> > top_results;
GetTopN(output->flat<float>(), kNumResults, kThreshold, &top_results);
```

In the rest of the book, we'll implement different versions of the `RunInferenceOnxxx` method to run different models with different inputs. So if you don't fully understand some of the previous code, don't worry; with a few more apps built, you'll feel comfortable writing your own inference logic for a new custom model.

Also, the complete iOS app, HelloTensorFlow, is in the book's source code repo.

Now, run the app in the Simulator or on an actual iOS device, first you'll see the following message box asking you to select a retrained model:

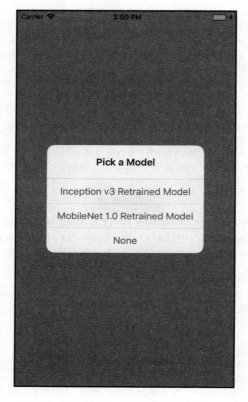

Figure 2.7 Selecting different retrained model for inference

Then you will see the inference results after selecting a model:

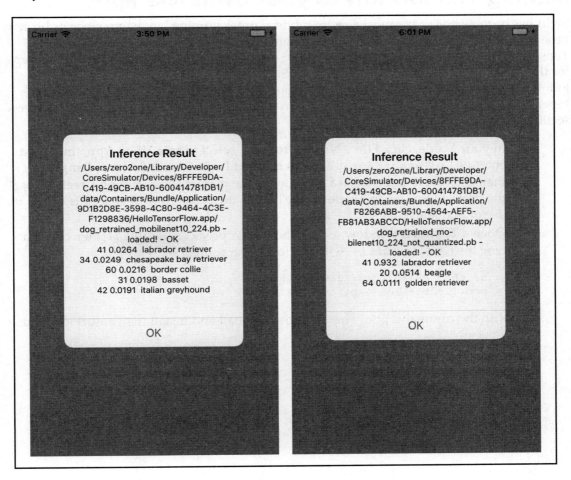

Figure 2.8 Inference results based on different retrained models

Notice that the MobileNet retrained model runs a lot faster, about one second on an iPhone 6, than the Inception v3 retrained model, runs about seven seconds on the same iPhone.

Adding TensorFlow to your Swift iOS app

Swift has become one of the most elegant modern programming languages since its birth in June 2014. So it'll be both fun and useful for some developers to integrate modern TensorFlow into their modern Swift-based iOS app. Steps to do that are similar to the steps for the Objective-C-based app but with some Swift-related trick. If you have already followed the steps for the Objective-C part, you may find some of the steps here are repetitive, but complete steps are provided anyway for those who may skip the Objective-C section and get to Swift directly:

1. In your Xcode, click **File | New | Project...**, select **Single View App**, then **Next**, enter `HelloTensorFlow_Swift` as **Product Name**, select **Swift** as **Language**, then click **Next** and choose a location for the project before hitting **Create**. Close the project window in Xcode (as we'll open the project's workspace file due to its use of pod later).
2. Open a Terminal Window, `cd` to where your project is located at, then create a new file named `Podfile`, with the following content:

    ```
    target 'HelloTensorFlow_Swift'
    pod 'TensorFlow-experimental'
    ```

3. Run the command `pod install` to download and install the TensorFlow pod;

4. Open the `HelloTensorFlow_Swift.xcworkspace` file in Xcode, then drag and drop the two files (`ios_image_load.mm` and `ios_image_load.h`) that handle image loading from the TensorFlow iOS sample directory `tensorflow/examples/ios/simple` to the `HelloTensorFlow_Swift` project folder. When you add the two files to the project, you'll see a message box, as shown in the following screenshot, asking you if you would like to configure an Objective-C bridging header, which is needed for Swift code to call C++ or Objective-C code. So click the **Create Bridging Header** button:

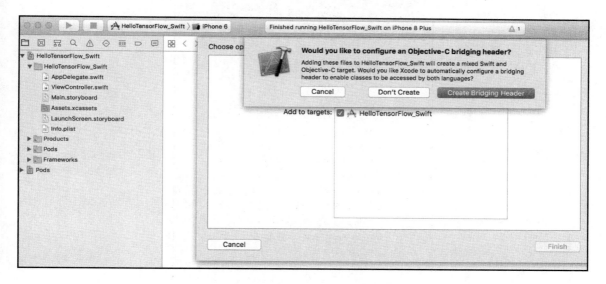

Figure 2.9 Creating Bridging Header when adding C++ file

Classifying Images with Transfer Learning

5. Also drag and drop the two models, `quantized_stripped_dogs_retrained.pb` and `dog_retrained_mobilenet10_224.pb`, the label file `dog_retrained_labels.txt` and a couple of test image files to the project folder – after that you should see something like this:

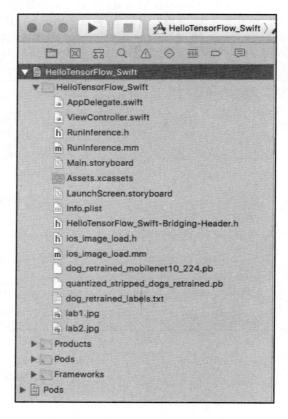

Figure 2.10 Adding utility files, model files, label file and image files

6. Create a new file named `RunInference.h` with the following code (one trick is that we have to use an Objective-C class as a wrapper to the `RunInferenceOnImage` method in the next step for our Swift code to make an indirect call to it. Otherwise, a build error will occur):

```
#import <Foundation/Foundation.h>
@interface RunInference_Wrapper : NSObject
    - (NSString *)run_inference_wrapper:(NSString *)name;
@end
```

7. Create another file named `RunInference.mm` which starts with the following `include` objects and prototype:

   ```
   #include <fstream>
   #include <queue>
   #include "tensorflow/core/framework/op_kernel.h"
   #include "tensorflow/core/public/session.h"
   #include "ios_image_load.h"

   NSString* RunInferenceOnImage(int wanted_width, int wanted_height,
   std::string input_layer, NSString *model);
   ```

8. Add `RunInference.mm` in the following code to implement the `RunInference_Wrapper` defined in its `.h` file:

   ```
   @implementation RunInference_Wrapper
   - (NSString *)run_inference_wrapper:(NSString *)name {
       if ([name isEqualToString:@"Inceptionv3"])
           return RunInferenceOnImage(299, 299, "Mul",
   @"quantized_stripped_dogs_retrained");
       else
           return RunInferenceOnImage(224, 224, "input",
   @"dog_retrained_mobilenet10_224");
   }
   @end
   ```

9. At the end of `RunInference.mm`, add the exact same methods as in the `ViewController.mm` in the Objective-C section, slightly different from the methods in the `tensorflow/example/ios/simple/RunModelViewController.mm`:

   ```
   class IfstreamInputStream : public namespace {
       class IfstreamInputStream : public
   ::google::protobuf::io::CopyingInputStream {
   ...
   static void GetTopN(
   ...
   bool PortableReadFileToProto(const std::string& file_name,
   ...
   NSString* FilePathForResourceName(NSString* name, NSString*
   extension) {
   ...
   NSString* RunInferenceOnImage(int wanted_width, int wanted_height,
   std::string input_layer, NSString *model) {
   ```

10. Now open the `ViewController.swift`, at the end of the `viewDidLoad` method, first add the code that adds a label to let users know what they can do with the app:

    ```
    let lbl = UILabel()
    lbl.translatesAutoresizingMaskIntoConstraints = false
    lbl.text = "Tap Anywhere"
    self.view.addSubview(lbl)
    ```

    Then the constraints to center the label in the screen:

    ```
    let horizontal = NSLayoutConstraint(item: lbl, attribute: .centerX, relatedBy: .equal, toItem: self.view, attribute: .centerX, multiplier: 1, constant: 0)

    let vertical = NSLayoutConstraint(item: lbl, attribute: .centerY, relatedBy: .equal, toItem: self.view, attribute: .centerY, multiplier: 1, constant: 0)

    self.view.addConstraint(horizontal)
    self.view.addConstraint(vertical)
    ```

    Finally, add a tap gesture recognizer there:

    ```
    let recognizer = UITapGestureRecognizer(target: self, action: #selector(ViewController.tapped(_:)))
    self.view.addGestureRecognizer(recognizer)
    ```

11. In the tap handler, we first add an `alert` action to allow the user to select the Inception v3 retrained model:

    ```
    let alert = UIAlertController(title: "Pick a Model", message: nil, preferredStyle: .actionSheet)
    alert.addAction(UIAlertAction(title: "Inception v3 Retrained Model", style: .default) { action in
        let result = RunInference_Wrapper().run_inference_wrapper("Inceptionv3")
        let alert2 = UIAlertController(title: "Inference Result", message: result, preferredStyle: .actionSheet)
        alert2.addAction(UIAlertAction(title: "OK", style: .default) { action2 in
        })
        self.present(alert2, animated: true, completion: nil)
    })
    ```

Then create another action for the MobileNet retrained model, as well as a `none` action, before presenting it:

```
alert.addAction(UIAlertAction(title: "MobileNet 1.0 Retrained Model", style: .default) { action in
    let result = RunInference_Wrapper().run_inference_wrapper("MobileNet")
    let alert2 = UIAlertController(title: "Inference Result", message: result, preferredStyle: .actionSheet)
    alert2.addAction(UIAlertAction(title: "OK", style: .default) { action2 in
    })
    self.present(alert2, animated: true, completion: nil)
})
alert.addAction(UIAlertAction(title: "None", style: .default) {
action in
})

self.present(alert, animated: true, completion: nil)
```

12. Open the `HelloTensorFlow_Swift-Bridging-Header.h` file, and add one line of code to it: `#include "RunInference.h"`.

Now run the app in the simulator, you'll see an alert controller asking you to select a model:

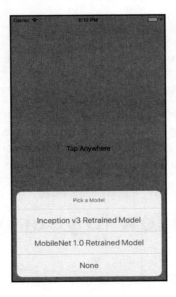

Figure 2.11 Selecting a retrained model for inference

And the inference results for different retrained models:

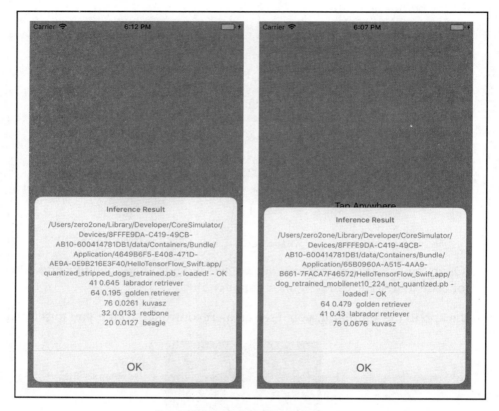

Figure 2.12 Inference results for different retrained models

There you go. Now that you know what it takes to add powerful TensorFlow models to your iOS apps, whether it's written in Objective-C or Swift, there's no reason to stop you from adding AI to your mobile apps, unless Android is your thing. But you know we'll certainly take care of Android too.

Adding TensorFlow to your own Android app

It turns out that adding TensorFlow to your own Android app is easier than iOS. Let's jump right to the steps:

Chapter 2

1. If you have an existing Android app, skip this. Otherwise, in Android Studio, select **File | New | New Project...** and accept all the defaults before clicking **Finish**.
2. Open the `build.gradle (Module: app)` file, and add compile `'org.tensorflow:tensorflow-android:+'` inside and at the end of dependencies `{...};`.
3. Build the `gradle` file and you'll see `libtensorflow_inference.so`, the TensorFlow native library that Java code talks to, inside the subfolders of the location `app/build/intermediates/transforms/mergeJniLibs/debug/0/lib` of your `app` directory.
4. If this is a new project, you can create the `assets` folder by first switching to **Packages**, then by right-mouse clicking the app and selecting **New | Folder | Assets Folder**, as shown in the following screenshot, and switching back from **Packages** to **Android**:

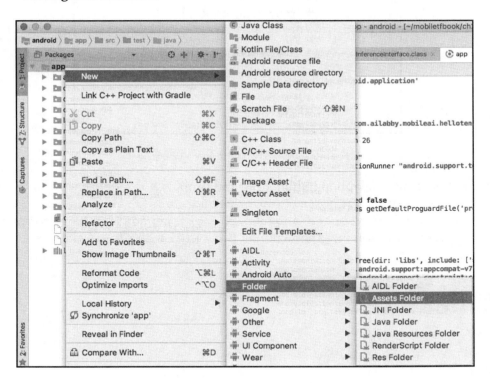

Figure 2.13 Adding Assets Folder to a new project

5. Drag and drop the two retrained model files and the label file, as well as a couple of test images, to the assets folder, as shown here:

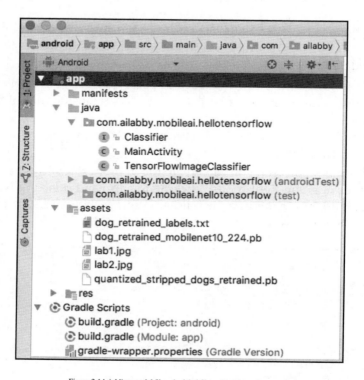

Figure 2.14 Adding model files, the label file and test images to assets

6. Hold down the option button, drag and drop `TensorFlowImageClassifier.java` and `Classifier.java` from `tensorflow/examples/android/src/org/tensorflow/demo` to your project's Java folder, as shown:

Chapter 2

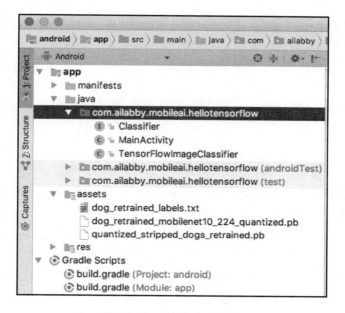

Figure 2.15 Adding TensorFlow classifier files to the project

7. Open `MainActivity`, first create the constants related to the retrained MobileNet model – the input image size, node names, model filename and label filename:

```
private static final int INPUT_SIZE = 224;
private static final int IMAGE_MEAN = 128;
private static final float IMAGE_STD = 128;
private static final String INPUT_NAME = "input";
private static final String OUTPUT_NAME = "final_result";

private static final String MODEL_FILE =
"file:///android_asset/dog_retrained_mobilenet10_224.pb";
private static final String LABEL_FILE =
"file:///android_asset/dog_retrained_labels.txt";
private static final String IMG_FILE = "lab1.jpg";
```

[61]

Classifying Images with Transfer Learning

8. Now, inside the `onCreate` method, first create a `Classifier` instance:

```
Classifier classifier = TensorFlowImageClassifier.create(
            getAssets(),
            MODEL_FILE,
            LABEL_FILE,
            INPUT_SIZE,
            IMAGE_MEAN,
            IMAGE_STD,
            INPUT_NAME,
            OUTPUT_NAME);
```

Then read our test image from the assets folder, resize it as specified by the model, and call the inference method, `recognizeImage`:

```
Bitmap bitmap = BitmapFactory.decodeStream(getAssets().open(IMG_FILE));
Bitmap croppedBitmap = Bitmap.createScaledBitmap(bitmap, INPUT_SIZE, INPUT_SIZE, true);
final List<Classifier.Recognition> results =
classifier.recognizeImage(croppedBitmap);
```

For simplicity, we don't have any UI-related code added to the Android app, but you can set a breakpoint at the line after you get the results, and debug run the app; you'll see the results as shown in the following screenshot:

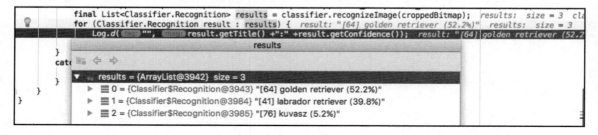

Figure 2.16 Recognition results using the MobileNet retrained model

If you switch to use the Inception v3 retrained model by changing `MODEL_FILE` to `quantized_stripped_dogs_retrained.pb`, `INPUT_SIZE` to `299`, and `INPUT_NAME` to `Mul`, then debug the app and you will get the results as shown here:

```
final long startTime = SystemClock.uptimeMillis();   startTime: 48300981
try {
    Bitmap bitmap = BitmapFactory.decodeStream(getAssets().open(IMG_FILE));   bitmap: ""
    croppedBitmap = Bitmap.createScaledBitmap(bitmap, INPUT_SIZE, INPUT_SIZE,  filter: true);
    final List<Classifier.Recognition> results = classifier.recognizeImage(croppedBitmap);
    for (Classifier.Recognition result : results) {   result: "[64] golden retriever (23.8%
        Log.d("", "", result.getTitle() +":" +result.getConfidence());   result:
    }
}
catch
```

results = {ArrayList@3941} size = 2
- 0 = {Classifier$Recognition@3942} "[41] labrador retriever (57.3%)"
- 1 = {Classifier$Recognition@3991} "[64] golden retriever (23.8%)"

Figure 2.17 Recognition results using the Inception v3 retrained model

Now that you see how to add TensorFlow and retrained models to your own iOS and Android apps, it shouldn't be that difficult if you want to add the non-TensorFlow related features, such as using your phone's camera to take a dog picture and recognize its breed.

Summary

In this chapter, we first gave a quick overview on what is transfer learning and why we can and should use it to retrain pretrained deep learning image classification models. Then we presented detailed information on how to retrain an Inception v3-based model and a MobileNet model so we can better understand and recognize our best friends. After that, we first showed how to use the retrained models in the TensorFlow sample iOS and Android apps, then gave step-by-step tutorials on how to add TensorFlow to your own iOS app, both Objective-C and Swift-based, and your own Android app.

Now that we have our best friends, with a few nice and clean tricks, covered, we know there are a lot of other guys out there, good or bad. In the next chapter, we'll learn how to be smarter, how to recognize all the interesting objects in a picture, and locate them, on your smart phone anytime, anywhere.

3
Detecting Objects and Their Locations

Object detection goes one step further than the image classification discussed in the previous chapter. Image classification just returns a label of class for an image, while object detection returns a list of objects identified in an image, along with a bounding box for each identified object. Modern object detection algorithms use deep learning to build models that can be used to detect and localize all kinds of objects in a single image. In the past few years, faster and more accurate object detection algorithms came one after another, and in June 2017, Google released the TensorFlow Object Detection API that incorporates several leading object detection algorithms.

In this chapter, we'll first give a quick overview of object detection: the process of creating an effective deep learning model for object detection and then using the model for inference. Then, we'll discus in detail how the TensorFlow Object Detection API works, how you can use several of its models for inference, and how you can retrain them with your own dataset. After that, we'll show you how you can use the pre-trained object detection models as well as the retrained models in your iOS app. We'll cover some powerful tips that let you build a custom TensorFlow iOS library manually to fix the problem with using the TensorFlow pod; this will help you get ready for dealing with any TensorFlow-supported models covered in the rest of the book. In this chapter, we won't provide an Android example app for object detection, as the TensorFlow source code already comes with a good example of doing so, using both a TensorFlow Object Detection pre-trained model as well as the YOLO model, which we'll cover last in this chapter. We'll show you how to use another leading object detection model, YOLO v2, in your iOS app. In summary, we're going to cover the following topics in this chapter:

- Object detection: a quick overview
- Setting up the TensorFlow Object Detection API

- Retraining SSD-MobileNet and Faster RCNN models
- Using object detection models in iOS
- Using YOLO2: another object detection model

Object detection–a quick overview

Since the breakthrough in neural network in 2012, when a deep CNN model called **AlexNet** won the annual ImageNet visual recognition challenge by dramatically reducing the error rate, many researchers in computer vision and natural language processing have started to take advantage of the power of deep learning models. Modern deep-learning-based object detections are all based on CNN and built on top of pre-trained models such as AlexNet, Google Inception, or another popular net VGG. These CNNs typically have trained millions of parameters and can convert an input image to a set of features that can be further used for tasks such as image classification, which we covered in the previous chapter, and object detection, among other computer-vision-related tasks.

In 2014, a state-of-the-art object detector that retrained AlexNet with a labeled object detection dataset, called RCNN (Regions with CNN features), was proposed, and it offered a big improvement in accuracy over traditional detection methods. RCNN combines a technique called region proposals, which generates about 2,000 possible region candidates, and runs a CNN on each of those regions for classification and bounding box predictions. It then merges those results to generate the detection result. The training process of RCNN is pretty complicated and takes several days, and the inference speed is also quite slow, taking almost a minute on an image on a GPU.

Since RCNN was proposed, better-performing object detection algorithms came one after another: Fast RCNN, Faster RCNN, YOLO (You Only Look Once), SSD (Single Shot MultiBox Detector), and YOLO v2.

Andrej Karpathy wrote a good introduction to RCNN, "Playing around with RCNN, State of the Art Object Detector" in 2014 (`https://cs.stanford.edu/people/karpathy/rcnn`). There's a nice video lecture, "Spatial Localization and Detection", as part of Stanford's CS231n course by Justin Johnson, on object detection, with details on RCNN, Fast RCNN, Faster RCNN, and YOLO. SSD is described in detail at `https://github.com/weiliu89/caffe/tree/ssd`. And the cool YOLO2 website is `https://pjreddie.com/darknet/yolo`.

Fast RCNN significantly improves both the training process and inference time (10 hours of training and 2.x seconds of inference) by first applying a CNN on the whole input image, instead of thousands of proposed regions, and then dealing with region proposals. Faster RCNN further improves the inference speed to real time (0.2 seconds) by using a region proposal network so that after training, the time-consuming region proposal process is no longer needed.

Unlike the RCNN family of detection, SSD and YOLO are both single-shot methods, meaning they apply a single CNN to the full input image without using region proposals and region classification. This makes both methods very fast, and their Mean Average Precision (mAPs) are about 80%, outperforming that of Faster RCNN.

If this is the first time you heard of these methods, you probably would feel a bit lost. But as a developer interested in powering up your mobile apps with AI, you don't need to understand all the details in setting up the deep neural network architectures and training the models for object detection; you should just be able to know how to use and, if needed, retrain pre-trained models and how to use the pre-trained or retrained models in your iOS and Android apps.

If you're really interested in deep learning research and want to know all the details of how each detector works to decide which one to use, you should definitely read the papers of each method and try to reproduce the training process on your own. It'll be a long but rewarding road. But if you want to take Andrej Karpathy's advice, "don't be a hero" (search on YouTube for "deep learning for computer vision Andrej"), then you can "take whatever works best, download a pre-trained model, potentially add/delete some parts of it, and fine-tune it on your app," which is also the approach we'll use here.

Before we start looking at what works best with TensorFlow, let's have a quick note on datasets. There are three main datasets used for training in object detection: PASCAL VOC (http://host.robots.ox.ac.uk/pascal/VOC), ImageNet (http://image-net.org), and Microsoft COCO (http://cocodataset.org), and the number of classes they have are 20, 200, and 80, respectively. Most of the pre-trained models the TensorFlow Object Detection API currently supports are trained on the 80-class MS COCO dataset (for a complete list of the pre-trained models and the datasets they're trained on, see https://github.com/tensorflow/models/blob/master/research/object_detection/g3doc/detection_model_zoo.md).

Although we won't do the training from scratch, you'll see the frequent mention of the PASCAL VOC or MS COCO data format, as well as the 20 or 80 common classes they cover, in retraining or use of the trained models. In the last section of this chapter, we'll try both a VOC-trained YOLO model and a COCO-trained one.

Setting up the TensorFlow Object Detection API

The TensorFlow Object Detection API is documented in detail at its official site https://github.com/tensorflow/models/tree/master/research/object_detection, and you should definitely check out its "Quick Start: Jupyter notebook for off-the-shelf inference" guide for a quick idea of how to use a good pre-trained model for detection in Python. But the documentation there is spread out over many different pages, making it sometimes difficult to follow. In this and the next sections, we'll streamline the official documentation by reorganizing the important details documented in many different places and adding more examples and code explanations, and offer two step-by-step tutorials on:

1. How to set up the API and use its pre-trained models for off-the-shelf inference
2. How to retrain pre-trained models with the API for more specific detection tasks

Quick installation and example

Perform the following steps to install and run an object detection inference:

1. In your TensorFlow source root you created in Chapter 1,*Getting Started with Mobile TensorFlow*, get the TensorFlow models repo, which contains the TensorFlow Object Detection API as one of its research models:

    ```
    git clone https://github.com/tensorflow/models
    ```

2. Install the `matplotlib`, `pillow`, `lxml`, and `jupyter` libraries. On Ubuntu or Mac, you can run:

    ```
    sudo pip install pillow
    sudo pip install lxml
    sudo pip install jupyter
    sudo pip install matplotlib
    ```

3. Go to the `models/research` directory, then run the following command:

   `protoc object_detection/protos/*.proto --python_out=.`

   This will compile all the Protobufs in the `object_detection/protos` directory to make the TensorFlow Object Detection API happy. Protobuf, or Protocol Buffer, is an automated way to serialize and retrieve structured data, and it's lightweight and more efficient than XML. All you need to do is write a .proto file that describes the structure of your data, then use protoc, the proto compiler, to generate code that automatically parses and encodes the protobuf data. Notice the `--python_out` parameter specifies the language of the generated code. In a later section of this chapter, when we discuss how to use a model in iOS, we'll use the protoc compiler with `--cpp_out` so the generated code is in C++. For complete documentation on Protocol Buffers, see `https://developers.google.com/protocol-buffers`.

4. Still inside models/research, run `export PYTHONPATH=$PYTHONPATH:`pwd`:`pwd`/slim` and then `python object_detection/builders/model_builder_test.py` to verify everything works.

5. Launch the `jupyter notebook` command and open `http://localhost:8888` in a browser. Click `object_detection` first, then select the `object_detection_tutorial.ipynb` notebook and run the demo cell by cell.

Using pre-trained models

Let's now see the main components of using pre-trained TensorFlow object detection models for inference in the Python notebook. First, some key constants are defined:

```
MODEL_NAME = 'ssd_mobilenet_v1_coco_2017_11_17'
MODEL_FILE = MODEL_NAME + '.tar.gz'
DOWNLOAD_BASE = 'http://download.tensorflow.org/models/object_detection/'
PATH_TO_CKPT = MODEL_NAME + '/frozen_inference_graph.pb'
PATH_TO_LABELS = os.path.join('data', 'mscoco_label_map.pbtxt')
NUM_CLASSES = 90
```

The notebook code downloads and uses a pre-trained object detection model, `ssd_mobilenet_v1_coco_2017_11_17` (built with the SSD method, which we talked briefly in the previous section, on top of the MobileNet CNN model, which we covered in the previous chapter). A complete list of pre-trained models supported by the TensorFlow Object Detection API is at the TensorFlow detection model zoo: https://github.com/tensorflow/models/blob/master/research/object_detection/g3doc/detection_model_zoo.md, and most of them are trained with the MS COCO dataset. The exact model used for inference is the `frozen_inference_graph.pb` file (in the downloaded `ssd_mobilenet_v1_coco_2017_11_17.tar.gz` file), which is used for off-the-shelf inference as well as retraining.

The `mscoco_label_map.pbtxt` label file, located in `models/research/object_detection/data/mscoco_label_map.pbtxt`, has 90 (NUM_CLASSES) items for the types of objects that the `ssd_mobilenet_v1_coco_2017_11_17` model can detect. The first and last two items of it are:

```
item {
  name: "/m/01g317"
  id: 1
  display_name: "person"
}
item {
  name: "/m/0199g"
  id: 2
  display_name: "bicycle"
}
...
item {
  name: "/m/03wvsk"
  id: 89
  display_name: "hair drier"
}
item {
  name: "/m/012xff"
  id: 90
  display_name: "toothbrush"
}
```

We talked about Protobuf in step 3 earlier, and the proto file that describes the data in `mscoco_label_map.pbtxt` is `string_int_label_map.proto`, located in `models/research/object_detection/protos`, with the following content:

```
syntax = "proto2";
package object_detection.protos;
message StringIntLabelMapItem {
  optional string name = 1;
  optional int32 id = 2;
  optional string display_name = 3;
};

message StringIntLabelMap {
  repeated StringIntLabelMapItem item = 1;
};
```

So basically, the protoc compiler creates code based on `string_int_label_map.proto` and the code can then be used to efficiently serialize the data in `mscoco_label_map.pbtxt`. Later, when a CNN detects an object and returns an integer ID, it can be converted to the `name` or `display_name` for humans to read.

After the model is downloaded, unzipped, and loaded into memory, the label map file also gets loaded, and some test images, located in `models/research/object_detection/test_images` where you can add any of your own test images for detection test, are ready to be used. Next, appropriate input and output tensors are defined:

```
with detection_graph.as_default():
  with tf.Session(graph=detection_graph) as sess:
    image_tensor = detection_graph.get_tensor_by_name('image_tensor:0')
    detection_boxes = detection_graph.get_tensor_by_name('detection_boxes:0')
    detection_scores = detection_graph.get_tensor_by_name('detection_scores:0')
    detection_classes = detection_graph.get_tensor_by_name('detection_classes:0')
    num_detections = detection_graph.get_tensor_by_name('num_detections:0')
```

Again, if you're wondering where those input and output tensor names come from in the SSD model downloaded and saved in `models/research/object_detection/ssd_mobilenet_v1_coco_2017_11_17/frozen_inference_graph.pb`, you can use the following code in iPython to find out:

```
import tensorflow as tf
g=tf.GraphDef()
```

```
g.ParseFromString(open("object_detection/ssd_mobilenet_v1_coco_2017_11_17/f
rozen_inference_graph.pb","rb").read())
x=[n.name for n in g.node]
x[-4:]
x[:5]
The last two statements will return:
[u'detection_boxes',
 u'detection_scores',
 u'detection_classes',
 u'num_detections']
and
[u'Const', u'Const_1', u'Const_2', u'image_tensor', u'ToFloat']
```

Another way is to use the summarize graph tool we described in the previous chapter:

```
bazel-bin/tensorflow/tools/graph_transforms/summarize_graph --in_graph=
models/research/object_detection/ssd_mobilenet_v1_coco_2017_11_17/frozen_in
ference_graph.pb
```

This will generate the following output:

```
Found 1 possible inputs: (name=image_tensor, type=uint8(4),
shape=[?,?,?,3])
No variables spotted.
Found 4 possible outputs: (name=detection_boxes, op=Identity)
(name=detection_scores, op=Identity (name=detection_classes, op=Identity)
(name=num_detections, op=Identity)
```

After each test image is loaded, the actual detection runs:

```
image = Image.open(image_path)
image_np = load_image_into_numpy_array(image)
image_np_expanded = np.expand_dims(image_np, axis=0)
(boxes, scores, classes, num) = sess.run(
    [detection_boxes, detection_scores, detection_classes, num_detections],
    feed_dict={image_tensor: image_np_expanded})
```

Finally, the detected results are visualized using the `matplotlib` library. If you use the default two test images that come with the `tensorflow/models` repo, you'll see the results in Figure 3.1:

Figure 3.1 Detected objects with bounding boxes and confidence scores

Detecting Objects and Their Locations

In the *Using object detection models in iOS* section, we'll see how to use the same model and draw the same detected result on an iOS device.

You can also test other pre-trained models in the Tensorflow detection model zoo mentioned earlier. For example, if you use the `faster_rcnn_inception_v2_coco` model by replacing `MODEL_NAME = 'ssd_mobilenet_v1_coco_2017_11_17'` in the `object_detection_tutorial.ipynb` notebook with `MODEL_NAME = 'faster_rcnn_inception_v2_coco_2017_11_08'`, which you get from the URL in the TensorFlow detection model zoo page, or `MODEL_NAME = 'faster_rcnn_resnet101_coco_2017_11_08'`, you can see similar detection results with two other Faster RCNN-based models, but they take longer.

Also, using the `summarize_graph` tool on the two `faster_rcnn` models generates the same info on input and output:

```
Found 1 possible inputs: (name=image_tensor, type=uint8(4),
shape=[?,?,?,3])
Found 4 possible outputs: (name=detection_boxes, op=Identity)
(name=detection_scores, op=Identity) (name=detection_classes, op=Identity)
(name=num_detections, op=Identity)
```

Generally, MobileNet-based models are the fastest, but less accurate (with smaller mAP value) than other large Inception- or Resnet-CNN-based models. By the way, the sizes of the downloaded `ssd_mobilenet_v1_coco`, `faster_rcnn_inception_v2_coco_2017_11_08`, and `faster_rcnn_resnet101_coco_2017_11_08` files are 76 MB, 149 MB, and 593 MB, respectively. As we'll see later, on mobile devices, MobileNet-based models, such as `ssd_mobilenet_v1_coco`, run a lot faster and sometimes a large model, such as `faster_rcnn_resnet101_coco_2017_11_08`, would simply crash on an older iPhone. Hopefully, the problems you have can be solved using a MobileNet-based model, or a retrained MobileNet model, or a future version of `ssd_mobilenet` that will certainly offer even better accuracy, although the v1 of `ssd_mobilenet` is already good enough for many use cases.

Retraining SSD-MobileNet and Faster RCNN models

The pre-trained TensorFlow Object Detection models certainly work well for some problems. But sometimes, you may need to use your own annotated dataset (with bounding boxes around objects or parts of objects that are of particular interest to you) and retrain an existing model so it can more accurately detect a different set of object classes.

We'll use the same Oxford-IIIT Pets dataset, as documented in the TensorFlow Object Detection API site, to retrain two existing models on your local machine, instead of using Google Cloud covered in the documentation. We'll also add an explanation for each step when needed. The following is the step-by-step guide on how to retrain a TensorFlow object detection model using the Oxford Pets dataset:

1. In a Terminal window, preferably on our GPU-powered Ubuntu to make the retraining faster, `cd models/research first`, then run the following commands to download the dataset (`images.tar.gz` is about 800MB and `annotations.tar.gz` is 38MB):

    ```
    wget http://www.robots.ox.ac.uk/~vgg/data/pets/data/images.tar.gz
    wget http://www.robots.ox.ac.uk/~vgg/data/pets/data/annotations.tar.gz
    tar -xvf images.tar.gz
    tar -xvf annotations.tar.gz
    ```

2. Run the following command to convert the dataset to the TFRecords format:

    ```
    python object_detection/dataset_tools/create_pet_tf_record.py \
        --label_map_path=object_detection/data/pet_label_map.pbtxt \
        --data_dir=`pwd` \
        --output_dir=`pwd`
    ```

    This command will generate two TFRecord files, named `pet_train_with_masks.record` (268MB) and `pet_val_with_masks.record` (110MB), in the `models/research` directory. TFRecords is an interesting binary format that includes all the data a TensorFlow app can use for training or validation, and is the required file format if you want to retrain your own dataset with the TensorFlow Object Detection API.

3. Download and unzip the `ssd_mobilenet_v1_coco` model and the `faster_rcnn_resnet101_coco` model to the `models/research` directory if you haven't done so in the previous section when testing the object detection notebook:

```
wget http://storage.googleapis.com/download.tensorflow.org/models/object_detection/ssd_mobilenet_v1_coco_2017_11_17.tar.gz
tar -xvf ssd_mobilenet_v1_coco_2017_11_17.tar.gz
wget http://storage.googleapis.com/download.tensorflow.org/models/object_detection/faster_rcnn_resnet101_coco_11_06_2017.tar.gz
tar -xvf faster_rcnn_resnet101_coco_11_06_2017.tar.gz
```

4. Replace five occurrences of PATH_TO_BE_CONFIGURED in the `object_detection/samples/configs/faster_rcnn_resnet101_pets.config` file, so they become:

```
fine_tune_checkpoint: "faster_rcnn_resnet101_coco_11_06_2017/model.ckpt"
...
train_input_reader: {
tf_record_input_reader {
input_path: "pet_train_with_masks.record"
}
label_map_path: "object_detection/data/pet_label_map.pbtxt"
}
eval_input_reader: {
tf_record_input_reader {
input_path: "pet_val_with_masks.record"
}
label_map_path: "object_detection/data/pet_label_map.pbtxt"
...
}
```

The `faster_rcnn_resnet101_pets.config` file is used to specify the locations of the model's checkpoint file, which contains the model's trained weights, the TFRecords files for training and validation, which are generated in step 2, and the labeled items of the 37 classes of pets to be detected. The first and last items of `object_detection/data/pet_label_map.pbtxt` are as follows:

```
item {
id: 1
name: 'Abyssinian'
}
...
```

```
item {
id: 37
name: 'yorkshire_terrier'
}
```

5. Similarly, change five occurrences of PATH_TO_BE_CONFIGURED in the `object_detection/samples/configs/ssd_mobilenet_v1_pets.config` file, so they become:

```
fine_tune_checkpoint: "object_detection/ssd_mobilenet_v1_coco_2017_11_17/model.ckpt"
train_input_reader: {
tf_record_input_reader {
input_path: "pet_train_with_masks.record"
}
label_map_path: "object_detection/data/pet_label_map.pbtxt"
}
eval_input_reader: {
tf_record_input_reader {
input_path: "pet_val_with_masks.record"
}
label_map_path: "object_detection/data/pet_label_map.pbtxt"
...
}
```

6. Create a new `train_dir_faster_rcnn` directory, then run the retraining command:

```
python object_detection/train.py \
    --logtostderr \
    --pipeline_config_path=object_detection/samples/configs/faster_rcnn_resnet101_pets.config \
    --train_dir=train_dir_faster_rcnn
```

On a GPU-powered system, it takes less than 25,000 steps of training to reach the loss of 0.2 or so from the initial loss of 5.0:

```
tensorflow/core/common_runtime/gpu/gpu_device.cc:1030] Found device 0 with properties:
 name: GeForce GTX 1070 major: 6 minor: 1 memoryClockRate(GHz): 1.7845
 pciBusID: 0000:01:00.0
 totalMemory: 7.92GiB freeMemory: 7.44GiB
INFO:tensorflow:global step 1: loss = 5.1661 (15.482 sec/step)
INFO:tensorflow:global step 2: loss = 4.6045 (0.927 sec/step)
INFO:tensorflow:global step 3: loss = 5.2665 (0.958 sec/step)
```

Detecting Objects and Their Locations

```
...
INFO:tensorflow:global step 25448: loss = 0.2042 (0.372 sec/step)
INFO:tensorflow:global step 25449: loss = 0.4230 (0.378 sec/step)
INFO:tensorflow:global step 25450: loss = 0.1240 (0.386 sec/step)
```

7. Press *Ctrl + C* to end the running of the retraining script above after about 20,000 steps (for about 2 hours). Create a new `train_dir_ssd_mobilenet` directory, then run:

```
python object_detection/train.py \
    --logtostderr \
    --
pipeline_config_path=object_detection/samples/configs/ssd_mobilenet_v1_pets.config \
    --train_dir=train_dir_ssd_mobilenet
```

The training results should look like the following:

```
INFO:tensorflow:global step 1: loss = 136.2856 (23.130 sec/step)
INFO:tensorflow:global step 2: loss = 126.9009 (0.633 sec/step)
INFO:tensorflow:global step 3: loss = 119.0644 (0.741 sec/step)
...
INFO:tensorflow:global step 22310: loss = 1.5473 (0.460 sec/step)
INFO:tensorflow:global step 22311: loss = 2.0510 (0.456 sec/step)
INFO:tensorflow:global step 22312: loss = 1.6745 (0.461 sec/step)
```

You can see the retraining of the `SSD_Mobilenet` model both starts and ends with a bigger loss than that of the `Faster_RCNN` model.

8. Terminate the preceding retraining script after about 20,000 training steps. Then create a new `eval_dir` directory and run the evaluation script:

```
python object_detection/eval.py \
    --logtostderr \
    --
pipeline_config_path=object_detection/samples/configs/faster_rcnn_resnet101_pets.config \
    --checkpoint_dir=train_dir_faster_rcnn \
    --eval_dir=eval_dir
```

9. Open another Terminal window, cd to the TensorFlow root, then `models/research`, and run `tensorboard --logdir=..` In a browser, open `http://localhost:6006`, and you'll see the loss graph, like in Figure 3.2:

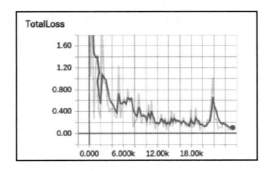

Figure 3.2 Total loss trend when retraining an object detection model

You'll also see some evaluation results, as in Figure 3.3:

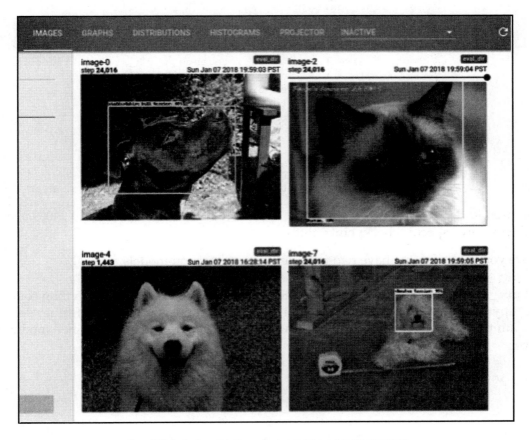

Figure 3.3 Evaluation image detection results when retraining an object detection model

10. Similarly, you can run the evaluation script for the `SSD_MobileNet` model and then use TensorBoard to view its loss trend and evaluation image results:

    ```
    python object_detection/eval.py \
        --logtostderr \
        --
    pipeline_config_path=object_detection/samples/configs/ssd_mobilenet_v1_pets.config \
        --checkpoint_dir=train_dir_ssd_mobilenet \
        --eval_dir=eval_dir_mobilenet
    ```

11. You can generate the retrained graphs using the following commands:

    ```
    python object_detection/export_inference_graph.py \
        --input_type image_tensor \
        --pipeline_config_path
    object_detection/samples/configs/ssd_mobilenet_v1_pets.config \
        --trained_checkpoint_prefix
    train_dir_ssd_mobilenet/model.ckpt-21817 \
        --output_directory output_inference_graph_ssd_mobilenet.pb

    python object_detection/export_inference_graph.py \
        --input_type image_tensor \
        --pipeline_config_path
    object_detection/samples/configs/faster_rcnn_resnet101_pets.config \
        --trained_checkpoint_prefix
    train_dir_faster_rcnn/model.ckpt-24009 \
        --output_directory output_inference_graph_faster_rcnn.pb
    ```

You need to replace the `--trained_checkpoint_prefix` value (21817 and 24009 above) with your own specific checkpoint values.

There you go—you now have two retrained object detection models, `output_inference_graph_ssd_mobilenet.pb` and `output_inference_graph_faster_rcnn.pb`, ready to be used in your Python code (for the Jupyter notebook in the last section) or your mobile apps. Without any further delay, let's jump to the mobile world and see how to use the pre-trained and retrained models we have.

Using object detection models in iOS

In the previous chapter, we showed you how to use the TensorFlow-experimental pod to quickly add TensorFlow to your iOS app. The TensorFlow-experimental pod works fine for models such as Inception and MobileNet or their retrained models. But if you use the TensorFlow-experimental pod, at least as of this writing (January 2018), with the `SSD_MobileNet` model, you're likely to get the following error message when loading the `ssd_mobilenet` graph file:

```
Could not create TensorFlow Graph: Not found: Op type not registered
'NonMaxSuppressionV2'
```

Unless the TensorFlow-experimental pod gets updated to include **op not registered here**, the only way to fix these problems is to create the custom TensorFlow iOS library by building it from the TensorFlow source, and that's why we showed you in Chapter 1, *Getting Started with Mobile TensorFlow*, how to get and set up TensorFlow from its source. Let's look at the steps to build your own TensorFlow iOS library and use it to create a new iOS app with TensorFlow support.

Building TensorFlow iOS libraries manually

Simply perform the following steps to build your own TensorFlow iOS libraries:

1. Open a new Terminal on your Mac, `cd` to your TensorFlow source root, which is `~/tensorflow-1.4.0`, if you unzipped the TensorFlow 1.4 source zip to your home directory.
2. Run the `tensorflow/contrib/makefile/build_all_ios.sh` command, which takes from 20 minutes to about an hour, depending on your Mac speed, to finish. After the build process completes successfully, you'll have three libraries created:

   ```
   tensorflow/contrib/makefile/gen/protobuf_ios/lib/libprotobuf-lite.a
   tensorflow/contrib/makefile/gen/protobuf_ios/lib/libprotobuf.a
   tensorflow/contrib/makefile/gen/lib/libtensorflow-core.a
   ```

The first two libraries handle the protobuf data we talked about earlier. The last library is the iOS universal static library.

> If you run the app, after completing the following steps, and encounter an error shown in your Xcode console, "Invalid argument: No OpKernel was registered to support Op 'Less' with these attrs. Registered devices: [CPU], Registered kernels: device='CPU'; T in [DT_FLOAT]", you'll need to change the `tensorflow/contrib/makefile/Makefile` file before taking step 2 here (see the section Building custom TensorFlow library for iOS in `Chapter 7`, *Recognizing Drawing with CNN and LSTM*, for more details). You may not see the error when using newer versions of TensorFlow.

Using TensorFlow iOS libraries in an app

To use the libraries in your own app, do the following:

1. In Xcode, click **File** | **New** | **Project...**, select **Single View App**, then enter **TFObjectDetectionAPI** as **Product Name** and choose **Objective-C** for Language (if you want to use Swift, see the previous chapter on how to add TensorFlow to a Swift-based iOS app and make the necessary changes pictured here), then select the location for the project and hit Create.

2. In the `TFObjectDetectionAPI` project, click the **PROJECT Name** and under **Build Settings**, hit **+** and **Add User-Defined Setting**, then enter TENSORFLOW_ROOT for the path of your TensorFlow source root (`$HOME/tensorflow-1.4` for example), as shown in Figure 3.4. This user-defined setting will be used in other settings to make it easy to change project settings later if you want to refer to a newer TensorFlow source:

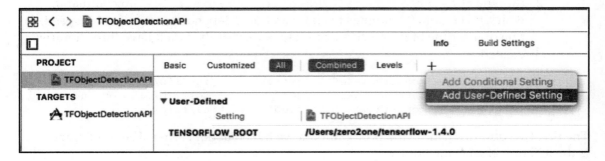

Figure 3.4 Adding a TENSORFLOW_ROOT user-defined setting

3. Click the target and under Build Settings, search for "other linker flags". Add the following value to it:

   ```
   -force_load
   $(TENSORFLOW_ROOT)/tensorflow/contrib/makefile/gen/lib/libtensorflow-core.a
   $(TENSORFLOW_ROOT)/tensorflow/contrib/makefile/gen/protobuf_ios/lib/libprotobuf.a
   $(TENSORFLOW_ROOT)/tensorflow/contrib/makefile/gen/protobuf_ios/lib/libprotobuf-lite.a
   $(TENSORFLOW_ROOT)/tensorflow/contrib/makefile/downloads/nsync/builds/lipo.ios.c++11/nsync.a
   ```

   The first –force_load is needed because it makes sure that the C++ constructors TensorFlow requires will be linked, otherwise you can still build and run the app but will encounter an error about sessions unregistered.

   The last library is for nsync, a C library that exports mutex and other synchronization methods (`https://github.com/google/nsync`). It's introduced in newer TensorFlow versions.

4. Search for "header search paths", add the following value to it:

   ```
   $(TENSORFLOW_ROOT)
   $(TENSORFLOW_ROOT)/tensorflow/contrib/makefile/downloads/protobuf/src  $(TENSORFLOW_ROOT)/tensorflow/contrib/makefile/downloads
   $(TENSORFLOW_ROOT)/tensorflow/contrib/makefile/downloads/eigen
   $(TENSORFLOW_ROOT)/tensorflow/contrib/makefile/gen/proto
   ```

Detecting Objects and Their Locations

After this, you'll see something like Figure 3.5:

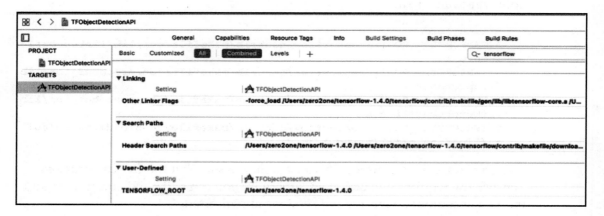

Figure 3.5 Adding all TensorFlow-related build settings for the target

5. In the target's Build Phases, add the Accelerate framework in the **Link Binary with Libraries**, as in Figure 3.6:

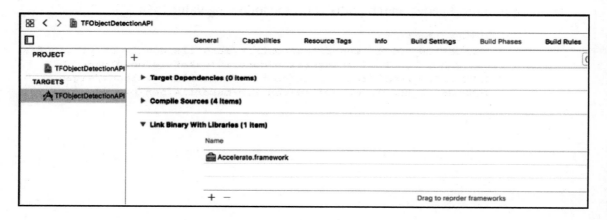

Figure 3.6 Adding the Accelerate framework

Chapter 3

6. Go back to the Terminal you used to build the TensorFlow iOS libraries, find the following two lines of code in `tensorflow/core/platform/default/mutex.h`:

   ```
   #include "nsync_cv.h"
   #include "nsync_mu.h"
   ```

   Then change them to this:

   ```
   #include "nsync/public/nsync_cv.h"
   #include "nsync/public/nsync_mu.h"
   ```

That's all it takes to add TensorFlow, your manually-built TensorFlow libraries, to your iOS app.

Loading the TensorFlow object detection models in your app with the manual TensorFlow libraries built from a later version of TensorFlow, such as 1.4, won't have the error you may see when using the TensorFlow-experimental pod or the manual libraries built from an earlier version. This is because a file named `tf_op_files.txt`, located in `tensorflow/contrib/makefile` and used to define what operations should be built and included for the TensorFlow library, has more operations defined in TensorFlow 1.4 than earlier versions. For example, the `tf_op_files.txt` file in TensorFlow 1.4 has a line, `tensorflow/core/kernels/non_max_suppression_op.cc`, that defines the `NonMaxSuppressionV2` operation, and that's why the library we manually built has the operation defined, preventing the error "Could not create TensorFlow Graph: Not found: Op type not registered 'NonMaxSuppressionV2'" that we'd see if using the TensorFlow pod from happening. In the future, if you encounter a similar "Op type not registered" error, you can fix it by adding the right source code file that defines the operation to the `tf_op_files.txt` file and then run the `build_all_ios.sh` again to create a new `libtensorflow-core.a` file.

[85]

Adding an object detection feature to an iOS app

Now perform the following steps to add the model files, label file, and code to the app and run to see the object detection in action:

1. Drag and drop the three object detection model graphs, `ssd_mobilenet_v1_frozen_inference_graph.pb`, `faster_rcnn_inceptionv2_frozen_inference_graph.pb`, and `faster_rcnn_resnet101_frozen_inference_graph.pb`, we downloaded (and renamed here for clarification) in the previous section, as well as the `mscoco_label_map.pbtxt` label map file and a couple of test images, to the `TFObjectDetectionAPI` project.
2. Add the `ios_image_load.mm` and its `.h` files from the TensorFlow iOS sample simple app or the iOS app we created in the last chapter to the project.
3. Download the Protocol Buffers release 3.4.0 at https://github.com/google/protobuf/releases (the `protoc-3.4.0-osx-x86_64.zip` file on Mac). The exact 3.4.0 version is required to work with the TensorFlow 1.4 library and later versions of TensorFlow may require a later version of the Protocol Buffers.
4. Assuming the downloaded file is unzipped into `~/Downloads` directory, open a Terminal window and run the following commands:

   ```
   cd <TENSORFLOW_ROOT>/models/research/object_detection/protos

   ~/Downloads/protoc-3.4.0-osx-x86_64/bin/protoc
   string_int_label_map.proto --
   cpp_out=<path_to_your_TFObjectDetectionAPI_project>, the same
   location as your code files and the three graph files.
   ```

5. After the protoc compiler command finishes, you'll see two files generated in your project's source directory: `string_int_label_map.pb.cc` and `string_int_label_map.pb.h`. Add the two files to the project in Xcode.
6. In Xcode, rename `ViewController.m` to `ViewController.mm`, as we did in the previous chapter, then similar to the `HelloTensorFlow` app's `ViewController.mm` in Chapter 2, *Classifying Images with Transfer Learning*, add in the tapped handler three UIAlertAction's for the three object detection models we have added to the project and will test. The complete project files should now look like Figure 3.7:

Chapter 3

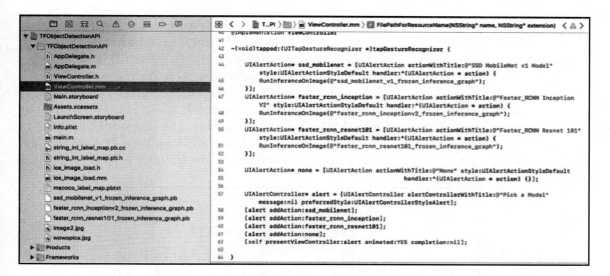

Figure 3.7 The TFObjectDetection API project files

7. Continue to add the rest of code in `ViewController.mm`. In `viewDidLoad`, add the code that programmatically creates a new `UIImageView` to display the test image first and the detected results after a specific model is selected to run on a the test image, then add the following function implementations:

```
NSString* FilePathForResourceName(NSString* name, NSString* extension)
int LoadLablesFile(const string pbtxtFileName,
object_detection::protos::StringIntLabelMap *imageLabels)
string GetDisplayName(const
object_detection::protos::StringIntLabelMap* labels, int index)
Status LoadGraph(const string& graph_file_name,
std::unique_ptr<tensorflow::Session>* session)
void DrawTopDetections(std::vector<Tensor>& outputs, int
image_width, int image_height)
void RunInferenceOnImage(NSString *model)
```

We'll explain these function implementations after the next step, and you can get all the source code in the ch3/ios folder of the book's source code repo.

[87]

8. Run the app in an iOS simulator or device. You'll first see an image on the screen. Tap anywhere and you'll see a dialog box asking you to select a model. Select the `SSD MobileNet` model and it takes about one second in the simulator and five seconds on iPhone 6 to get the detection results drawn on the image. The Faster RCNN Inception V2 takes longer (about 5 seconds in the simulator and over 20 on iPhone 6); this model also is more accurate than the `SSD MobileNet`, catching one dog object missed by the `SSD MobileNet` model. The last model, Faster RCNN Resnet 101, takes close to 20 seconds in the iOS simulator, but crashes on an iPhone 6 due to its size. Figure 3.8 summarizes the running results:

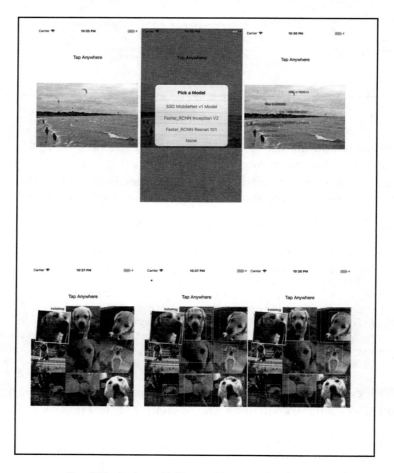

Figure 3.8 Running the app with different models and showing detection results

Back to the functions in step 7, the `FilePathForResourceName` function is a helper function used to return the file path for a resource: the `mscoco_label_map.pbtxt` file that defines the ID, internal name, and display name of 90 object classes to be detected, the model graph files, and the test images. Its implementation is the same as what we see in the `HelloTensorFlow` app in the previous chapter.

The `LoadLablesFile` and `GetDisplayName` functions use the Google Protobuf API to load and parse the `mscoco_label_map.pbtxt` file, and return the display name for the ID of a detected object.

LoadGraph attempts to load one of the three user-selected model files and returns the loading status.

The two key functions are `RunInferenceOnImage` and `DrawTopDetections`. As we saw in the *Setting up the TensorFlow Object Detection API* section, the summarize_graph tool shows the following info on the three pre-trained object detection models we used in the app (notice the uint8 type):

```
Found 1 possible inputs: (name=image_tensor, type=uint8(4),
shape=[?,?,?,3])
```

That's why we need to create an image tensor with `uint8`, instead of the `float` type to feed to our model, or you'll get an error when running the model. Also notice that when we use the for-loop to convert `image_data` to `image_tensor` of the `Tensor` type that's expected by the TensorFlow C++ API's `Run` method of `Session`, we don't use `input_mean` and `input_std` as we did when using image classification models (for a detailed comparison, see the `RunInferenceOnImage` implementation of the HelloTensorFlow app in Chapter 2, *Classifying Images with Transfer Learning*). And we know there are four output named detection_boxes, detection_scores, detection_classes, and num_detections, so `RunInferenceOnImage` has the following code to feed a model the image input and get the four output:

```
tensorflow::Tensor image_tensor(tensorflow::DT_UINT8,
tensorflow::TensorShape({1, image_height, image_width, wanted_channels}));

auto image_tensor_mapped = image_tensor.tensor<uint8, 4>();
tensorflow::uint8* in = image_data.data();
uint8* c_out = image_tensor_mapped.data();
for (int y = 0; y < image_height; ++y) {
    tensorflow::uint8* in_row = in + (y * image_width * image_channels);
    uint8* out_row = c_out + (y * image_width * wanted_channels);
    for (int x = 0; x < image_width; ++x) {
        tensorflow::uint8* in_pixel = in_row + (x * image_channels);
        uint8* out_pixel = out_row + (x * wanted_channels);
```

```
            for (int c = 0; c < wanted_channels; ++c) {
                out_pixel[c] = in_pixel[c];
            }
        }
    }
    std::vector<Tensor> outputs;
    Status run_status = session->Run({{"image_tensor", image_tensor}},
            {"detection_boxes", "detection_scores", "detection_classes",
    "num_detections"}, {}, &outputs);
```

To draw the bounding boxes on detected objects, we pass the `outputs` tensor vector to `DrawTopDetections`, which uses the following code to parse the `outputs` vector to get the values for the four output and loop through each detection to get the bounding box values (left, top, right, bottom) as well as display name for the detected object ID, so you can write code to draw the bounding box with the name:

```
auto detection_boxes = outputs[0].flat<float>();
auto detection_scores = outputs[1].flat<float>();
auto detection_classes = outputs[2].flat<float>();
auto num_detections = outputs[3].flat<float>()(0);

LOG(INFO) << "num_detections: " << num_detections << ", detection_scores
size: " << detection_scores.size() << ", detection_classes size: " <<
detection_classes.size() << ", detection_boxes size: " <<
detection_boxes.size();

for (int i = 0; i < num_detections; i++) {
    float left = detection_boxes(i * 4 + 1) * image_width;
    float top = detection_boxes(i * 4 + 0) * image_height;
    float right = detection_boxes(i * 4 + 3) * image_width;
    float bottom = detection_boxes((i * 4 + 2)) * image_height;
    string displayName = GetDisplayName(&imageLabels,
detection_classes(i));
    LOG(INFO) << "Detected " << i << ": " << displayName << ", " << score
<< ", (" << left << ", " << top << ", " << right << ", " << bottom << ")";
    ...
}
```

The preceding `LOG(INFO)` lines, when running with the second test image in Figure 3.1, and the demo image shown in the TensorFlow Object Detection API website, will output the following info:

```
num_detections: 100, detection_scores size: 100, detection_classes size: 100, detection_boxes size: 400
 Detected 0: person, 0.916851, (533.138, 498.37, 553.206, 533.727)
 Detected 1: kite, 0.828284, (467.467, 344.695, 485.3, 362.049)
 Detected 2: person, 0.779872, (78.2835, 516.831, 101.287, 560.955)
 Detected 3: kite, 0.769913, (591.238, 72.0729, 676.863, 149.322)
```

So that's what it takes to use an existing pre-trained object detection model in an iOS app. What about using our retrained object detection models in iOS? It turns out that it's almost the same as using the pre-trained models–there's no need to modify the `input_size`, `input_mean`, `input_std` and `input_name` as we did in the previous chapter when dealing with retrained image classification models. You just need to do the following things:

1. Add your retrained model, for example the `output_inference_graph_ssd_mobilenet.pb` file created in the previous section, the label map file used for the model's retraining, for example `pet_label_map.pbtxt`, and optionally some new test images specific to the retrained model to the `TFObjectDetectionAPI` project
2. In `ViewController.mm`, call `RunInferenceOnImage` with the retrained model
3. Still in `ViewController.mm`, call `LoadLablesFile([FilePathForResourceName(@"pet_label_map", @"pbtxt") UTF8String], &imageLabels);` inside the `DrawTopDetections` function

That's it. Run the app and you can see the detected results are more fine-tuned for the retrained model. For example, using the preceding retrained model, generated by retraining with the Oxford pet dataset, we expect to see the bounding boxes are around the head area instead of the whole body, and that's exactly what we see with the test image shown in Figure 3.9:

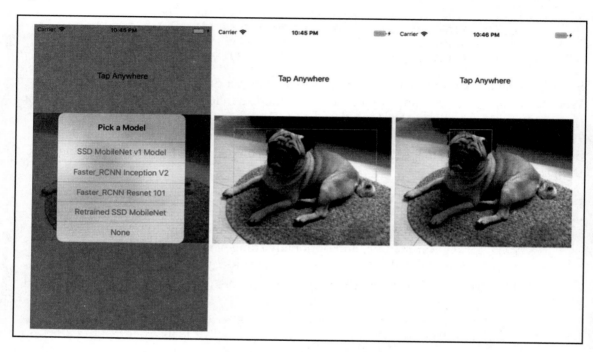

Figure 3.9 Comparing detected results of pre-trained and retrained models

Using YOLO2–another object-detection model

As we mentioned in the first section, YOLO2 (https://pjreddie.com/darknet/yolo) is another cool object-detection model using a different approach from the RCNN family. It uses a single neural network to divide an input image into fixed-size regions (but no region proposals as in the RCNN-family methods) and predict bounding boxes, classes, and probabilities for each region.

The TensorFlow Android example app has sample code for using a pre-trained YOLO model, but there's no iOS example. As YOLO2 is one of the fastest object-detection models and also pretty accurate (see the mAP comparison of it with SSD models at its website), it's worth taking a look at what it takes to use it in an iOS app.

YOLO uses a unique open source neural network framework called Darknet (https://pjreddie.com/darknet) to train its models. There's another library called darkflow (https://github.com/thtrieu/darkflow) that can convert the neural network weights of YOLO models trained with Darknet to the TensorFlow graph format, as well as retraining pre-trained models.

To build the YOLO2 models in the TensorFlow format, first get the darkflow repo at https://github.com/thtrieu/darkflow. Because it requires Python 3 and TensorFlow 1.0 (it's possible that Python 2.7 and TensorFlow 1.4 or later also work), we'll use Anaconda to set up a new TensorFlow 1.0 environment with Python 3 support:

```
conda create --name tf1.0_p35 python=3.5
source activate tf1.0_p35
conda install -c derickl tensorflow
```

Also run `conda install -c menpo opencv3` to install OpenCV 3, another dependency of darkflow. Now `cd` to the darkflow directory and run `pip install .` to install darkflow.

Next, we need to download the weights of pre-trained YOLO models – we'll try two Tiny-YOLO models, which are super fast but less accurate than the full YOLO models. The iOS code to run both Tiny-YOLO models and YOLO models are about the same, so we'll just show you how to run the Tiny-YOLO models.

You can download the weights and config files for tiny-yolo-voc (trained with the 20-object-class PASCAL VOC dataset) and tiny-yolo (trained with the 80-object-class MS COCO dataset) at the official YOLO2 web site or the darkflow repo. Now, run the following commands to convert the weights to the TensorFlow graph files:

```
flow --model cfg/tiny-yolo-voc.cfg --load bin/tiny-yolo-voc.weights --savepb
flow --model cfg/tiny-yolo.cfg --load bin/tiny-yolo.weights --savepb
```

The two generated files `tiny-yolo-voc.pb` and `tiny-yolo.pb` will be located in the `built_graph` directory. Now, go to your TensorFlow source root directory, and run the following commands to create the quantized models as we did in the previous chapter:

```
python tensorflow/tools/quantization/quantize_graph.py --input=darkflow/built_graph/tiny-yolo.pb --output_node_names=output --output=quantized_tiny-yolo.pb --mode=weights
```

Detecting Objects and Their Locations

```
python tensorflow/tools/quantization/quantize_graph.py --
input=darkflow/built_graph/tiny-yolo-voc.pb --output_node_names=output --
output=quantized_tiny-yolo-voc.pb --mode=weights
```

Now follow these steps to see how to use the two YOLO models in our iOS app:

1. Drag both `quantized_tiny-yolo-voc.pb` and `quantized_tiny-yolo.pb` to the `TFObjectDetectionAPI` project
2. Add two new alert actions in `ViewController.mm`, so when running the app you'll see the models available for running, as shown in Figure 3.10:

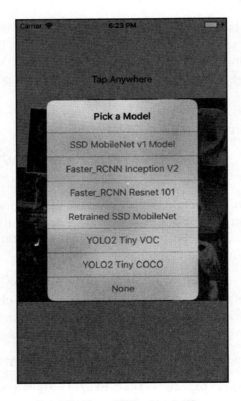

Figure 3.10 Adding two YOLO models to the iOS app

3. Add the following code to process an input image into a tensor to be fed to the input node and run a TensorFlow session with a YOLO model graph loaded to generate the detection outputs:

```
tensorflow::Tensor
image_tensor(tensorflow::DT_FLOAT,tensorflow::TensorShape({1,
wanted_height, wanted_width, wanted_channels}));
auto image_tensor_mapped = image_tensor.tensor<float, 4>();
tensorflow::uint8* in = image_data.data();
float* out = image_tensor_mapped.data();
for (int y = 0; y < wanted_height; ++y) {
   ...
   out_pixel[c] = in_pixel[c] / 255.0f;
}
std::vector<tensorflow::Tensor> outputs;
tensorflow::Status run_status = session->Run({{"input",
image_tensor}}, {"output"}, {}, &outputs);
```

Notice the slight but important difference in `for-loop` and `session->Run` here from the code used for image classification in the previous chapter and for object detection using other models shown earlier in this chapter (we didn't show the code snippet in ... as it's the same as in those two examples). To make image data conversion correct, you need to either understand the model detail or learn from a working example in Python, Android, or iOS, plus of course the necessary debugging. To make the input and output node names right, you can use the `summarize_graph` tool or the Python code snippet we showed several times.

4. Pass the output result to a function called YoloPostProcess, which is similar to the post processing code in the `tensorflow/examples/android/src/org/tensorflow/demo/TensorFlowYoloDetector.java` Android example file:

```
tensorflow::Tensor* output = &outputs[0];
std::vector<std::pair<float, int> > top_results;
YoloPostProcess(model, output->flat<float>(), &top_results);
```

We won't show the rest of the code here. You can check out the complete iOS app in `ch3/ios` of the source code repo.

5. Run the app and select YOLO2 Tiny VOC or YOLO2 Tiny COCO, you'll see similar speed but less accurate detection results than using the SSD MobileNet v1 model.

Both TensorFlow Object Detection models and YOLO2 models run pretty fast on mobile devices, although the MobileNet-based TensorFlow models and the Tiny YOLO2 models are less accurate. The larger Faster RNN models and full YOLO2 models are more accurate, but they take much longer or won't even run on mobile devices. So the best way to add fast object detection to your mobile apps is to use the SSD MobileNet or Tiny-YOLO2 models, or their retrained and fine-tuned models. Future releases of the models are very likely to have even better performance and accuracy. With the knowledge covered in this chapter, you should be able to enable object detection in your iOS apps quickly.

Summary

In this chapter, we first gave a quick overview of different deep-learning-based object detection methods. Then we covered in detail how to use the TensorFlow Object Detection API to make off-the-shelf inference with pre-trained models and how to retrain pre-trained TensorFlow Object Detection models in Python. We also provided detailed tutorials on how to manually build the TensorFlow iOS library, create a new iOS app using the library, and how to use the pre-existing and retrained SSD MobileNet and Faster RCNN models in iOS. Finally, we showed what it takes to use another powerful object detection model, YOLO2, in your iOS app.

In the next chapter, our third computer-vision-related task, we'll take a closer look at how to train and build a fun deep learning model in Python and TensorFlow, and how to use it in your iOS and Android apps to add amazing art styles to an image.

4
Transforming Pictures with Amazing Art Styles

Since deep neural networks took off in 2012 with the winning ImageNet challenge of AlexNet, AI researchers have been applying deep learning technology, including pre-trained deep CNN models, to more and more problem domains. What can be more creative than creating art? One idea that has been proposed and implemented, called neural style transfer, allows you to take advantage of a pre-trained deep neural network model and to transfer the style of an image, or any Van Gogh or Monet masterpiece, for example, to another image such as your profile picture or a picture of your favorite dog, thereby creating an image that mixes the content of your picture and the style of a masterpiece. There's actually an iOS app called Prisma that won Best App of the Year in 2016 that does just that; in just a few seconds, it transforms your pictures with any of the styles you choose.

In this chapter, we'll first give an overview of three neural style transfer methods, the original one, the much-improved one, and the further-improved one. Then we'll take a detailed look at how to use the second method to train a fast neural style transfer model that can be used in your iOS and Android smartphones to achieve what Prisma does. Next, we'll actually use the model in both an iOS app and an Android app, walking you through the whole process of creating such apps from scratch. Finally, we'll give you a quick introduction to the TensorFlow Magenta open source project, which you can use to build more music and art generation apps based on deep learning, and show you how to use a single pre-trained style transfer model, created on the basis of the latest research advance in neural style transfer and including 26 cool art styles, in your iOS and Android apps to achieve even faster performance and results. In summary, we'll cover the following topics in this chapter:

- Neural Style Transfer – a quick overview
- Training fast neural - style transfer models
- Using fast neural - style transfer models in iOS
- Using fast neural - style transfer models in Android

- Using the TensorFlow Magenta multi-style model in iOS
- Using the TensorFlow Magenta multi-style model in Android

Neural Style Transfer – a quick overview

The original idea and algorithm of using a deep neural network to merge the content of an image with the style of another was published in a paper titled *A Neural Algorithm of Artistic Style* (https://arxiv.org/abs/1508.06576) in the summer of 2015. It was based on a pre-trained deep CNN model called VGG-19 (https://arxiv.org/pdf/1409.1556.pdf), the winner of the 2014 ImageNet image recognition challenge, which has 16 convolutional layers, or feature maps, representing different levels of the image content. In this original method, the final transferred image is first initialized as a white noise image merged with the content image. The content loss function is defined as the squared error loss of a specific set of feature representations on the convolutional layer, conv4_2, of the content image and the result image after both being fed into the VGG-19 network. The style loss function calculates the total error difference of the style image and the resulting image on five different convolutional layers. The total loss is then defined as the sum of the content loss and the style loss. During training, the loss gets minimized, and a resulting image that mixes the content of one image and the style of another is generated.

Although the result of the original neural style transfer algorithm was pretty amazing, its performance was very poor—training was part of the style-transferred image generation process and it usually took several minutes on a GPU and about an hour on a CPU to generate good results.

If you're interested in the details of the original algorithm, you can read the paper along with a well-documented Python implementation, at https://github.com/log0/neural-style-painting/blob/master/art.py. We won't discuss this original algorithm as it's not feasible to run on mobile phone, but it's fun and instrumental to try it to get a better understanding of how to use a pre-trained deep CNN model for different computer vision tasks.

Naturally, in 2016, a new algorithm that is "three orders of magnitude faster" was published in the paper, *Perceptual Losses for Real-Time Style Transfer and Super-Resolution* (https://cs.stanford.edu/people/jcjohns/eccv16/) by *Justin Johnson*, et al. It uses a separate training process and defines better loss functions that are themselves deep neural networks. After the training, which can take a few hours on a GPU as we'll see in the next section, using the trained model to generate a style-transferred image is almost real time on computer and only takes a few seconds on a smartphone.

There's still one shortcoming of using this fast neural transfer algorithm: a model can only be trained for a specific style, so in order to use different styles in your app, you have to train those styles one by one to generate one model for each style. A new paper titled *A Learned Representation For Artistic Style* (https://arxiv.org/abs/1610.07629) was published in 2017 and it found that a single deep neural network model can generalize many different styles. The TensorFlow Magenta project (https://github.com/tensorflow/magenta/tree/master/magenta/models/image_stylization) includes pre-trained models with multiple styles, and we'll see in the last two sections of this chapter how easy it is to use such a model in iOS and Android apps to produce powerful and amazing art effects.

Training fast neural-style transfer models

In this section, we'll show you how to train models using the fast neural-style transfer algorithm with TensorFlow. Perform the following steps to train such a model:

1. On a Terminal of your Mac or preferably GPU-powered Ubuntu, run `git clone https://github.com/jeffxtang/fast-style-transfer`, which is a fork of a nice TensorFlow implementation of Johnson's fast-style transfer, modified to allow the trained model to be used in iOS or Android apps.
2. `cd` to the fast-style-transfer directory, then run the `setup.sh` script to download the pre-trained VGG-19 model file as well as the MS COCO training dataset, which we mentioned in the previous chapter – note that it can take several hours to download the large files.
3. Run the following commands to create checkpoint files with training using a style image named `starry_night.jpg` and a content image named `ww1.jpg`:

   ```
   mkdir checkpoints
   mkdir test_dir
   python style.py --style images/starry_night.jpg --test
   images/ww1.jpg --test-dir test_dir --content-weight 1.5e1 --
   checkpoint-dir checkpoints --checkpoint-iterations 1000 --batch-
   size 10
   ```

There are a few other style images in the images directory that you can use to create different checkpoint files. The `starry_night.jpg` style image used here is a famous painting by Vincent van Gogh, as shown in Figure 4.1:

Figure 4.1 Using Van Gogh's painting as the style image

The whole training takes about five hours on the NVIDIA GTX 1070 GPU-powered Ubuntu we set up in Chapter 1, *Getting Started with Mobile TensorFlow*, and would certainly take much longer on a CPU.

 The script was originally written for TensorFlow 0.12 but modified later for TensorFlow 1.1, and has been verified to also run OK in the TensorFlow 1.4 with Python 2.7 environment we set up earlier.

4. Open the `evaluate.py` file in a text editor, and uncomment the following two lines of code (on lines 158 and 159):

```
# saver = tf.train.Saver()
# saver.save(sess, "checkpoints_ios/fns.ckpt")
```

5. Run the following command to create a new checkpoint with the input image named `img_placeholder` and the transferred image named `preds`:

```
python evaluate.py --checkpoint checkpoints \
    --in-path examples/content/dog.jpg \
    --out-path examples/content/dog-output.jpg
```

6. Run the following command to build a TensorFlow graph file that combines the graph definition and the weights in the checkpoint. This will create a .pb file of about 6.7 MB:

   ```
   python freeze.py --model_folder=checkpoints_ios --output_graph fst_frozen.pb
   ```

7. Assuming you have a /tf_files directory, copy the fst_frozen.pb file generated to /tf_files, cd to your TensorFlow source root directly, likely ~/tensorflow-1.4.0, then run the following command to generate the quantized model of the .pb file (we covered quantization in Chapter 2, *Classifying Images with Transfer Learning*):

   ```
   bazel-bin/tensorflow/tools/quantization/quantize_graph \
   --input=/tf_files/fst_frozen.pb \
   --output_node_names=preds \
   --output=/tf_files/fst_frozen_quantized.pb \
   --mode=weights
   ```

This will reduce the frozen graph file size from 6.7MB to about 1.7MB, meaning if you put 50 models for 50 different styles in your app, the added size will be about 85MB. Apple announced in September 2017 that the cellular over-the-air app download limit has been increased to 150MB, so users should still be able to download an app with over 50 different styles over the cellular network.

That's all it takes to train and quantize a fast neural transfer model using a style image and an input image. You can check out the generated images in the test_dir directory, generated in step 3, to see the effects of the style transfer. If needed, you can play with hyper-parameters, documented at https://github.com/jeffxtang/fast-style-transfer/blob/master/docs.md#style, to see different, and hopefully better, style transfer effects.

One important note before we see how to use these models in our iOS and Android apps is that you need to write down the exact image width and height of the image specified as the value for the --in-path parameter in step 5 and use the image width and height values in your iOS or Android code (you'll see how soon), otherwise you'll get an Conv2DCustomBackpropInput: Size of out_backprop doesn't match computed error when running your model in your app.

Using fast neural-style transfer models in iOS

It turns out that we'll have no problem using the `fst_frozen_quantized.pb` model file, generated in step 7, in an iOS app built with the TensorFlow-experimental pod as shown in Chapter 2, *Classifying Images with Transfer Learning*, but the pre-trained multi-style model file from the TensorFlow Magenta project, which we'll use in a later section of this chapter, won't load with the TensorFlow pod (as of January 2018)—it'll throw the following error when trying to load the multi-style model file:

```
Could not create TensorFlow Graph: Invalid argument: No OpKernel was
registered to support Op 'Mul' with these attrs. Registered devices: [CPU],
Registered kernels:
  device='CPU'; T in [DT_FLOAT]
    [[Node: transformer/expand/conv1/mul_1 =
Mul[T=DT_INT32](transformer/expand/conv1/mul_1/x,
transformer/expand/conv1/strided_slice_1)]]
```

We talked about the reason and how to use manually-built TensorFlow libraries to fix this error in Chapter 3, *Detecting Objects and Their Locations*. As we'll use both models in the same iOS app, we'll create a new iOS app using the more powerful manually-built TensorFlow libraries.

Adding and testing with fast neural transfer models

If you haven't manually built TensorFlow libraries, you need to go back to the previous chapter to do that first. Then perform the following steps to add TensorFlow support and fast neural-style transfer model files to your iOS app and test run the app:

1. If you already have an iOS app with TensorFlow manual libraries added, you can skip this step. Otherwise, similar to what we did in the previous chapter, create a new Objective-C-based iOS app named, for example, `NeuralStyleTransfer`, or in your existing app, create a new user-defined setting under your PROJECT's **Build Settings** named `TENSORFLOW_ROOT` with the `$HOME/tensorflow-1.4.0` value, assuming that's where you have your TensorFlow 1.4.0 installed, and then in your TARGET's **Build Settings**, set Other Linker Flags to be:

```
-force_load
$(TENSORFLOW_ROOT)/tensorflow/contrib/makefile/gen/lib/libtensorflo
w-core.a
$(TENSORFLOW_ROOT)/tensorflow/contrib/makefile/gen/protobuf_ios/lib
/libprotobuf.a
$(TENSORFLOW_ROOT)/tensorflow/contrib/makefile/gen/protobuf_ios/lib
/libprotobuf-lite.a
$(TENSORFLOW_ROOT)/tensorflow/contrib/makefile/downloads/nsync/buil
ds/lipo.ios.c++11/nsync.a
```

Then set Header Search Paths to be:

```
$(TENSORFLOW_ROOT)
$(TENSORFLOW_ROOT)/tensorflow/contrib/makefile/downloads/protobuf/s
rc $(TENSORFLOW_ROOT)/tensorflow/contrib/makefile/downloads
$(TENSORFLOW_ROOT)/tensorflow/contrib/makefile/downloads/eigen
$(TENSORFLOW_ROOT)/tensorflow/contrib/makefile/gen/proto
```

2. Drag and drop the `fst_frozen_quantized.pb` file and a few test images to your project's folder. Copy the same `ios_image_load.mm` and `.h` file we used in the previous chapters from the previous iOS apps, or from the `NeuralStyleTransfer` app folder under `Ch4/ios` of the book's source repo to the project.

3. Rename `ViewController.m` to `ViewController.mm` and replace it and `ViewController.h` with the `ViewController.h` and `.mm` files from `Ch4/ios/NeuralStyleTransfer`. We'll take a detailed look at the core code snippet after test-running the app.

4. Run the app on the iOS simulator or your iOS device, and you'll see a dog picture, like in Figure 4.2:

Figure 4.2 The original dog picture before style applied

5. Tap to select **Fast Style Transfer**, and after a few seconds, you'll see a new picture in Figure 4.3 with the starry night style transferred:

Figure 4.3 Like having Van Gogh draw your favorite dog

You can easily build other models with different styles by just choosing your favorite pictures as your style images and following the steps in the previous section. Then you can follow the steps in this section to use the models in your iOS app. If you're interested in understanding how the model gets trained, you should check out the code in the GitHub repo in the previous section. Let's take a detailed look at the iOS code that uses the model to do the magic.

Looking back at the iOS code using fast neural transfer models

There are several key code snippets in `ViewController.mm` that are unique in the pre-processing of the input image and post-processing of the transferred image:

1. Two constants, `wanted_width` and `wanted_height`, are defined to be the same values as the image width and height of the image `examples/content/dog.jpg` of the repo in step 5:

    ```
    const int wanted_width = 300;
    const int wanted_height = 400;
    ```

2. The iOS's dispatch queue is used to load and run our fast neural transfer model in a non-UI thread and, after the style transferred image is generated, to send the image to the UI thread for display:

    ```
    dispatch_async(dispatch_get_global_queue(0, 0), ^{
        UIImage *img = imageStyleTransfer(@"fst_frozen_quantized");
        dispatch_async(dispatch_get_main_queue(), ^{
            _lbl.text = @"Tap Anywhere";
            _iv.image = img;
        });
    });
    ```

3. A 3-dimensional tensor of floating numbers is defined and used to convert the input image data to:

    ```
    tensorflow::Tensor image_tensor(tensorflow::DT_FLOAT,
    tensorflow::TensorShape({wanted_height, wanted_width,
    wanted_channels}));
    auto image_tensor_mapped = image_tensor.tensor<float, 3>();
    ```

4. The input node name and output node name sent to the TensorFlow `Session->Run` method are defined to be the same as when the model is trained:

    ```
    std::string input_layer = "img_placeholder";
    std::string output_layer = "preds";
    std::vector<tensorflow::Tensor> outputs;
    tensorflow::Status run_status = session->Run({{input_layer,
    image_tensor}} {output_layer}, {}, &outputs);
    ```

5. After the model finishes running and sends back the output tensor, which contains RGB values in the range of 0 to 255, we need to call a utility function called `tensorToUIImage` to convert the tensor data to an RGB buffer first:

```
UIImage *imgScaled = tensorToUIImage(model, output->flat<float>(),
image_width, image_height);

static UIImage* tensorToUIImage(NSString *model, const
Eigen::TensorMap<Eigen::Tensor<float, 1, Eigen::RowMajor>,
Eigen::Aligned>& outputTensor, int image_width, int image_height) {
    const int count = outputTensor.size();
    unsigned char* buffer = (unsigned char*)malloc(count);

    for (int i = 0; i < count; ++i) {
        const float value = outputTensor(i);
        int n;
        if (value < 0) n = 0;
        else if (value > 255) n = 255;
        else n = (int)value;
        buffer[i] = n;
    }
}
```

6. Then, we convert the buffer to a `UIImage` instance before resizing it and returning it for display:

```
UIImage *img = [ViewController convertRGBBufferToUIImage:buffer
withWidth:wanted_width withHeight:wanted_height];
UIImage *imgScaled = [img scaleToSize:CGSizeMake(image_width,
image_height)];
return imgScaled;
```

The complete code and app are in the `Ch4/ios/NeuralStyleTransfer` folder.

Using fast neural-style transfer models in Android

In Chapter 2, *Classifying Images with Transfer Learning*, we described how to add TensorFlow to your own Android app, but without any UI. Let's create a new Android app to use the fast-style transfer models we trained earlier and used in iOS.

Because this Android app offers a good opportunity to use the minimal TensorFlow-related code and Android UI and threaded code to run a complete TensorFlow model-powered app, we'll go through each line of the code added from scratch to help you further understand what it takes to develop an Android TensorFlow app from scratch:

1. In Android Studio, select **File** | **New** | **New Project...** and enter `FastNeuralTransfer` as the **Application Name**; accept all the defaults before clicking Finish.

2. Create a new `assets` folder, as shown in Figure 2.13, and drag the fast neural transfer models you have trained, from the iOS app if you have tried it in the previous section, or from the folder `/tf_files` as in step 7 of the *Training fast neural-style transfer models* section, along with some test images to the `assets` folder.

3. In the app's `build.gradle` file, add a line, `compile 'org.tensorflow:tensorflow-android:+'`, to the end of `dependencies`.

4. Open the `res/layout/activity_main.xml` file, remove the default TextView there, and first add an ImageView to show the image before and after the style transfer:

   ```
   <ImageView
       android:id="@+id/imageview"
       android:layout_width="match_parent"
       android:layout_height="match_parent"
       app:layout_constraintBottom_toBottomOf="parent"
       app:layout_constraintLeft_toLeftOf="parent"
       app:layout_constraintRight_toRightOf="parent"
       app:layout_constraintTop_toTopOf="parent"/>
   ```

5. Add a Button to initiate the style transfer action:

   ```
   <Button
       android:id="@+id/button"
       android:layout_width="wrap_content"
       android:layout_height="wrap_content"
       android:text="Style Transfer"
       app:layout_constraintBottom_toBottomOf="parent"
       app:layout_constraintHorizontal_bias="0.502"
       app:layout_constraintLeft_toLeftOf="parent"
       app:layout_constraintRight_toRightOf="parent"
       app:layout_constraintTop_toTopOf="parent"
       app:layout_constraintVertical_bias="0.965" />
   ```

6. In the app's `MainActivity.java` file, first enter our most important import:

   ```
   import org.tensorflow.contrib.android.TensorFlowInferenceInterface;
   ```

   `TensorFlowInferenceInterface` provides the JAVA interface to access the native TensorFlow inference APIs. Then make sure the `MainActivity` class implements the `Runnable` interface as we need to keep our app responsive and load and run the TensorFlow model on a worker thread.

7. At the beginning of the class, define six constants, as follows:

   ```
   private static final String MODEL_FILE =
   "file:///android_asset/fst_frozen_quantized.pb";
   private static final String INPUT_NODE = "img_placeholder";
   private static final String OUTPUT_NODE = "preds";
   private static final String IMAGE_NAME = "pug1.jpg";
   private static final int WANTED_WIDTH = 300;
   private static final int WANTED_HEIGHT = 400;
   ```

   You can use any of your trained model files for `MODEL_FILE`. The values of `INPUT_NODE` and `OUTPUT_NODE` are the same as what we set in the Python training script and used in the iOS app. The `WANTED_WIDTH` and `WANTED_HEIGHT` are, again, the same values as the image width and height of the image for `--in-path` that we used in step 5 of the *Training fast neural-style transfer models* section.

8. Declare four instance variables:

   ```
   private ImageView mImageView;
   private Button mButton;
   private Bitmap mTransferredBitmap;

   private TensorFlowInferenceInterface mInferenceInterface;
   ```

   `mImageView` and `mButton` will be set using the simple `findViewById` method in the `onCreate` method. `mTransferredBitmap` will hold the bitmap of a transferred image so `mImageView` can show it. `mInferenceInterface` is used to load our TensorFlow model, feed the input image to the model, run the model, and return the inference result.

9. Create a `Handler` instance to handle the task of showing the resulting transferred image in the main thread after our TensorFlow inference thread sends a message to the `Handler` instance, we also create a handy `Toast` message:

```
Handler mHandler = new Handler() {
    @Override
    public void handleMessage(Message msg) {
        mButton.setText("Style Transfer");
        String text = (String)msg.obj;
        Toast.makeText(MainActivity.this, text,
                Toast.LENGTH_SHORT).show();
        mImageView.setImageBitmap(mTransferredBitmap);
    } };
```

10. Inside the `onCreate` method, we'll bind the `ImageView` in the layout xml file with the `mImageView` instance variable, load the bitmap of our test image in the `assets` folder, and show it in the `ImageView`:

```
mImageView = findViewById(R.id.imageview);
try {
    AssetManager am = getAssets();
    InputStream is = am.open(IMAGE_NAME);
    Bitmap bitmap = BitmapFactory.decodeStream(is);
    mImageView.setImageBitmap(bitmap);
} catch (IOException e) {
    e.printStackTrace();
}
```

11. Set `mButton` similarly and also set up a click listener so when the button is tapped, a new thread is created and started, calling the `run` method:

```
mButton = findViewById(R.id.button);
mButton.setOnClickListener(new View.OnClickListener() {
    @Override
    public void onClick(View v) {
        mButton.setText("Processing...");
        Thread thread = new Thread(MainActivity.this);
        thread.start();
    }
});
```

12. In the thread's `run` method, we first declare three arrays and allocate appropriate memories for them: the `intValues` array holds the pixel values of the test image, with each pixel value representing a 32-bit ARGB (Alpha, Red, Green, Blue) value; the `floatValues` array holds the Red, Green, and Blue values of each pixel separately, as expected by the model, so it has a size three times bigger than `intValues`, and the `outputValues` has the same size as `floatValues`, but holds the output values of the model:

```
public void run() {
    int[] intValues = new int[WANTED_WIDTH * WANTED_HEIGHT];
    float[] floatValues = new float[WANTED_WIDTH * WANTED_HEIGHT * 3];
    float[] outputValues = new float[WANTED_WIDTH * WANTED_HEIGHT * 3];
```

Then we get the bitmap data of the test image, scale it to match the size of the image used in training, and load the scaled bitmap's pixels to the `intValues` array and convert it to the `floatValues`:

```
Bitmap bitmap = BitmapFactory.decodeStream(getAssets().open(IMAGE_NAME));
Bitmap scaledBitmap = Bitmap.createScaledBitmap(bitmap, WANTED_WIDTH, WANTED_HEIGHT, true);
scaledBitmap.getPixels(intValues, 0, scaledBitmap.getWidth(), 0, 0, scaledBitmap.getWidth(), scaledBitmap.getHeight());

for (int i = 0; i < intValues.length; i++) {
    final int val = intValues[i];
    floatValues[i*3] = ((val >> 16) & 0xFF);
    floatValues[i*3+1] = ((val >> 8) & 0xFF);
    floatValues[i*3+2] = (val & 0xFF);
}
```

Note that `val`, or each element of the `intValues` pixel array, is a 32-bit integer holding ARGB in each of its 8-bit areas. We use the right bit shift (for Red and Green) and bitwise AND operations to extract the Red, Green, and Blue values of each pixel, ignoring the Alpha value of the leftmost 8-bit in an `intValues` element. So `floatValues[i*3]`, `floatValues[i*3+1]`, and `floatValues[i*3+2]` hold the Red, Green, and Blue value of a pixel, respectively.

Now we create a new `TensorFlowInferenceInterface` instance, passing it with an `AssetManager` instance and the model file name in the `assets` folder, and use the `TensorFlowInferenceInterface` instance to feed the `INPUT_NODE` with the converted `floatValues` array. If more than one input node is expected by a model, you can just call multiple `feed` methods. We then run the model by passing a string array of output node names. Here for our fast-style transfer model, we only have one input node and one output node. Finally, we fetch the model's output values by passing the output node name. You can call multiple fetches if you expect to receive more than one output node:

```
AssetManager assetManager = getAssets();
mInferenceInterface = new TensorFlowInferenceInterface(assetManager, MODEL_FILE);
mInferenceInterface.feed(INPUT_NODE, floatValues, WANTED_HEIGHT, WANTED_WIDTH, 3);
mInferenceInterface.run(new String[] {OUTPUT_NODE}, false);
mInferenceInterface.fetch(OUTPUT_NODE, outputValues);
```

The `outputValues` generated by the model hold one of the 8-bit Red, Green, and Blue values in the range of 0 to 255 in each element, and we first use the left bit shift operation on the Red and Green values, but with different shift sizes (16 and 8), then use the bitwise OR operation to combine the 8-bit Alpha value (0xFF) with 8-bit RGB values, saving the result in the `intValues` array:

```
for (int i=0; i < intValues.length; ++i) {
    intValues[i] = 0xFF000000
                | (((int) outputValues[i*3]) << 16)
                | (((int) outputValues[i*3+1]) << 8)
                | ((int) outputValues[i*3+2]);
```

We then create a new Bitmap instance and set its pixel values using the `intValues` array, scale the bitmap to the test image's original size, and save the scaled bitmap to `mTransferredBitmap`:

```
Bitmap outputBitmap = scaledBitmap.copy( scaledBitmap.getConfig() , true);
outputBitmap.setPixels(intValues, 0, outputBitmap.getWidth(), 0, 0,
outputBitmap.getWidth(), outputBitmap.getHeight());
mTransferredBitmap = Bitmap.createScaledBitmap(outputBitmap,
bitmap.getWidth(), bitmap.getHeight(), true);
```

Finally, we send our main thread's handler a message to let it know it's time to show the style-transferred image:

```
Message msg = new Message();
msg.obj = "Tranfer Processing Done";
mHandler.sendMessage(msg);
```

So in a total of fewer than 100 lines of code, you have a complete Android app that does the amazing style transfer on your image. Run the app on your Android device or virtual device, you'll first see your test image with a button, tap the button and a few seconds later, you'll see the style transferred image, as shown in Figure 4.4:

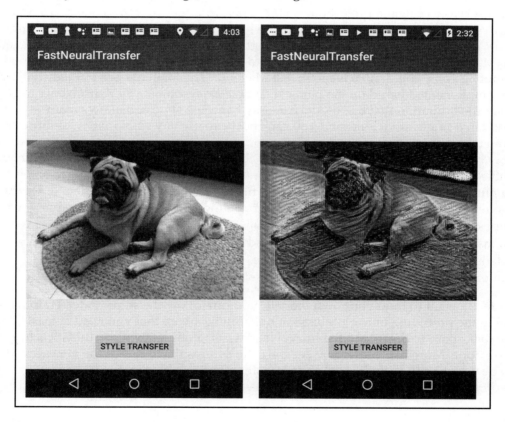

Figure 4.4 The original image and style transferred image on Android

Transforming Pictures with Amazing Art Styles

One problem we have with the fast neural-style models is that even though after quantization each model is only 1.7MB, we still need to train separately for each style, and each trained model can only have one style transfer supported. Fortunately, there's a great solution to this problem.

Using the TensorFlow Magenta multi-style model in iOS

The TensorFlow Magenta project (`https://github.com/tensorflow/magenta`) allows you to use over 10 pre-trained models to generate new music and images. We'll focus on using Magenta's image stylization models in this and the next sections. You can follow the link to install Magenta on your computer, although to use their cool image-style transfer models in your mobile apps, you don't have to install Magenta. The Magenta pre-trained style transfer model, implemented based on the paper *A Learned Representation for Artistic Style* in 2017, removes the limitation that one model can only have one style and allows many styles to be included in a single model file, and you can choose to use any combination of those styles. You can take a quick look at the demo at `https://github.com/tensorflow/magenta/tree/master/magenta/models/image_stylization`, but the two pre-trained checkpoint models available for download there can't be used directly in your mobile apps due to some NaNs (not a number) error saved in the checkpoint files. We won't go through detailed steps on how to remove those numbers and generate a `.pb` model file that can be used in your apps (if interested, you can check out `https://github.com/tensorflow/tensorflow/issues/9678`), instead, we'll simply use the pre-trained `stylize_quantized.pb` model file included in the TensorFlow Android example at `tensorflow/examples/android/assets` to see how it works.

If you really want to train your own models, you can follow the steps under Training a Model in the preceding image_stylization link. But be aware that you need at least 500 GB of free disk space to download the ImageNet dataset, along with powerful GPUs to get the training done. More likely after you see the code and results in this or the next section, you'd be very happy with the cool style transfer effects enabled by the pre-trained `stylize_quantized.pb` model.

Perform the following steps to use and run the multi-style model in our iOS app created earlier in this chapter:

1. Drag and drop the `stylize_quantized.pb` file from `tensorflow/examples/android/assets` to your iOS apps folder in Xcode.
2. Add a new `UIAlertAction` to the tap handler, using the same `dispatch_async` we used for loading and processing fast transfer-style models:

```
UIAlertAction* multi_style_transfer = [UIAlertAction
actionWithTitle:@"Multistyle Transfer"
style:UIAlertActionStyleDefault handler:^(UIAlertAction * action) {
    _lbl.text = @"Processing...";
    _iv.image = [UIImage imageNamed:image_name];
    dispatch_async(dispatch_get_global_queue(0, 0), ^{
        UIImage *img = imageStyleTransfer(@"stylize_quantized");
        dispatch_async(dispatch_get_main_queue(), ^{
            _lbl.text = @"Tap Anywhere";
            _iv.image = img;
        });
    });
}];
```

3. Replace the `input_layer` and `output_layer` values with the correct ones for the new model and add a new input node name called `style_num` (these values are from the example Android code in `StylizeActivity.java`, but you can also use the `summarize_graph` tool, TensorBoard, or the code snippet we showed in the previous chapters to find them out):

```
std::string input_layer = "input";
std::string style_layer = "style_num";
std::string output_layer = "transformer/expand/conv3/conv/Sigmoid";
```

4. Different from the fast-style transfer models, the multi-style model here expects a 4-dimensional tensor of floating numbers as the image input:

```
tensorflow::Tensor image_tensor(tensorflow::DT_FLOAT,
tensorflow::TensorShape({1, wanted_height, wanted_width,
wanted_channels}));
auto image_tensor_mapped = image_tensor.tensor<float, 4>();
```

[115]

5. We also need to define `style_tensor` as another Tensor with the shape (NUM_STYLES*1), where NUM_STYLES is defined in the beginning of `ViewController.mm` as `const int NUM_STYLES = 26;`. The number 26 is the number of styles built into the `stylize_quantized.pb` model file, where you can run the Android TF Stylize app and see the 26 results, as shown in Figure 4.5. Notice that the 20th image, the one on the bottom left, is the familiar starry night by Van Gogh:

Figure 4.5 The 26 style images in the multi-style model

```
tensorflow::Tensor style_tensor(tensorflow::DT_FLOAT,
tensorflow::TensorShape({ NUM_STYLES, 1}));
auto style_tensor_mapped = style_tensor.tensor<float, 2>();
float* out_style = style_tensor_mapped.data();
for (int i = 0; i < NUM_STYLES; i++) {
    out_style[i] = 0.0 / NUM_STYLES;
}
out_style[19] = 1.0;
```

The sum of all the values in the `out_style` array needs to be 1, and the final style-transferred image will be the mix of the styles weighted by the values specified in the `out_style` array. For example, the preceding code will only use the starry night style (array index 19 corresponds to the 20th image in the style image list in Figure 4.5).

If you want an equal mix of the starry night image and the top-right image, you need to replace the last line in the preceding block of code with the following:

```
out_style[4] = 0.5;
out_style[19] = 0.5;
```

If you want an equal mix of all 26 styles, change the preceding for loop to the following and don't set other values to any specific `out_style` element:

```
for (int i = 0; i < NUM_STYLES; i++) {
    out_style[i] = 1.0 / NUM_STYLES;
}
```

You'll see in Figures 4.8 and 4.9 later the style transferred effects of all three of these settings.

6. Change the `session->Run` call to the following line to send both the image tensor and the style tensor to the model:

```
tensorflow::Status run_status = session->Run({{input_layer, image_tensor}, {style_layer, style_tensor}}, {output_layer}, {}, &outputs);
```

Those are all the changes you need to make to run your iOS app with the multi-style model. Run your app now and you'll see something like Figure 4.6 first:

Figure 4.6 Showing the original content image

Tap anywhere and you'll see two style choices, as shown in Figure 4.7:

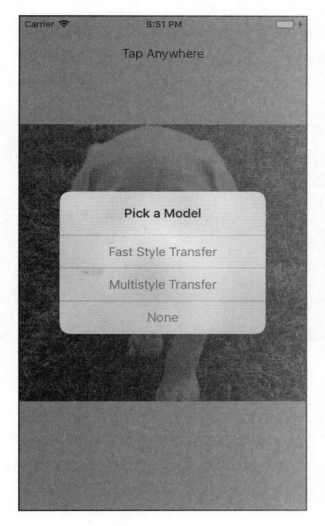

Figure 4.7 Showing the choice of two style models

The results of the two transferred images, with out_style[19] = 1.0; are shown in Figure 4.8:

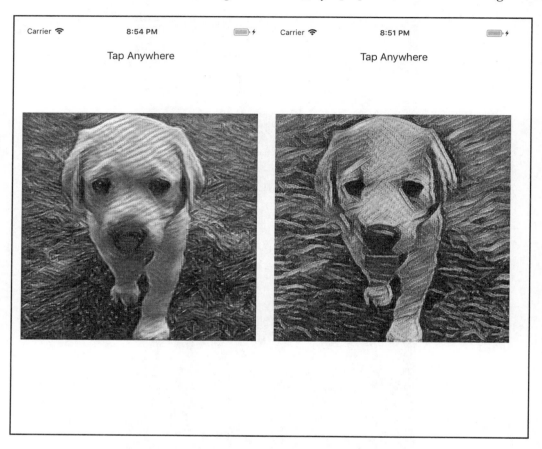

Figure 4.8 The style-transferred results with two different models (fast style transfer on the left, multi-style on the right)

The results of using an equal mix of the starry night image and the top right image in Figure 4.5, as well as an equal mix of all the 26 styles, are shown in Figure 4.9:

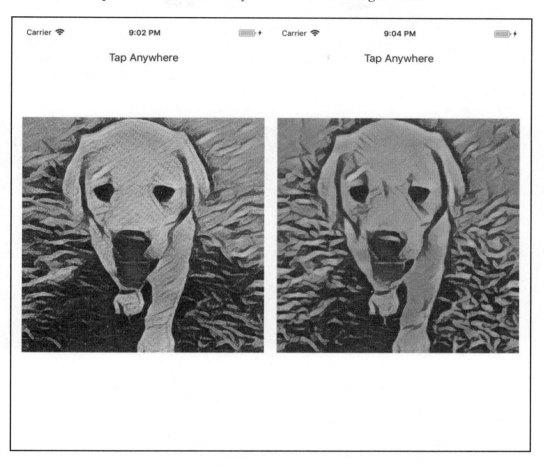

Figure 4.9 Results of different mixes of multiple styles (half starry night half the other on the left, mix of all 26 styles on the right)

The multi-style model runs on an iPhone 6 in about 5 seconds, about 2-3 times faster than the fast-style transfer model runs.

Using the TensorFlow Magenta multi-style model in Android

Although the TensorFlow Android example app already has code that uses the multi-style model (actually we used the model from the Android example app in our iOS app in the previous section), the TensorFlow-related code in the example app is mixed with a lot of UI code in the over 600-line `StylizeActivity.java` file. There's also a Codelab for TensorFlow Android Style Transfer (https://codelabs.developers.google.com/codelabs/tensorflow-style-transfer-android/index.html) you can go through, but the code there is about the same as the TensorFlow Android example app. Since we already have a minimalist implementation of an Android app using a TensorFlow fast-style transfer model, it'd be interesting to see how we can change just a few lines of code and have a powerful multi-style style-transfer app. This should also be a more intuitive way to see how you can add a great TensorFlow model to your existing Android app.

So this is what takes to use the multi-style transfer model in the Android app we built earlier:

1. Drag and drop the `stylize_quantized.pb` file from `tensorflow/examples/android/assets` to our Android app's `assets` folder.

2. In Android Studio, open `MainActivity.java`, locate these three lines of code:

   ```
   private static final String MODEL_FILE = "file:///android_asset/fst_frozen_quantized.pb";
   private static final String INPUT_NODE = "img_placeholder";
   private static final String OUTPUT_NODE = "preds";
   ```

   And then replace them with the following four lines:

   ```
   private static final int NUM_STYLES = 26;
   private static final String MODEL_FILE = "file:///android_asset/stylize_quantized.pb";
   private static final String INPUT_NODE = "input";
   private static final String OUTPUT_NODE = "transformer/expand/conv3/conv/Sigmoid";
   ```

   The values are the same as the iOS app we built in the previous section. If you only do Android app development and skipped the previous iOS section, just do a quick read on the explanation of step 3 in the previous iOS section.

3. Replace the following code snippet that feeds the input image to the fast-style transfer model and processes the output image:

```
mInferenceInterface.feed(INPUT_NODE, floatValues, WANTED_HEIGHT,
    WANTED_WIDTH, 3);
mInferenceInterface.run(new String[] {OUTPUT_NODE}, false);
mInferenceInterface.fetch(OUTPUT_NODE, outputValues);
for (int i = 0; i < intValues.length; ++i) {
    intValues[i] = 0xFF000000
                 | (((int) outputValues[i * 3]) << 16)
                 | (((int) outputValues[i * 3 + 1]) << 8)
                 | ((int) outputValues[i * 3 + 2]);
}
```

with the code snippet that first sets a `styleVals` array (if you're confused about the `styleVals` and how the values in the array can be set, check out the notes in step 5 of the previous section):

```
final float[] styleVals = new float[NUM_STYLES];
for (int i = 0; i < NUM_STYLES; ++i) {
    styleVals[i] = 0.0f / NUM_STYLES;
}
styleVals[19] = 0.5f;
styleVals[4] = 0.5f;
```

It then feeds both the input image tensor and the style value tensor to the model and runs the model to fetch the transferred image:

```
mInferenceInterface.feed(INPUT_NODE, floatValues, 1, WANTED_HEIGHT,
    WANTED_WIDTH, 3);
mInferenceInterface.feed("style_num", styleVals, NUM_STYLES);
mInferenceInterface.run(new String[] {OUTPUT_NODE}, false);
mInferenceInterface.fetch(OUTPUT_NODE, outputValues);
```

Finally, it processes the output:

```
for (int i=0; i < intValues.length; ++i) {
    intValues[i] = 0xFF000000
                 | (((int) (outputValues[i*3] * 255)) << 16)
                 | (((int) (outputValues[i*3+1] * 255)) << 8)
                 | ((int) (outputValues[i*3+2] * 255));
}
```

Notice the multi-style model returns an array of floating numbers to `outputValues`, each of which is in the range of 0.0 to 1.0, so we need to multiply each by 255 before applying the left bit shift operations to get the Red and Green values and then applying the bitwise OR to set the final ARGB value to each element of the `intValues` array.

That's all it takes to add the cool multi-style model to a standalone Android app. Now let's run the app and play with a different test image but the same three combinations of style values as in the iOS app.

With both the 20th and 5th style images equally mixed as in the code snippet in step 3, the original image and the transferred image are shown in Figure 4.10:

Figure 4.10 The original content image and the style transferred image with a mix of the 5th image and the starry night image

If you replace the following two lines of code:

```
styleVals[19] = 0.5f;
styleVals[4] = 0.5f;
```

with a single line of code, `styleVals[19] = 1.5f;`, or replace the following code snippet:

```
for (int i = 0; i < NUM_STYLES; ++i) {
    styleVals[i] = 0.0f / NUM_STYLES;
}
styleVals[19] = 0.5f;
styleVals[4] = 0.5f;
```

with this code snippet:

```
for (int i = 0; i < NUM_STYLES; ++i) {
    styleVals[i] = 1.0f / NUM_STYLES;
}
```

Then you'll see the effects in Figure 4.11:

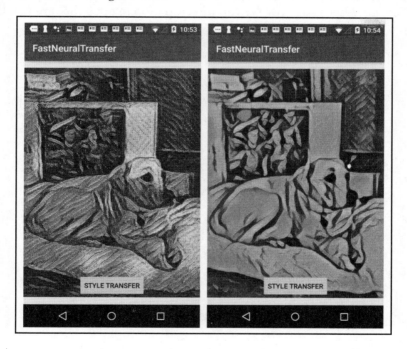

Figure 4.11 Images stylized with the starry night style only and with all the 26 styles mixed equally

Looks like we mobile developers can also be great artists, with the help of some powerful TensorFlow models and the knowledge of how to use them in mobile apps, as covered here.

Summary

In this chapter, we first gave an overview of different neural-style transfer methods developed since 2015. Then we showed how to train a second-generation style transfer model that's fast enough to run on mobile devices in a few seconds. After that, we covered how to use the model in both an iOS app and an Android app, built from scratch with a minimalist approach with fewer than 100 total lines of code. Finally, we talked about how to use a TensorFlow Magenta multi-style neural transfer model, which includes 26 amazing art styles in a single small model, in both iOS and Android apps.

In the next chapter, we'll explore another task deemed as intelligent when demonstrated by us humans or our best friends: to be able to recognize voice commands. Who wouldn't want our dogs to understand commands such as "sit," "come," "no," or our babies to respond to "yes," "stop," or "go"? Let's see how we can develop mobile apps that are just like them.

Understanding Simple Speech Commands

Voice services, such as Apple Siri, Amazon Alexa, Google Assistant, and Google Translate, have become more and more popular these days, as voice is the most natural and effective way for us to find information or accomplish tasks in certain scenarios. Many of those voice services are cloud-based because user speech can be pretty long and freeform, and **automatic speech recognition** (**ASR**) is very complicated and requires a lot of computing power. In fact, only in recent years has ASR in natural and noisy environments become feasible thanks to the breakthrough in deep learning.

But in some cases, it makes sense to be able to recognize simple speech commands offline on a device. For example, to control the movement of a Raspberry-Pi-driven robot, you don't need complicated voice commands, and not only is on-device ASR faster than a cloud-based solution, but it is also always available, even in an environment with no network access. Simple speech command recognition on a device can also save network bandwidth by only sending complicated user speech to the server when a certain clear user command is uttered.

In this chapter, we'll first give an overview of ASR technologies covering both state-of-the-art deep-learning-based systems and top open source projects. Then we'll discuss how to train and retrain a TensorFlow model to recognize simple speech commands such as "left," "right," "up," "down," "stop," and "go." Next, we'll use the trained model to build a simple Android app, followed by two complete iOS apps, one implemented with Objective-C and the other with Swift. We haven't covered Swift-based iOS app using TensorFlow models in the previous two chapters, and this chapter is a good place to review and strengthen our understanding of building Swift-based TensorFlow iOS apps.

In summary, we'll cover the following topics in this chapter:

- Speech recognition - a quick overview
- Training a simple commands-recognition model
- Using a simple speech recognition model in Android
- Using a simple speech recognition model in iOS with Objective-C
- Using a simple speech recognition model in iOS with Swift

Speech recognition – a quick overview

The first practical speaker-independent, large-vocabulary, and continuous speech recognition systems emerged in the 1990s. In the early 2000s, speech recognition engines offered by leading startups Nuance and SpeechWorks powered many of the first-generation web-based voice services, such as TellMe, AOL by Phone, and BeVocal. Speech recognition systems built then were mainly based on the traditional **Hidden Markov Models** (**HMM**) and required manually-written grammar and quiet environments to help the recognition engine work more accurately.

Modern speech recognition engines can pretty much understand any utterance by people under noisy environments and are based on end-to-end deep learning, especially another type of deep neural network more suitable for natural language processing, called **recurrent neural network** (**RNN**). Unlike traditional HMM-based speech recognition, where human expertise is required to build and fine-tune hand-designed features and acoustic and language models, end-to-end RNN-based speech recognition systems transcribe audio input directly to text, without the need to convert the audio input to a phonetic representation for further processing.

RNN allows us to handle sequences of input and/or output, because the network, by design, has memory of previous items in an input sequence or can generate a sequence of output. This makes RNN more appropriate for speech recognition, (where the input is a sequence of words uttered by users), image captioning, (where the output is a natural language sentence consisting of a series of words), text generation, and time series prediction. If you're unfamiliar with RNN, you should definitely check out *Andrey Karpathy's* blog, *The Unreasonable Effectiveness of Recurrent Neural Networks* (http://karpathy.github.io/2015/05/21/rnn-effectiveness). We'll also cover some detailed RNN models later in the book.

The first research paper on end-to-end speech recognition with RNN was published in 2014 (http://proceedings.mlr.press/v32/graves14.pdf) using a **Connectionist Temporal Classification** (**CTC**) layer. Later in 2014, Baidu released Deep Speech (https://arxiv.org/abs/1412.5567), one of the first commercial systems built with CTC-based end-to-end RNN but powered with huge datasets, and achieved significantly lower error rates in a noisy environment than traditional ASR systems. If you're interested, you can check out a TensorFlow implementation of Deep Speech (https://github.com/mozilla/DeepSpeech), but the generated model would require too many resources to run on a mobile phone, due to a problem with such CTC-based systems. It needs a large language model during deployment to correct generated text errors partially caused by the nature of RNN (read the RNN blog linked earlier to gain some insight if you wonder why).

Newer speech recognition systems in 2015 and 2016 used similar end-to-end RNN methods but replaced the CTC layer with an attention-based model (https://arxiv.org/pdf/1508.01211.pdf) so there's no need for a large language model when running the model, making it possible to deploy on mobile devices with limited memory. We won't explore such a possibility and cover how to use the latest advanced ASR models in mobile apps in this version of the book. Instead, we'll start with a simpler speech recognition model that we know will certainly work well on mobile devices.

To add offline speech-recognition capability to mobile apps, you can also use one of the two leading open source speech recognition projects:

- CMU Sphinx (https://cmusphinx.github.io), which started about 20 years ago but is still under active development. To build Android apps with speech recognition, you can use its PocketSphinx built for Android (https://github.com/cmusphinx/pocketsphinx-android). To build iOS apps with speech recognition, you can use the OpenEars framework (https://www.politepix.com/openears), a free SDK that uses CMU PocketSphinx to build offline speech recognition and text-to-speech in iOS apps.
- Kaldi (https://github.com/kaldi-asr/kaldi), which started in 2009 and has been very active recently, with 165 contributors as of January 2018. To try it on Android, you can check out this blog article: http://jcsilva.github.io/2017/03/18/compile-kaldi-android. For iOS, check out this prototype of using Kaldi on iOS: https://github.com/keenresearch/keenasr-ios-poc.

As this is a book about using TensorFlow on mobile devices, and TensorFlow can be used to build powerful models for image processing, speech processing, and text processing, among other intelligent tasks, in the rest of the chapter, we'll focus on how to use TensorFlow to train a simple speech recognition model and use it in mobile apps.

Training a simple commands recognition model

In this section, we'll summarize the steps used in the well-written TensorFlow Simple Audio Recognition tutorial (https://www.tensorflow.org/versions/master/tutorials/audio_recognition) and also add a few tips that can be of help to you while training the model.

The simple speech commands recognition model we'll build will be able to recognize 10 words: "yes," "no," "up," "down," "left," "right," "on," "off," "stop," and "go"; it can also detect silence. If it finds no silence and none of the 10 words, it'll generate "unknown." The speech commands dataset we'll download and use for training the model, when we run the `tensorflow/example/speech_commands/train.py` script later, actually contains 20 more words in addition to those 10 words: "zero," "two," "three," ..., "ten" (the 20 words you've seen so far are called the core words), and 10 auxiliary words: "bed," "bird," "cat," "dog," "happy," "house," "marvin," "sheila," "tree," and "wow." Core words just have more recorded .wav files (about 2350) than auxiliary words (about 1750).

The speech commands dataset is collected from an Open Speech Recording site (https://aiyprojects.withgoogle.com/open_speech_recording). You should give it a try and maybe contribute a few minutes of your own recordings to help it improve and also get a sense of how you can collect your own speech commands dataset if needed. There's also a Kaggle competition (https://www.kaggle.com/c/tensorflow-speech-recognition-challenge) on using the dataset to build a model and you can learn more about speech models and tips there.

The model to be trained and used in mobile apps is based on the paper Convolutional Neural Networks for Small-footprint Keyword Spotting (http://www.isca-speech.org/archive/interspeech_2015/papers/i15_1478.pdf), which is different from most other RNN-based large-scale speech recognition models. That a CNN-based model for speech recognition is possible is interesting but feasible, because for simple speech commands recognition, we can convert the audio signal in a short period of time to an image, or more accurately, a spectrogram, which is the frequency distribution of the audio signal during that time window (see the TensorFlow tutorial link at the beginning of this section for a sample spectrogram image generated using the `wav_to_spectrogram` script). In other words, we can transform an audio signal from its original time-domain representation to its frequency domain representation. The best algorithm for doing this conversion is **Discrete Fourier Transform (DFT)**, and **Fast Fourier Transform (FFT)** is just an efficient algorithm for DFT implementation.

As mobile developers, you probably don't need to understand DFT and FFT. But you'd better appreciate how all this model training works when used in mobile apps by knowing that behind the scenes of the TensorFlow simple speech commands model training that we're about to cover, it's the use of FFT, one of the top 10 algorithms in the 20th century, among other things of course, that makes the CNN-based speech command recognition model training possible. For a fun and intuitive tutorial on DFT, you can read this article: http://practicalcryptography.com/miscellaneous/machine-learning/intuitive-guide-discrete-fourier-transform.

Let's now perform the following steps to train a model of simple speech commands recognition:

1. On a Terminal, `cd` to your TensorFlow source root, likely `~/tensorflow-1.4.0`.
2. Simply run the following command to download the speech commands dataset we talked about earlier:

   ```
   python tensorflow/examples/speech_commands/train.py
   ```

   There are many arguments you can use: `--wanted_words` by default are the 10 core words starting with "yes"; you can add more words that can be recognized by your model with this argument. To train your own speech commands dataset, use `--data_url --data_dir=<path_to_your_dataset>` to disable the downloading of the speech commands dataset and access your own dataset, where each command should be named as its own folder, which should contain 1000-2000 audio clips all about 1 second long; if the audio clips are all longer, you can change the `--clip_duration_ms` argument value accordingly. See the `train.py` source code and the TensorFlow Simple Audio Recognition tutorial for more details.

3. If you accept all the default arguments for `train.py`, after downloading the 1.48 GB speech commands dataset, the whole training of 18,000 steps takes about 90 minutes on a GTX-1070 GPU-powered Ubuntu. After the training is done, you should see a list of checkpoint files inside the `/tmp/speech_commands_train` folder, as well as the `conv.pbtxt` graph definition file and a label file named `conv_labels.txt` that contains the list of commands (same as what the `--wanted_words` argument is as default or set to, plus two additional words, "_silence" and "_unknown", in the beginning of the file):

   ```
   -rw-rw-r-- 1 jeff jeff   75437 Dec  9 21:08 conv.ckpt-18000.meta
   -rw-rw-r-- 1 jeff jeff     433 Dec  9 21:08 checkpoint
   -rw-rw-r-- 1 jeff jeff 3707448 Dec  9 21:08
   ```

```
conv.ckpt-18000.data-00000-of-00001
-rw-rw-r-- 1 jeff jeff 315 Dec 9 21:08 conv.ckpt-18000.index
-rw-rw-r-- 1 jeff jeff 75437 Dec 9 21:08 conv.ckpt-17900.meta
-rw-rw-r-- 1 jeff jeff 3707448 Dec 9 21:08
conv.ckpt-17900.data-00000-of-00001
-rw-rw-r-- 1 jeff jeff 315 Dec 9 21:08 conv.ckpt-17900.index
-rw-rw-r-- 1 jeff jeff 75437 Dec 9 21:07 conv.ckpt-17800.meta
-rw-rw-r-- 1 jeff jeff 3707448 Dec 9 21:07
conv.ckpt-17800.data-00000-of-00001
-rw-rw-r-- 1 jeff jeff 315 Dec 9 21:07 conv.ckpt-17800.index
-rw-rw-r-- 1 jeff jeff 75437 Dec 9 21:07 conv.ckpt-17700.meta
-rw-rw-r-- 1 jeff jeff 3707448 Dec 9 21:07
conv.ckpt-17700.data-00000-of-00001
-rw-rw-r-- 1 jeff jeff 315 Dec 9 21:07 conv.ckpt-17700.index
-rw-rw-r-- 1 jeff jeff 75437 Dec 9 21:06 conv.ckpt-17600.meta
-rw-rw-r-- 1 jeff jeff 3707448 Dec 9 21:06
conv.ckpt-17600.data-00000-of-00001
-rw-rw-r-- 1 jeff jeff 315 Dec 9 21:06 conv.ckpt-17600.index
-rw-rw-r-- 1 jeff jeff 60 Dec 9 19:41 conv_labels.txt
-rw-rw-r-- 1 jeff jeff 121649 Dec 9 19:41 conv.pbtxt
```

The `conv_labels.txt` contains the following commands:

```
_silence_
_unknown_
yes
no
up
down
left
right
on
off
stop
go
```

Now run the following command to combine the graph definition file and the checkpoint file into a single model file that we can use in mobile apps:

```
python tensorflow/examples/speech_commands/freeze.py \
--start_checkpoint=/tmp/speech_commands_train/conv.ckpt-18000 \
--output_file=/tmp/speech_commands_graph.pb
```

4. Optionally, before you deploy the `speech_commands_graph.pb` model file in mobile apps, you can give it a quick test using the following command:

```
python tensorflow/examples/speech_commands/label_wav.py \
--graph=/tmp/speech_commands_graph.pb \
--labels=/tmp/speech_commands_train/conv_labels.txt \
--wav=/tmp/speech_dataset/go/9d171fee_nohash_1.wav
```

You'll see an output like this:

```
go (score = 0.48427)
no (score = 0.17657)
_unknown_ (score = 0.08560)
```

5. Use the `summarize_graph` tool to find out the names of input nodes and output nodes:

```
bazel-bin/tensorflow/tools/graph_transforms/summarize_graph --
in_graph=/tmp/speech_commands_graph.pb
```

The output should be as follows:

```
Found 1 possible inputs: (name=wav_data, type=string(7), shape=[])
No variables spotted.
Found 1 possible outputs: (name=labels_softmax, op=Softmax)
```

Unfortunately, it's only correct for the output name, and doesn't show other possible inputs. Using `tensorboard --logdir /tmp/retrain_logs` and then opening `http://localhost:6006` on a browser to interact with the graph doesn't help either. But our little code snippet shown in the previous chapters to find out about the input and output names helps—the following is the interaction with iPython:

```
In [1]: import tensorflow as tf
In [2]: g=tf.GraphDef()
In [3]:
g.ParseFromString(open("/tmp/speech_commands_graph.pb","rb").read())
In [4]: x=[n.name for n in g.node]
In [5]: x
Out[5]:
[u'wav_data',
 u'decoded_sample_data',
 u'AudioSpectrogram',
 ...
 u'MatMul',
 u'add_2',
 u'labels_softmax']
```

So we see both `wav_data` and `decoded_sample_data` are possible input. We'd have to dig deep into the model training code to find out exactly which input names we should use if we don't see a comment in the `freeze.py` file: "The resulting graph has an input for WAV-encoded data named `wav_data`, one for raw PCM data (as floats in the range -1.0 to 1.0) called `decoded_sample_data`, and the output is called `labels_softmax`." Actually, in the case of this model, there's a TensorFlow Android example app, part of what we saw in `Chapter 1`, *Getting Started with Mobile TensorFlow*, called TF Speech that specifically defines those input names and output names. In some chapters later in the book, you'll see how to look into the source code for model training when needed to find out the crucial input and output node names, with or without the help of our three methods. Or hopefully, by the time you read this book, the TensorFlow `summarize_graph` tool will be improved to give us accurate input and output node names.

Now it's time to use our hot new model in mobile apps.

Using a simple speech recognition model in Android

The TensorFlow Android example app for simple speech commands recognition, located at `tensorflow/example/android`, has code that does audio recording and recognition in the `SpeechActivity.java` file, which assumes the app needs to be always ready for new audio commands. While this certainly makes sense in some cases, it also results in code that's more complicated than the code that would do recording and recognition only after the user presses a button, like the way Apple's Siri works. In this section, we'll show you how to create a new Android app and add the minimum possible code to record users' speech commands and display recognition results. This should help you more easily integrate the model to your own Android app. But if you need to deal with the case where speech commands should always be automatically recorded and recognized, you should check out the TensorFlow example Android app.

Building a new app using the model

Perform the following steps to build a complete new Android app that uses the `speech_commands_graph.pb` model we built in the last section:

1. Create a new Android app named `AudioRecognition` by accepting all the defaults as in the previous chapters, then add the `compile 'org.tensorflow:tensorflow-android:+'` line to the end of the app's `build.gradle` file's dependencies.
2. Add `<uses-permission android:name="android.permission.RECORD_AUDIO" />` to the app's `AndroidManifest.xml` file so the app can be allowed to record audio.
3. Create a new assets folder, then drag and drop the `speech_commands_graph.pb` and `conv_actions_labels.txt` files, generated in steps 2 and 3 in the previous section, to the `assets` folder.
4. Change the `activity_main.xml` file to hold three UI elements. The first one is a `TextView` for recognition result display:

   ```
   <TextView
       android:id="@+id/textview"
       android:layout_width="wrap_content"
       android:layout_height="wrap_content"
       android:text=""
       android:textSize="24sp"
       android:textStyle="bold"
       app:layout_constraintBottom_toBottomOf="parent"
       app:layout_constraintLeft_toLeftOf="parent"
       app:layout_constraintRight_toRightOf="parent"
       app:layout_constraintTop_toTopOf="parent" />
   ```

   The second `TextView` is to display the 10 default commands we have trained using the `train.py` Python program in step 2 of the last section:

   ```
   <TextView
       android:layout_width="wrap_content"
       android:layout_height="wrap_content"
       android:text="yes no up down left right on off stop go"
       app:layout_constraintBottom_toBottomOf="parent"
       app:layout_constraintHorizontal_bias="0.50"
       app:layout_constraintLeft_toLeftOf="parent"
       app:layout_constraintRight_toRightOf="parent"
       app:layout_constraintTop_toTopOf="parent"
       app:layout_constraintVertical_bias="0.25" />
   ```

The last UI element is a Button that, when tapped, starts recording audio for one second and then send the recording to our model for recognition:

```xml
<Button
    android:id="@+id/button"
    android:layout_width="wrap_content"
    android:layout_height="wrap_content"
    android:text="Start"
    app:layout_constraintBottom_toBottomOf="parent"
    app:layout_constraintHorizontal_bias="0.50"
    app:layout_constraintLeft_toLeftOf="parent"
    app:layout_constraintRight_toRightOf="parent"
    app:layout_constraintTop_toTopOf="parent"
    app:layout_constraintVertical_bias="0.8" />
```

5. Open `MainActivity.java`, first make the `MainActivity` implements `Runnable` class. Then add the following constants defining the model name, label name, input names, and output name:

```java
private static final String MODEL_FILENAME = "file:///android_asset/speech_commands_graph.pb";
private static final String LABEL_FILENAME = "file:///android_asset/conv_actions_labels.txt";
private static final String INPUT_DATA_NAME = "decoded_sample_data:0";
private static final String INPUT_SAMPLE_RATE_NAME = "decoded_sample_data:1";
private static final String OUTPUT_NODE_NAME = "labels_softmax";
```

6. Declare four instance variables:

```java
private TensorFlowInferenceInterface mInferenceInterface;
private List<String> mLabels = new ArrayList<String>();
private Button mButton;
private TextView mTextView;
```

7. In the `onCreate` method, we first instantiate `mButton` and `mTextView` then set up the button click event handler, which first changes the button title, then launches a thread to do recording and recognition:

```java
mButton = findViewById(R.id.button);
mTextView = findViewById(R.id.textview);
mButton.setOnClickListener(new View.OnClickListener() {
    @Override
    public void onClick(View v) {
        mButton.setText("Listening...");
        Thread thread = new Thread(MainActivity.this);
```

```
            thread.start();
        }
});
```

At the end of the `onCreate` method, we read the content of the label file line by line and save each line in the `mLabels` array list.

8. In the beginning of the `public void run()` method, started when the Start button is clicked, add the code that first gets the minimum buffer size for creating an Android `AudioRecord` object, then uses `buffersize` to create a new `AudioRecord` instance with a 16,000 `SAMPLE_RATE` and 16-bit mono format, the type of raw audio expected by our model, and finally starts recording from the `AudioRecord` instance:

```
int bufferSize = AudioRecord.getMinBufferSize(SAMPLE_RATE,
AudioFormat.CHANNEL_IN_MONO, AudioFormat.ENCODING_PCM_16BIT);
AudioRecord record = new
AudioRecord(MediaRecorder.AudioSource.DEFAULT, SAMPLE_RATE,
AudioFormat.CHANNEL_IN_MONO, AudioFormat.ENCODING_PCM_16BIT,
bufferSize);

if (record.getState() != AudioRecord.STATE_INITIALIZED) return;
record.startRecording();
```

There are two classes in Android for recording audio: `MediaRecorder` and `AudioRecord`. `MediaRecorder` is easier to use than `AudioRecord`, but it saves compressed audio files until Android API Level 24 (Android 7.0), which supports recording raw, unprocessed audio. According to https://developer.android.com/about/dashboards/index.html, as of January 2018, there are more than 70% of Android devices in the market that still run Android versions older than 7.0. You probably would prefer not to target your app to Android 7.0 or above. In addition, to decode the compressed audio recorded by `MediaRecorder`, you have to use `MediaCodec`, which is pretty complicated to use. `AudioRecord`, albeit a low-level API, is actually perfect for recording raw unprocessed data which is then sent to the speech commands recognition model for processing.

Understanding Simple Speech Commands

9. Create two arrays of 16-bit short integers, `audioBuffer` and `recordingBuffer`, and for 1-second recording, every time after the `AudioRecord` object reads and fills the `audioBuffer` array, the actual data read gets appended to the `recordingBuffer`:

```
long shortsRead = 0;
int recordingOffset = 0;
short[] audioBuffer = new short[bufferSize / 2];
short[] recordingBuffer = new short[RECORDING_LENGTH];
while (shortsRead < RECORDING_LENGTH) { // 1 second of recording
    int numberOfShort = record.read(audioBuffer, 0, audioBuffer.length);
    shortsRead += numberOfShort;
    System.arraycopy(audioBuffer, 0, recordingBuffer, recordingOffset, numberOfShort);
    recordingOffset += numberOfShort;
}
record.stop();
record.release();
```

10. After the recording is done, we first change the button title to `Recognizing`:

```
runOnUiThread(new Runnable() {
    @Override
    public void run() {
        mButton.setText("Recognizing...");
    }
});
```

Then convert the `recordingBuffer` short array to a `float` array, also making each element of the `float` array in the range of -1.0 and 1.0, as our model expects floats between -1.0 and 1.0:

```
float[] floatInputBuffer = new float[RECORDING_LENGTH];
for (int i = 0; i < RECORDING_LENGTH; ++i) {
    floatInputBuffer[i] = recordingBuffer[i] / 32767.0f;
}
```

11. Create a new `TensorFlowInferenceInterface` as we did in the previous chapters, then call its `feed` method with two input nodes' names and values, one of which is the sample rate and the other is the raw audio data stored in the `floatInputBuffer` array:

```
AssetManager assetManager = getAssets();
mInferenceInterface = new TensorFlowInferenceInterface(assetManager, MODEL_FILENAME);
```

```
int[] sampleRate = new int[] {SAMPLE_RATE};
mInferenceInterface.feed(INPUT_SAMPLE_RATE_NAME, sampleRate);

mInferenceInterface.feed(INPUT_DATA_NAME, floatInputBuffer,
RECORDING_LENGTH, 1);
```

After that, we call the `run` method to run the recognition inference on our model and then `fetch` the output scores for each of the 10 speech commands and the "unknown" and "silence" output:

```
String[] outputScoresNames = new String[] {OUTPUT_NODE_NAME};
mInferenceInterface.run(outputScoresNames);

float[] outputScores = new float[mLabels.size()];
mInferenceInterface.fetch(OUTPUT_NODE_NAME, outputScores);
```

12. The `outputScores` array matches the `mLabels` list so we can easily find the top score and get its command name:

```
float max = outputScores[0];
int idx = 0;
for (int i=1; i<outputScores.length; i++) {
    if (outputScores[i] > max) {
        max = outputScores[i];
        idx = i;
    }
}
final String result = mLabels.get(idx);
```

Finally, we show the result in a `TextView` and change the button title back to `"Start"` so users can start to record and recognize speech commands again:

```
runOnUiThread(new Runnable() {
    @Override
    public void run() {
        mButton.setText("Start");
        mTextView.setText(result);
    }
});
```

Showing model-powered recognition results

Now run the app on your Android device. You'll see the initial screen as in Figure 5.1:

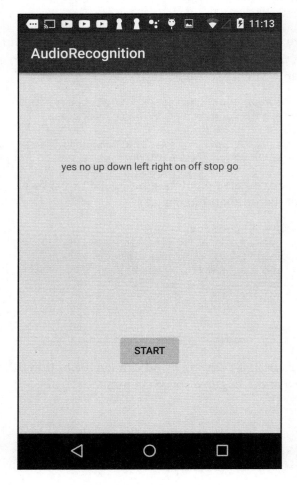

Figure 5.1 Showing the initial screen after app starts

Tap on the **START** button, and start saying one of the 10 commands shown on top. You'll see the button title changed to **Listening...** then **Recognizing...**, as shown in Figure 5.2:

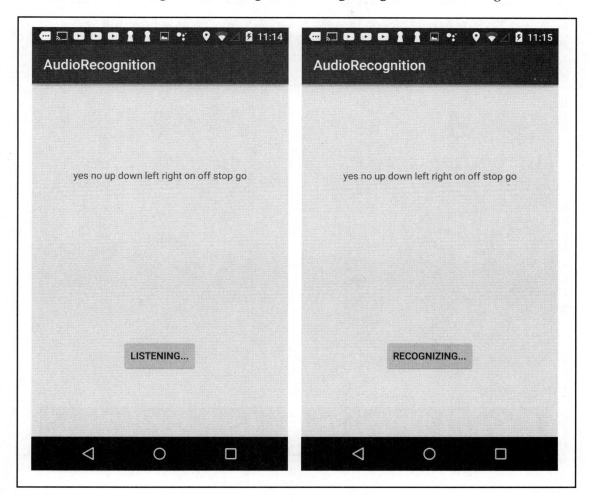

Figure 5.2 Listening to record audio and recognizing the recorded audio

Understanding Simple Speech Commands

Almost in real time, the recognized result shows in the middle of screen, like Figure 5.3:

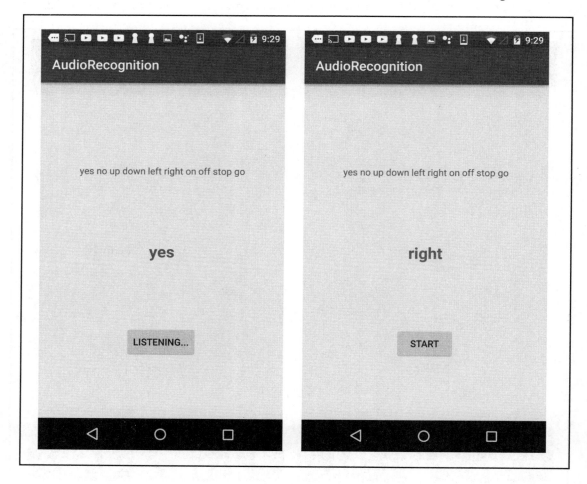

Figure 5.3 Showing the recognized speech commands

The whole recognition process completes almost instantly and the `speech_commands_graph.pb` model used for the recognition is only 3.7 MB. Sure, it only supports 10 speech commands, but the size won't change much even if dozens of commands are supported using the `train.py` script's `--wanted_words` argument or your own dataset, as we discussed in the training section.

Admittedly, the app screenshots here are not as colorful and attractive as those in the previous chapter (a picture is worth a thousand words), but speech recognition can certainly do something an artist can't do, such as issuing voice commands to control a robot's movement.

The complete source code of the app is in the `Ch5/android` folder of the book's source code repository on Github. Now let's see how we can build an iOS app using the model, which involves some tricky TensorFlow iOS library building and audio data preparation steps for the model to run correctly.

Using a simple speech recognition model in iOS with Objective-C

If you have gone through the iOS apps in the previous three chapters, you'd probably prefer using the manually-built TensorFlow iOS library to the TensorFlow experimental pod, as with the manual library method, you have more control on what TensorFlow operations can be added to make your models happy, and this better control is also one of the reasons we decided to focus on TensorFlow Mobile instead of TensorFlow Lite in `Chapter 1`, *Getting Started with Mobile TensorFlow*.

So although you can try to use the TensorFlow pod by the time you read the book to see whether the pod has been updated to support all the ops used in the model, we'll from now on always use the manually-built TensorFlow libraries (see steps 1 and 2 of the *Using object detection models in iOS* section in `Chapter 3`, *Detecting Objects and Their Locations*, for how) in our iOS apps.

Building a new app using the model

Now perform the following steps to create a new iOS app to use the speech commands recognition model:

1. Create a new Objective-C app named AudioRecognition in Xcode, and set the project to use TensorFlow manually-built libraries as summarized in step 1 of the *Using fast neural style transfer models in iOS* section of `Chapter 4`, *Transforming Pictures with Amazing Art Styles*. Also add `AudioToolbox.framework`, `AVFoundation.framework`, and `Accelerate.framework` to the Target's Link Binary With Libraries.

Understanding Simple Speech Commands

2. Drag and drop the `speech_commands_graph.pb` model file to the project.
3. Change the extension of `ViewController.m` to `mm`, then add the following headers used by audio recording and processing:

```
#import <AVFoundation/AVAudioRecorder.h>
#import <AVFoundation/AVAudioSettings.h>
#import <AVFoundation/AVAudioSession.h>
#import <AudioToolbox/AudioToolbox.h>
```

Also add the headers for TensorFlow:

```
#include <fstream>
#include "tensorflow/core/framework/op_kernel.h"
#include "tensorflow/core/framework/tensor.h"
#include "tensorflow/core/public/session.h"
```

Now, define an audio `SAMPLE_RATE` constant, a C-pointer to a float array holding the audio data to be sent to the model, our key `audioRecognition` function signature, and two properties holding the recorded file path and an iOS `AVAudioRecorder` instance. We also need to let `ViewController` implement the `AudioRecorderDelegate` so it knows when the recording is finished:

```
const int SAMPLE_RATE = 16000;
float *floatInputBuffer;
std::string audioRecognition(float* floatInputBuffer, int length);
@interface ViewController () <AVAudioRecorderDelegate>
@property (nonatomic, strong) NSString *recorderFilePath;
@property (nonatomic, strong) AVAudioRecorder *recorder;
@end
```

We won't show here the code snippet that programmatically creates two UI elements: a button that when tapped starts recording for 1 second of audio and then sends the audio to our model for recognition, and a label that shows the recognition results. But we'll show some UI code in Swift in the next section for a refresher.

4. Inside the button's `UIControlEventTouchUpInside` handler, we first create an `AVAudioSession` instance and set its category as record and make it active:

```
AVAudioSession *audioSession = [AVAudioSession sharedInstance];
NSError *err = nil;
[audioSession setCategory:AVAudioSessionCategoryPlayAndRecord error:&err];
if(err){
    NSLog(@"audioSession: %@", [[err userInfo] description]);
```

```
        return;
    }

    [audioSession setActive:YES error:&err];
    if(err){
        NSLog(@"audioSession: %@", [[err userInfo] description]);
        return;
    }
```

Then create a dictionary of record settings:

```
NSMutableDictionary *recordSetting = [[NSMutableDictionary alloc]
init];
[recordSetting setValue:[NSNumber
numberWithInt:kAudioFormatLinearPCM] forKey:AVFormatIDKey];
[recordSetting setValue:[NSNumber numberWithFloat:SAMPLE_RATE]
forKey:AVSampleRateKey];
[recordSetting setValue:[NSNumber numberWithInt: 1]
forKey:AVNumberOfChannelsKey];
[recordSetting setValue :[NSNumber numberWithInt:16]
forKey:AVLinearPCMBitDepthKey];
[recordSetting setValue :[NSNumber numberWithBool:NO]
forKey:AVLinearPCMIsBigEndianKey];
[recordSetting setValue :[NSNumber numberWithBool:NO]
forKey:AVLinearPCMIsFloatKey];
[recordSetting setValue:[NSNumber numberWithInt:AVAudioQualityMax]
forKey:AVEncoderAudioQualityKey];
```

Finally, in the button tap handler, we define where to save the recorded audio, create an `AVAudioRecorder` instance, set its delegate and start recording for 1 second:

```
self.recorderFilePath = [NSString
stringWithFormat:@"%@/recorded_file.wav", [NSHomeDirectory()
stringByAppendingPathComponent:@"tmp"]];
NSURL *url = [NSURL fileURLWithPath:_recorderFilePath];
err = nil;
_recorder = [[ AVAudioRecorder alloc] initWithURL:url
settings:recordSetting error:&err];
if(!_recorder){
    NSLog(@"recorder: %@", [[err userInfo] description]);
    return;
}
[_recorder setDelegate:self];
[_recorder prepareToRecord];
[_recorder recordForDuration:1];
```

5. In the delegate method of `AVAudioRecorderDelegate`, `audioRecorderDidFinishRecording`, we use Apple's Extended Audio File Services, which is for reading and writing compressed and linear PCM audio files, to load the recorded audio, convert it to the format expected by the model, and read the audio data into the memory. We won't show this part of the code here, which is mainly based on this blog: https://batmobile.blogs.ilrt.org/loading-audio-file-on-an-iphone/. After this processing, `floatInputBuffer` points to the raw audio samples. Now we can pass the data to our `audioRecognition` method in a worker thread and show the result in the UI thread:

```
dispatch_async(dispatch_get_global_queue(0, 0), ^{
    std::string command = audioRecognition(floatInputBuffer, totalRead);
    delete [] floatInputBuffer;
    dispatch_async(dispatch_get_main_queue(), ^{
        NSString *cmd = [NSString stringWithCString:command.c_str() encoding:[NSString defaultCStringEncoding]];
        [_lbl setText:cmd];
        [_btn setTitle:@"Start" forState:UIControlStateNormal];
    });
});
```

6. Inside the `audioRecognition` method, we first define a C++ `string` array with the 10 commands to be recognized as well as the two special values, "_silence_" and "_unknown_":

```
std::string commands[] = {"_silence_", "_unknown_", "yes", "no", "up", "down", "left", "right", "on", "off", "stop", "go"};
```

After the standard TensorFlow `Session`, `Status`, and `GraphDef` setup, as we have done in the iOS apps in the previous chapters, we read out model file and try to create a TensorFlow `Session` with it:

```
NSString* network_path = FilePathForResourceName(@"speech_commands_graph", @"pb");

PortableReadFileToProto([network_path UTF8String], &tensorflow_graph);

tensorflow::Status s = session->Create(tensorflow_graph);
if (!s.ok()) {
    LOG(ERROR) << "Could not create TensorFlow Graph: " << s;
    return "";
}
```

If the session gets created successfully, we define two input node names and one output node name for the model:

```
std::string input_name1 = "decoded_sample_data:0";
std::string input_name2 = "decoded_sample_data:1";
std::string output_name = "labels_softmax";
```

7. For `"decoded_sample_data:0"`, we need to send the sample rate value as a scalar (or you'll get an error when calling the TensorFlow Session's `run` method), and the tensor is defined in the TensorFlow C++ API as follows:

```
tensorflow::Tensor samplerate_tensor(tensorflow::DT_INT32,
tensorflow::TensorShape());
samplerate_tensor.scalar<int>()() = SAMPLE_RATE;
```

For `"decoded_sample_data:1"`, the audio data in float numbers needs to be converted from the `floatInputBuffer` array to a TensorFlow `audio_tensor` tensor, in a way similar to how the `image_tensor` was defined and set in the previous chapters:

```
tensorflow::Tensor audio_tensor(tensorflow::DT_FLOAT,
tensorflow::TensorShape({length, 1}));
auto audio_tensor_mapped = audio_tensor.tensor<float, 2>();
float* out = audio_tensor_mapped.data();
for (int i = 0; i < length; i++) {
    out[i] = floatInputBuffer[i];
}
```

Now we can run the model like before with the inputs and get the output:

```
std::vector<tensorflow::Tensor> outputScores;
tensorflow::Status run_status = session->Run({{input_name1,
audio_tensor}, {input_name2, samplerate_tensor}},{output_name}, {},
&outputScores);
if (!run_status.ok()) {
    LOG(ERROR) << "Running model failed: " << run_status;
    return "";
}
```

Understanding Simple Speech Commands

8. We do a simple parsing on the model's `outputScores` output and return the top score. The `outputScores` is a vector of the TensorFlow tensor, with its first element holding the 12 score values for the 12 possible recognition results. Those 12 score values can be accessed via the `flat` method and checked for the max score:

```
tensorflow::Tensor* output = &outputScores[0];
const Eigen::TensorMap<Eigen::Tensor<float, 1, Eigen::RowMajor>,
Eigen::Aligned>& prediction = output->flat<float>();
const long count = prediction.size();
int idx = 0;
float max = prediction(0);
for (int i = 1; i < count; i++) {
    const float value = prediction(i);
    printf("%d: %f", i, value);
    if (value > max) {
        max = value;
        idx = i;
    }
}

return commands[idx];
```

One other thing you need to do before the app can record any audio is create a new *Privacy - Microphone Usage Description* property in the app's `Info.plist` file and set the property's value to something like "to hear and recognize your voice commands".

Now run the app on either the iOS simulator (if your Xcode version is older than 9.2 and iOS simulator version older than 10.0, you may have to run the app on your actual iOS device as chances are you can't record audio in the iOS simulator before 10.0) or an iPhone, you'll first see the initial screen with a **Start** button in the center, then tap the button and say one of the 10 commands, the recognition result should appear on the top, as in Figure 5.4:

Chapter 5

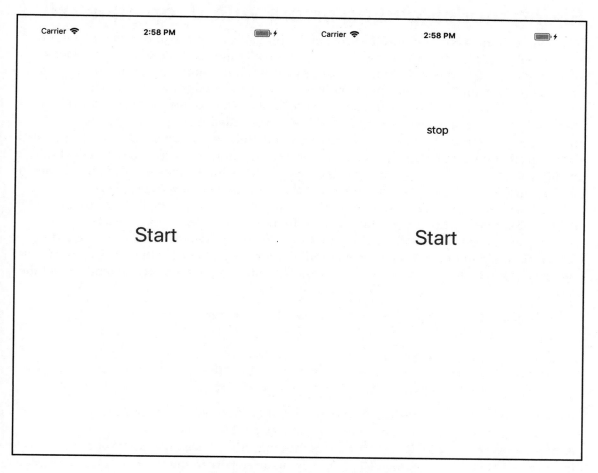

Figure 5.4 Showing the initial screen and recognition result

Yes, the recognition result should appear, but actually it won't, because you'll get an error in your Xcode output pane:

```
Could not create TensorFlow Graph: Not found: Op type not registered
'DecodeWav' in binary running on XXX's-MacBook-Pro.local. Make sure the Op
and Kernel are registered in the binary running in this process.
```

Fixing model-loading errors with tf_op_files.txt

We have seen this kind of notorious error in the previous chapters, and it could take a lot of pain to figure out the fix unless you know what it really means. TensorFlow operations consist of two parts: the definition, called ops (this is a little confusing, as an op can mean either both its definition and its implementation, or just its definition), located in the `tensorflow/core/ops` folder, and the implementation, called kernels, located in the `tensorflow/core/kernels` folder. There's a file called `tf_op_files.txt` in the `tensorflow/contrib/makefile` folder that lists the definitions and implementations of the ops that need to be built into the TensorFlow iOS library when you manually build the library. The `tf_op_files.txt` file is supposed to include all the ops definitions files, as documented in the *TensorFlow Preparing models for mobile deployment* (https://www.tensorflow.org/mobile/prepare_models), because they take up very little space. But not all the operations' op definitions are included in the `tf_op_files.txt` file as of TensorFlow 1.4 or 1.5 . So when we see an "Op type not registered" error, we need to find out which op definition and implementation files are responsible for that op. In our case, the op type is named `DecodeWav`. We can run the following two shell commands to get the info:

```
$ grep 'REGISTER.*"DecodeWav"' tensorflow/core/ops/*.cc
tensorflow/core/ops/audio_ops.cc:REGISTER_OP("DecodeWav")
```

```
$ grep 'REGISTER.*"DecodeWav"' tensorflow/core/kernels/*.cc
tensorflow/core/kernels/decode_wav_op.cc:REGISTER_KERNEL_BUILDER(Name("DecodeWav").Device(DEVICE_CPU), DecodeWavOp);
```

In the `tf_op_files.txt` file of TensorFlow 1.4, there's already a line of text, `tensorflow/core/kernels/decode_wav_op.cc`, but sure enough, `tensorflow/core/ops/audio_ops.cc` is missing. All we need to do is to add a line, `tensorflow/core/ops/audio_ops.cc`, to anywhere in the `tf_op_files.txt` file, and run `tensorflow/contrib/makefile/build_all_ios.sh` as we did in Chapter 3, *Detecting Objects and Their Locations*, to rebuild the TensorFlow iOS library. Then run the iOS app again and keep tapping the **Start** button and speaking voice commands to be recognized or misunderstood till you're bored.

 The process on how to fix the "Not found: Op type not registered" error is a big takeaway from this chapter, as it can possibly save you a lot of time when working on other TensorFlow models in the future.

But before you jump ahead to see what another new TensorFlow AI model will be covered and used in the next chapter, let's give other iOS developers who prefer to use the newer and, at least to them, cooler Swift language some consideration.

Using a simple speech recognition model in iOS with Swift

We created a Swift-based iOS app that uses the TensorFlow pod in Chapter 2, *Classifying Images with Transfer Learning*. Let's now create a new Swift app that uses the TensorFlow iOS libraries we manually built in the last section and use the speech commands model in our Swift app:

1. Create a new Single View iOS project from Xcode, and set up the project in the same way as steps 1 and 2 in the previous section, except set the **Language** as Swift.
2. Select Xcode **File | New | File** ... and select **Objective-C File**. Enter the name `RunInference`. You'll see a message box asking you "Would you like to configure an Objective-C bridging header?" Click the **Create Bridging Header**. Rename the file `RunInference.m` to `RunInfence.mm` as we'll mix our C, C++, and Objective-C code to do post-recording audio processing and recognition. We're still using Objective-C in the Swift app because to call the TensorFlow C++ code from Swift, we need to have an Objective-C class as a wrapper to the C++ code.
3. Create a header file called `RunInference.h`, and add this code to it:

```
@interface RunInference_Wrapper : NSObject
- (NSString *)run_inference_wrapper:(NSString*)recorderFilePath;
@end
```

Your app in Xcode should look like Figure 5.5 now:

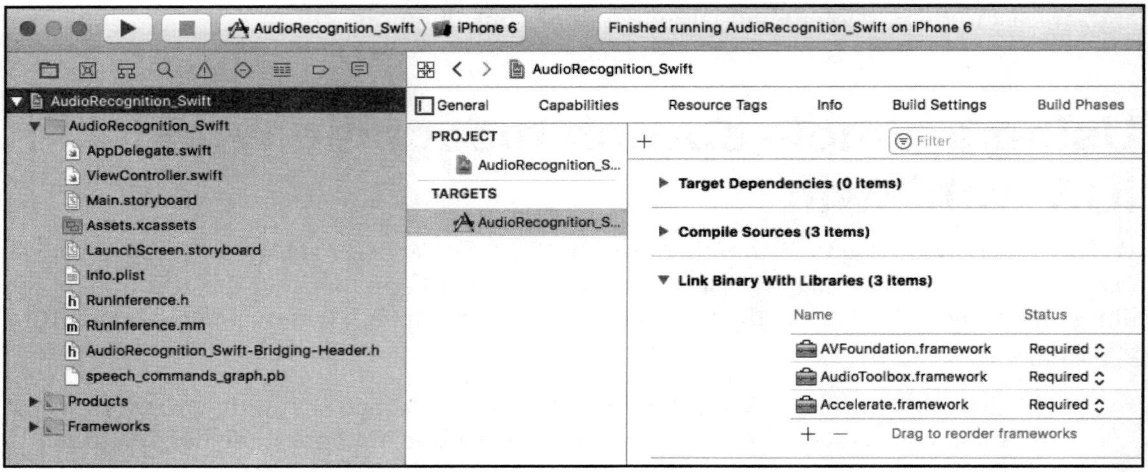

Figure 5.5 The Swift-based iOS app project

4. Open `ViewController.swift`. Add the following code at the top after `import UIKit`:

```
import AVFoundation

let _lbl = UILabel()
let _btn = UIButton(type: .system)
var _recorderFilePath: String!
```

Then make the `ViewController` look like this (the code snippet that defines `NSLayoutConstraint` for the _btn and _lbl and calls `addConstraint` is not showing):

```
class ViewController: UIViewController, AVAudioRecorderDelegate {
    var audioRecorder: AVAudioRecorder!
override func viewDidLoad() {
    super.viewDidLoad()
    _btn.translatesAutoresizingMaskIntoConstraints = false
    _btn.titleLabel?.font = UIFont.systemFont(ofSize:32)
    _btn.setTitle("Start", for: .normal)
    self.view.addSubview(_btn)
    _btn.addTarget(self, action:#selector(btnTapped), for: .touchUpInside)

    _lbl.translatesAutoresizingMaskIntoConstraints = false
```

```
        self.view.addSubview(_lbl)
```

5. Add a button tap handler and inside it, first request the user's permission for recording:

```
@objc func btnTapped() {
    _lbl.text = "..."
    _btn.setTitle("Listening...", for: .normal)
    AVAudioSession.sharedInstance().requestRecordPermission () {
        [unowned self] allowed in
        if allowed {
            print("mic allowed")
        } else {
            print("denied by user")
            return
        }
    }
```

Then create an `AudioSession` instance and set its category to record and status to active, just like what we did in the Objective-C version:

```
let audioSession = AVAudioSession.sharedInstance()

do {
    try audioSession.setCategory(AVAudioSessionCategoryRecord)
    try audioSession.setActive(true)
} catch {
    print("recording exception")
    return
}
```

Now define the settings to be used by `AVAudioRecorder`:

```
let settings = [
    AVFormatIDKey: Int(kAudioFormatLinearPCM),
    AVSampleRateKey: 16000,
    AVNumberOfChannelsKey: 1,
    AVLinearPCMBitDepthKey: 16,
    AVLinearPCMIsBigEndianKey: false,
    AVLinearPCMIsFloatKey: false,
    AVEncoderAudioQualityKey: AVAudioQuality.high.rawValue
    ] as [String : Any]
```

Set the file path to save the recorded audio, create an `AVAudioRecorder` instance, set its delegate and start recording for 1 second:

```
do {
    _recorderFilePath =
NSHomeDirectory().stringByAppendingPathComponent(path:
"tmp").stringByAppendingPathComponent(path: "recorded_file.wav")
    audioRecorder = try AVAudioRecorder(url:
NSURL.fileURL(withPath: _recorderFilePath), settings: settings)
    audioRecorder.delegate = self
    audioRecorder.record(forDuration: 1)
} catch let error {
    print("error:" + error.localizedDescription)
}
```

6. At the end of `ViewController.swift`, add an `AVAudioRecorderDelegate` method, `audioRecorderDidFinishRecording`, with the following implementation, which mainly calls `run_inference_wrapper` to do audio post-processing and recognition:

```
func audioRecorderDidFinishRecording(_ recorder: AVAudioRecorder,
successfully flag: Bool) {
    _btn.setTitle("Recognizing...", for: .normal)
    if flag {
        let result =
RunInference_Wrapper().run_inference_wrapper(_recorderFilePath)
        _lbl.text = result
    }
    else {
        _lbl.text = "Recording error"
    }
    _btn.setTitle("Start", for: .normal)
}
```

In the `AudioRecognition_Swift-Bridging-Header.h` file, add `#include "RunInference.h"` so the preceding Swift code, `RunInference_Wrapper().run_inference_wrapper(_recorderFilePath)`, works.

7. In the `RunInference.mm`, inside the `run_inference_wrapper` method, copy the code from `ViewController.mm` in the Objective-C `AudioRecognition` app, described in steps 5-8 of the last section, that converts the saved recorded audio to the format the TensorFlow model accepts and then sends it along with sample rate to the model to get the recognition result:

```
@implementation RunInference_Wrapper
- (NSString *)run_inference_wrapper:(NSString*)recorderFilePath {
...
}
```

If you really want to port as much code as possible to Swift, you can replace the audio file conversion code in C with Swift (see https://developer.apple.com/documentation/audiotoolbox/extended_audio_file_services for details). There are also some unofficial open source projects that provide the Swift wrapper of the official TensorFlow C++ API. But for simplicity and the right balance, we'll keep the TensorFlow model inference, and in this example, the audio file reading and conversion as well, in C++ and Objective-C, working together with the Swift code, which controls the UI and audio recording, and initiates the call to do audio processing and recognition.

That's all it takes to build a Swift iOS app that uses the speech commands recognition model. Now you can run it on your iOS simulator or actual device and see the exact same results as the Objective-C version.

Summary

In this chapter, we first gave a quick overview of speech recognition and how modern ASR systems were built using end-to-end deep learning methods. Then we covered how to train a TensorFlow model to recognize simple speech commands, and presented step-by-step tutorials on how to use the model in an Android app, as well as in both Objective-C- and Swift-based iOS apps. We also discussed how to fix a common model-loading error in iOS by finding out the missing TensorFlow op or kernel file, adding it, and rebuilding the TensorFlow iOS library.

ASR is for converting speech to text. In the next chapter, we'll explore another model that has text as the output, and the text there will be full, natural-language sentences instead of the simple commands in this chapter. We'll cover how to build a model to convert an image, our old friend, to text, and how to use the model in mobile apps. Observing and describing what you see in natural language requires true human intelligence. One of the best for this task is Sherlock Holmes; we certainly won't be as good as Holmes yet, but let's see how we can start.

6
Describing Images in Natural Language

If image classification and object detection are intelligent tasks, describing an image in natural language is definitely a more challenging task that requires more intelligence – just think for a moment about how everyone grows from a newborn (who learns to recognize objects and detect their locations) to a three-year old (who learns to tell a story about a picture). The official term for the task of describing an image in natural language is image captioning. Unlike speech recognition, which has a long history of research and development, image captioning (with full natural language, not just keyword output) has only had a short but exciting history of research due to its complexityand the deep learning breakthrough in 2012.

In this chapter, we'll first review how a deep learning-based image captioning model that won the 2015 Microsoft COCO (a large-scale object detection, segmentation, and captioning dataset, which we covered briefly in `Chapter 3`, *Detecting Objects and Their Locations*), Image Captioning Challenge, works. Then we'll summarize the steps for training the model in TensorFlow, and go through in detail how to prepare and optimize such a complicated model to be deployed on mobile devices. After that, we'll show you step-by-step tutorials on how to build iOS and Android apps to use the model to generate a natural language sentence describing an image. As the model involves both computer vision and natural language processing, you'll see for the first time how CNN and RNN, the two main deep neural network architectures, work together and how you can write iOS and Android code to access trained networks and make multiple inferences. In summary, we'll cover the following topics in this chapter:

- Image captioning – how it works
- Training and freezing an image captioning model
- Transforming and optimizing the image captioning model
- Using the image captioning model in iOS
- Using the image captioning model in Android

Image captioning – how it works

The model that won the first MSCOCO Image Captioning Challenge in 2015 is described in the paper, *Show and Tell: Lessons learned from the 2015 MSCOCO Image Captioning Challenge* (`https://arxiv.org/pdf/1609.06647.pdf`). Before we talk about the training process, which is also covered pretty well in TensorFlow's im2txt model documentation website at `https://github.com/tensorflow/models/tree/master/research/im2txt`, let's first get a basic understanding of how the model works. This will also help you understand training and inference code in Python, as well as the inference code in iOS and Android you'll see later in the chapter.

The winning Show and Tell model is trained using an end-to-end method, similar to the latest deep learning-based speech recognition models we covered briefly in the previous chapter. It uses the MSCOCO image captioning 2014 dataset, available for download at `http://cocodataset.org/#download`, which has more than 82k training images and target natural language sentences describing them. The model is trained to maximize the probability of outputting the target natural language sentence for each input image. Different from other more complicated training methods using multiple subsystems, the end-to-end method is elegant, simpler, and can achieve state-of-the-art results.

To process and represent an input image, the Show and Tell model uses a pretrained Inception v3 model, the same model we used for retraining our dog breed image classification model in `Chapter 2`, *Classifying Images with Transfer Learning*. The last hidden layer of the Inception v3 CNN network is used as the representation of the input image. Because of the nature of the CNN model, earlier layers capture more basic image info, while later layers capture the more high-level concepts of an image; thus by using the last hidden layer of an input image to represent the image, we can better prepare natural language output with high-level concepts. After all, we normally would start describing a picture with words such as "person" or "train", instead of "something with a sharp edge."

To represent each word in a target natural language output, a word embedding method is used. Word embedding is just the vector representations of words. There's a nice tutorial (`https://www.tensorflow.org/tutorials/word2vec`) on the TensorFlow website on how you can build a model to get the vector representations of words.

Now, with both the input image and output words represented (each such pair forms a training example), the best training model that can be used to maximize the probability of generating each word w in the target output, given the input image and the previous words before that word w, is an RNN sequence model, or more specifically, a **Long Short Term Memory** (**LSTM**) type of RNN model. LSTM is known for solving the vanishing and exploding gradients problem inherent in regular RNN models; to better understand LSTM, you should check out this popular blog, http://colah.github.io/posts/2015-08-Understanding-LSTMs

The gradient concept is used in the back propagation process to update network weights so it can learn to generate better outputs. If you're not familiar with the back propagation process, one of the most fundamental and powerful algorithms in neural networks, you should definitely spend some time understanding it – just Google "backprop" and the top five results won't disappoint. Vanishing gradient means that, during the deep neural network back propagation learning process, network weights in earlier layers barely get updated so the network never converges; exploding gradient means that those weights get updated too wildly, causing the network to diverge a lot. So, if someone has a closed mind and never learns, or if someone gets crazy about new things as fast as he loses interest, you know what kind of gradient problem they seem to have.

After training, the CNN and LSTM models can then be used together for inference: given an input image, the model can estimate the probability of each word, thus predicting which n-best first words are most likely to generate for the output sentence; then, given the input image and the n-best first words, the n-best next words can be generated, and the process continues till we have a complete sentence when the model returns a specific end of sentence word, or the designated word length for the generated sentence has been reached (to prevent the model from being too wordy).

Using n-best words (meaning having n-best sentences at the end) at each word generation is called a beam search. When n, the so-called beam size, is 1, it becomes a greedy search or best search, based solely on the top probability value among all possible words returned by the model. The training and inference process in the next section from the official TensorFlow im2txt model uses the beam search with beam size set to 3, implemented in Python; for comparison, the iOS and Android apps we'll develop use the simpler greedy or best search. You'll see which method generates better captions.

Training and freezing an image captioning model

In this section, we'll first summarize the process of training the Show and Tell model called im2txt, documented at `https://github.com/tensorflow/models/tree/master/research/im2txt`, with some tips to help you better understand the process. Then we'll show some key changes to the Python code that comes with the im2txt model project, in order to freeze the model to be ready for use on mobile devices.

Training and testing caption generation

If you have followed the *Setting up the TensorFlow Object Detection API* section in Chapter 3, *Detecting Objects and Their Locations,* then you already have the `im2txt` folder installed; otherwise simply `cd` to your TensorFlow source root directory then run:

```
git clone https://github.com/tensorflow/models
```

One Python library you probably haven't installed is the **Natural Language ToolKit (NLTK)**, one of the most popular Python libraries for natural language processing. Just go to its website at `http://www.nltk.org` for installation instructions.

Now, follow these steps to train the model:

1. Set up a location to save the 2014 MSCOCO image captioning training and validation dataset by opening a Terminal and running:

    ```
    MSCOCO_DIR="${HOME}/im2txt/data/mscoco"
    ```

    Note that although the 2014 raw dataset to be downloaded and saved is about 20 GB, the dataset will be converted to the TFRecord format (we also used the TFRecord format in Chapter 3, *Detecting Objects and Their Locations*, to convert an object detection dataset), which is required to run the following training script and adds about 100 GB data. So you need a total of about 140 GB to train your own image captioning model using the TensorFlow im2txt project.

2. Go to where your im2txt source code is and download and process the MSCOCO dataset:

    ```
    cd <your_tensorflow_root>/models/research/im2txt
    bazel build //im2txt:download_and_preprocess_mscoco
    bazel-bin/im2txt/download_and_preprocess_mscoco "${MSCOCO_DIR}"
    ```

After the `download_and_preprocess_mscoco` script finishes, you'll see in the `$MSCOCO_DIR` folder all the training, validation, and testing data files in the TFRecord format.

There's also a file named `word_counts.txt` generated in the `$MSCOCO_DIR` folder. It has a total of 11,518 words and each line consists of a word, a space, and the number of times the word appears in the dataset. Only words with a count equal to or larger than 4 are saved in the file. Special words such as the beginning and end of a sentence are also saved (represented as `<S>` and `</S>`, respectively). You'll see later how we specifically use and parse the file in our iOS and Android apps for caption generation.

3. Get the Inception v3 checkpoint file by running the following commands:

```
INCEPTION_DIR="${HOME}/im2txt/data"
mkdir -p ${INCEPTION_DIR}
cd ${INCEPTION_DIR}
wget
"http://download.tensorflow.org/models/inception_v3_2016_08_28.tar.gz"
tar -xvf inception_v3_2016_08_28.tar.gz -C ${INCEPTION_DIR}
rm inception_v3_2016_08_28.tar.gz
```

After this, you'll see in your `${HOME}/im2txt/data` folder a file named `inception_v3.ckpt`, like this:

```
jeff@AiLabby:~/im2txt/data$ ls -lt inception_v3.ckpt
-rw-r----- 1 jeff jeff 108816380 Aug 28  2016 inception_v3.ckpt
```

4. Now we're ready to train our model with the following commands:

```
INCEPTION_CHECKPOINT="${HOME}/im2txt/data/inception_v3.ckpt"
MODEL_DIR="${HOME}/im2txt/model"
cd <your_tensorflow_root>/models/research/im2txt
bazel build -c opt //im2txt/...
bazel-bin/im2txt/train \
  --input_file_pattern="${MSCOCO_DIR}/train-?????-of-00256" \
  --inception_checkpoint_file="${INCEPTION_CHECKPOINT}" \
  --train_dir="${MODEL_DIR}/train" \
  --train_inception=false \
  --number_of_steps=1000000
```

Describing Images in Natural Language

Even on a GPU, such as our Nvidia GTX 1070 set up in `Chapter 1, Getting Started with Mobile TensorFlow`, the whole 1M steps, specified in the preceding `--number_of_steps` parameter, would take more than 5 days and nights, as it takes about 6.5 hours to run 50K steps. Fortunately, as you'll see soon, even with about 50K steps, the image captioning results are already pretty good. Also notice that you can cancel the `train` script at anytime, then later rerun it and the script will start from the last saved checkpoint; checkpoints get saved every 10 minutes by default, so at worst you'd lose only 10 minutes of training.

After several hours of training, cancel the preceding `train` script and then take a look at where `--train_dir` points to. You'll see something like this (by default five sets of checkpoint files are saved, but we only show three here):

```
ls -lt $MODEL_DIR/train
-rw-rw-r-- 1 jeff jeff 2171543 Feb 6 22:17 model.ckpt-109587.meta
-rw-rw-r-- 1 jeff jeff 463 Feb 6 22:17 checkpoint
-rw-rw-r-- 1 jeff jeff 149002244 Feb 6 22:17 model.ckpt-109587.data-00000-of-00001
-rw-rw-r-- 1 jeff jeff 16873 Feb 6 22:17 model.ckpt-109587.index
-rw-rw-r-- 1 jeff jeff 2171543 Feb 6 22:07 model.ckpt-109332.meta
-rw-rw-r-- 1 jeff jeff 16873 Feb 6 22:07 model.ckpt-109332.index
-rw-rw-r-- 1 jeff jeff 149002244 Feb 6 22:07 model.ckpt-109332.data-00000-of-00001
-rw-rw-r-- 1 jeff jeff 2171543 Feb 6 21:57 model.ckpt-109068.meta
-rw-rw-r-- 1 jeff jeff 149002244 Feb 6 21:57 model.ckpt-109068.data-00000-of-00001
-rw-rw-r-- 1 jeff jeff 16873 Feb 6 21:57 model.ckpt-109068.index
-rw-rw-r-- 1 jeff jeff 4812699 Feb 6 14:27 graph.pbtxt
```

You can tell each set of checkpoint files (`model.ckpt-109068.*`, and `model.ckpt-109332.*`, `model.ckpt-109587.*`) gets generated every 10 minutes. `graph.pbtxt` is the graph definition file of the model (in text format), and the `model.ckpt-??????.meta` file also contains the graph definition of the model in addition to some other metadata for a specific checkpoint, such as `model.ckpt-109587.data-00000-of-00001` (notice its size is almost 150 MB because all the network parameters are saved there).

5. Test caption generation like this:

```
CHECKPOINT_PATH="${HOME}/im2txt/model/train"
VOCAB_FILE="${HOME}/im2txt/data/mscoco/word_counts.txt"
IMAGE_FILE="${HOME}/im2txt/data/mscoco/raw-data/val2014/COCO_val2014_000000224477.jpg"
bazel build -c opt //im2txt:run_inference
```

```
bazel-bin/im2txt/run_inference \
  --checkpoint_path=${CHECKPOINT_PATH} \
  --vocab_file=${VOCAB_FILE} \
  --input_files=${IMAGE_FILE}
```

The `CHECKPOINT_PATH` is set to be the same path as `--train_dir` is set to. The `run_inference` script will generate something like this (not exactly the same, depending on how many steps of training have been run):

```
Captions for image COCO_val2014_000000224477.jpg:
  0) a man on a surfboard riding a wave . (p=0.015135)
  1) a person on a surfboard riding a wave . (p=0.011918)
  2) a man riding a surfboard on top of a wave . (p=0.009856)
```

This is pretty cool. Wouldn't it be even cooler if we could run this model on our smartphones? But before we can do that, there are a few extra steps we need to take, due to the relatively greater complexity of the model and also the way the `train` and `run_inference` scripts in Python were written.

Freezing the image captioning model

In Chapter 4, *Transforming Pictures with Amazing Art Styles*, and Chapter 5, *Understanding Simple Speech Commands*, we used two slightly different versions of a script called `freeze.py` to merge trained network weights with the network graph definition to a self-sufficient model file, a boon we can use on mobile devices. TensorFlow also comes with a more universal version of the `freeze` script, called `freeze_graph.py`, located in the `tensorflow/python/tools` folder, that you can use to build a model file. To make it work, you need to provide it with at least four parameters (to see all the available parameters, check out `tensorflow/python/tools/freeze_graph.py`):

- `--input_graph` or `--input_meta_graph`: a graph definition file of the model. For example, in the output of the command `ls -lt $MODEL_DIR/train` in Step 4 of the last section, `model.ckpt-109587.meta` is a meta graph file that contains the model's graph definition and other checkpoint-related metadata, and `graph.pbtxt` is just the model's graph definition.
- `--input_checkpoint`: a specific checkpoint file, for example, `model.ckpt-109587`. Notice you don't specify the full filename of the large sized checkpoint file, `model.ckpt-109587.data-00000-of-00001`.

- `--output_graph`: the path to the frozen model file – this is the one used on mobile devices.
- `--output_node_names`: the list of output node names, separated with a comma, that tells the `freeze_graph` tool what part of the model and weights are to be included in the frozen model, so nodes and weights not required for generating the specific output node names will be left.

So for this model, how do we figure out the must-have output node names, as well as the input node names, also essential for inference as we have seen in iOS and Android apps in the previous chapters? Because we already used the `run_inference` script to generate our test image's caption, we can see how it makes the inference.

Go to your im2txt source code folder, `models/research/im2txt/im2txt`: you may want to open this in a nice editor such as Atom or Sublime Text, or a Python IDE such as PyCharm, or just open it (https://github.com/tensorflow/models/tree/master/research/im2txt/im2txt) from your browser. In `run_inference.py`, there's a call to `build_graph_from_config` in `inference_utils/inference_wrapper_base.py`, which calls `build_model` in `inference_wrapper.py`, which further calls a `build` method in `show_and_tell_model.py`. The `build` method finally calls, among other things, a `build_input` method, which has the following code:

```
if self.mode == "inference":
    image_feed = tf.placeholder(dtype=tf.string, shape=[],
name="image_feed")
    input_feed = tf.placeholder(dtype=tf.int64,
        shape=[None], # batch_size
        name="input_feed")
```

And a `build_model` method, which has:

```
if self.mode == "inference":
    tf.concat(axis=1, values=initial_state, name="initial_state")
    state_feed = tf.placeholder(dtype=tf.float32,
        shape=[None, sum(lstm_cell.state_size)],
        name="state_feed")
...
tf.concat(axis=1, values=state_tuple, name="state")
...
tf.nn.softmax(logits, name="softmax")
```

So, the three placeholders named `image_feed`, `input_feed`, and `state_feed` should be the input node names, while `initial_state`, `state`, and `softmax` should be the output node names. Furthermore, two methods defined in `inference_wrapper.py` confirm our detective work – the first one is:

```
def feed_image(self, sess, encoded_image):
    initial_state = sess.run(fetches="lstm/initial_state:0",
                             feed_dict={"image_feed:0": encoded_image})
    return initial_state
```

So, we provide `image_feed` and get `initial_state` back (the `lstm/` prefix just means that the node is under the `lstm` scope). The second method is:

```
def inference_step(self, sess, input_feed, state_feed):
    softmax_output, state_output = sess.run(
        fetches=["softmax:0", "lstm/state:0"],
        feed_dict={
            "input_feed:0": input_feed,
            "lstm/state_feed:0": state_feed,
        })
    return softmax_output, state_output, None
```

We feed in `input_feed` and `state_feed`, and get back `softmax` and `state`. In total, three input node names and three output names.

Note that these nodes are created only if the `mode` is "inference", as `show_and_tell_model.py` is used by both `train.py` and `run_inference.py`. This means that the model's graph definition file and weights located in `--checkpoint_path`, generated with `train` in Step 5, will be modified after running the `run_inference.py` script. So, how do we save the updated graph definition and checkpoint files?

It turns out that, in `run_inference.py`, after a TensorFlow session is created, there's also a call `restore_fn(sess)` to load the checkpoint file, and the call is defined in `inference_utils/inference_wrapper_base.py`:

```
def _restore_fn(sess):
    saver.restore(sess, checkpoint_path)
```

When reaching the `saver.restore` call after starting `run_inference.py`, the updated graph definition has been made so we can just save a new checkpoint and graph file there, making the `_restore_fn` function the following:

```
def _restore_fn(sess):
    saver.restore(sess, checkpoint_path)
```

```
saver.save(sess, "model/image2text")
tf.train.write_graph(sess.graph_def, "model", 'im2txt4.pbtxt')
tf.summary.FileWriter("logdir", sess.graph_def)
```

The line `tf.train.write_graph(sess.graph_def, "model", 'im2txt4.pbtxt')` is optional as, when saving a new checkpoint file by calling `saver.save`, a meta file also gets generated, which can be used along with the checkpoint file by `freeze_graph.py`. But it's generated here for those who'd love to see everything in the plain text format or who prefer to use a graph definition file with the `--in_graph` parameter when freezing a model. The last line `tf.summary.FileWriter("logdir", sess.graph_def)` is also optional, but it generates an event file that can be visualized by TensorBoard. So with these changes, after running `run_inference.py` again (remember to run `bazel build -c opt //im2txt:run_inference` first unless you run the `run_inference.py` directly with Python), you'll see in your `model` directory the following new checkpoint files and a new graph definition file:

```
jeff@AiLabby:~/tensorflow-1.5.0/models/research/im2txt$ ls -lt model
-rw-rw-r-- 1 jeff jeff 2076964 Feb 7 12:33 image2text.pbtxt
-rw-rw-r-- 1 jeff jeff 1343049 Feb 7 12:33 image2text.meta
-rw-rw-r-- 1 jeff jeff 77 Feb 7 12:33 checkpoint
-rw-rw-r-- 1 jeff jeff 149002244 Feb 7 12:33 image2text.data-00000-of-00001
-rw-rw-r-- 1 jeff jeff 16873 Feb 7 12:33 image2text.index
```

And in your `logdir` directory:

```
jeff@AiLabby:~/tensorflow-1.5.0/models/research/im2txt$ ls -lt logdir
total 2124
-rw-rw-r-- 1 jeff jeff 2171623 Feb 7 12:33 events.out.tfevents.1518035604.AiLabby
```

Running the `bazel build` command to build a TensorFlow Python script is optional. You can just run the Python script directly. For example, we can run `python tensorflow/python/tools/freeze_graph.py` without building it first with `bazel build tensorflow/python/tools:freeze_graph` then running `bazel-bin/tensorflow/python/tools/freeze_graph`. But be aware that running the Python script directly will use the version of TensorFlow you've installed via pip, which may be different from the version you've downloaded as source and built by the `bazel build` command. This can be the cause of some confusing errors so be sure you know the TensorFlow version used to run a script. In addition, for a C++ based tool, you have to build it first with bazel before you can run it. For example, the `transform_graph` tool, which we'll see soon, is implemented in `transform_graph.cc` located at `tensorflow/tools/graph_transforms`; another important tool called `convert_graphdef_memmapped_format`, which we'll use for our iOS app later, is also implemented in C++ located at `tensorflow/contrib/util`.

Now that we're here, let's quickly use TensorBoard to take a look at what our graph looks like – simply run `tensorboard --logdir logdir`, and open `http://localhost:6006` from a browser. Figure 6.1 shows three output node names (**softmax** at the top, and **lstm/initial_state** and **lstm/state** at the top of the highlighted red rectangle) and one input node name (**state_feed** at the bottom):

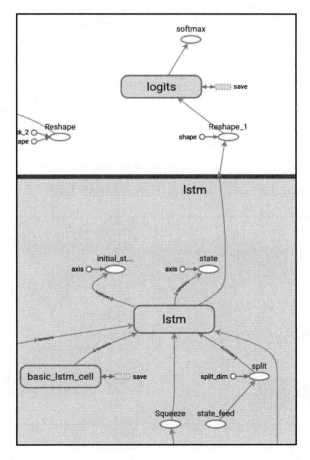

Figure 6.1: Diagram showing the three output node names and one input node name

[167]

Describing Images in Natural Language

Figure 6.2 shows one additional input node name, **image_feed**:

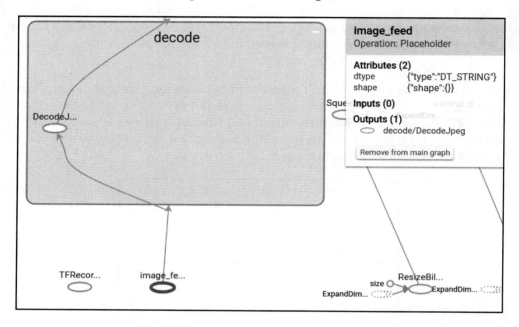

Figure 6.2: Diagram showing one additional input node name image_feed

Finally, Figure 6.3 shows the last input node name, **input_feed**:

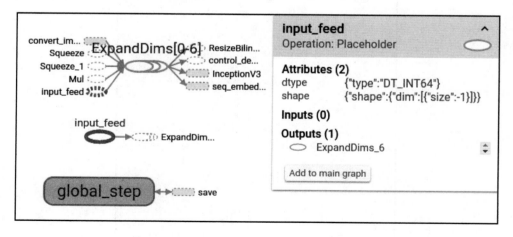

Figure 6.3: Diagram showing the last input node name input_feed

There are certainly a lot of details we can't and won't cover here. But you get the big picture and, equally important, enough details to move forward. Now running `freeze_graph.py` should be like a breeze (pun intended):

```
python tensorflow/python/tools/freeze_graph.py --
input_meta_graph=/home/jeff/tensorflow-1.5.0/models/research/im2txt/model/i
mage2text.meta --
input_checkpoint=/home/jeff/tensorflow-1.5.0/models/research/im2txt/model/i
mage2text --output_graph=/tmp/image2text_frozen.pb --
output_node_names="softmax,lstm/initial_state,lstm/state" --
input_binary=true
```

Notice we use the meta graph file here, along with the `--input_binary` parameter set to `true`, as by default it's false, meaning the `freeze_graph` tool expects the input graph or meta graph file to be in text format.

You can use the text-formatted graph file as input, in which case there's no need to provide the `--input_binary` parameter:

```
python tensorflow/python/tools/freeze_graph.py  --
input_graph=/home/jeff/tensorflow-1.5.0/models/research/im2txt/model/image2
text.pbtxt --
input_checkpoint=/home/jeff/tensorflow-1.5.0/models/research/im2txt/model/i
mage2text --output_graph=/tmp/image2text_frozen2.pb --
output_node_names="softmax,lstm/initial_state,lstm/state"
```

The sizes of the two output graph files, `image2text_frozen.pb` and `image2text_frozen2.pb`, will be slightly different, but they behave exactly the same when, after being transformed and possibly optimized, they're used on mobile devices.

Transforming and optimizing the image captioning model

If you really can't wait any longer and decide to try the freshly frozen hot model on your iOS or Android app now, you certainly can, but you'll be shown a fatal error, `No OpKernel was registered to support Op 'DecodeJpeg' with these attrs`, to force you to reconsider your decision.

Fixing errors with transformed models

Normally, you can use a tool called `strip_unused.py`, located at the same location as `freeze_graph.py` at `tensorflow/python/tools`, to remove the `DecodeJpeg` operation that is not included in the TensorFlow core library (see https://www.tensorflow.org/mobile/prepare_models#removing_training-only_nodes for more details), but since the input node image_feed requires the decode operation (Figure 6.2), a tool such as `strip_unused` won't treat the `DecodeJpeg` as unused so it won't be stripped. You can verify this by first running the `strip_unused` command as follows:

```
bazel-bin/tensorflow/python/tools/strip_unused --
input_graph=/tmp/image2text_frozen.pb --
output_graph=/tmp/image2text_frozen_stripped.pb --
input_node_names="image_feed,input_feed,lstm/state_feed" --
output_node_names="softmax,lstm/initial_state,lstm/state" --
input_binary=True
```

Then loadg the output graph in iPython and list the first several nodes like this:

```
import tensorflow as tf
g=tf.GraphDef()
g.ParseFromString(open("/tmp/image2text_frozen_stripped", "rb").read())
x=[n.name for n in g.node]
x[:6]
```

The output will be as follows:

```
[u'image_feed',
 u'input_feed',
 u'decode/DecodeJpeg',
 u'convert_image/Cast',
 u'convert_image/y',
 u'convert_image']
```

A second possible solution to fix the error for your iOS app is to add the unregistered op implementation to the `tf_op_files` file and rebuild the TensorFlow iOS library, as we did in Chapter 5, *Understanding Simple Speech Commands*. The bad news is that because there's no implementation of the `DecodeJpeg` functionality in TensorFlow, there's no way to add a TensorFlow implementation of `DecodeJpeg` to `tf_op_files`.

The fix to this annoyance is actually hinted at also in Figure 6.2, where a `convert_image` node is used as the decoded version of the `image_feed` input. To be more accurate, click on the **Cast** and **decode** nodes in the TensorBoard graph as shown in Figure 6.4, and you'll see from the TensorBoard info cards on the right that the input and output of Cast (named `convert_image/Cast`) are `decode/DecodeJpeg` and `convert_image`, and the input and output of decode are `image_feed` and `convert_image/Cast`:

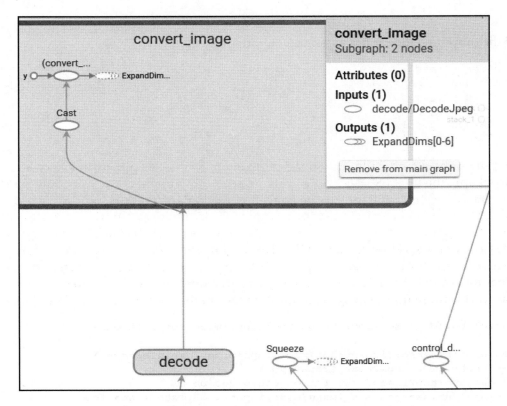

Figure 6.4: Looking into the decode and conver_image nodes

In fact, in `im2txt/ops/image_processing.py`, there's a line `image = tf.image.convert_image_dtype(image, dtype=tf.float32)` that converts a decoded image to floats. Let's replace `image_feed` with `convert_image/Cast`, the name shown in TensorBoard, as well as the output of the preceding code snippet, and run `strip_unused` again:

```
bazel-bin/tensorflow/python/tools/strip_unused --
input_graph=/tmp/image2text_frozen.pb   --
output_graph=/tmp/image2text_frozen_stripped.pb --
```

```
input_node_names="convert_image/Cast,input_feed,lstm/state_feed" --
output_node_names="softmax,lstm/initial_state,lstm/state"   --
input_binary=True
```

Now rerun the code snippet as follows:

```
g.ParseFromString(open("/tmp/image2text_frozen_stripped", "rb").read())
x=[n.name for n in g.node]
x[:6]
```

And the output no longer has a `decode/DecodeJpeg` node:

```
[u'input_feed',
 u'convert_image/Cast',
 u'convert_image/y',
 u'convert_image',
 u'ExpandDims_1/dim',
 u'ExpandDims_1']
```

If we use our new model file, `image2text_frozen_stripped.pb`, in an iOS or an Android app, the `No OpKernel was registered to support Op 'DecodeJpeg' with these attrs.` will for sure be gone. But another error occurs, `Not a valid TensorFlow Graph serialization: Input 0 of node ExpandDims_6 was passed float from input_feed:0 incompatible with expected int64`. If you go through a nice Google TensorFlow codelab called TensorFlow for Poets 2 (https://codelabs.developers.google.com/codelabs/tensorflow-for-poets-2), you may recall there's another tool called `optimize_for_inference` that does things similar to `strip_unused`, and it works nicely for the image classification task in the codelab. You can run it like this:

```
bazel build tensorflow/python/tools:optimize_for_inference

bazel-bin/tensorflow/python/tools/optimize_for_inference \
--input=/tmp/image2text_frozen.pb \
--output=/tmp/image2text_frozen_optimized.pb \
--input_names="convert_image/Cast,input_feed,lstm/state_feed" \
--output_names="softmax,lstm/initial_state,lstm/state"
```

But loading the output model file `image2text_frozen_optimized.pb` on iOS or Android app results in the same `Input 0 of node ExpandDims_6 was passed float from input_feed:0 incompatible with expected int64` error. Looks like, while we're trying to achieve, to some humble extent at least, what Holmes can do in this chapter, someone wants us to be like Holmes first.

If you have tried the `strip_unused` or `optimize_for_inference` tools on other models, such as those we have seen in the previous chapters, they work just fine. It turns out the two Python-based tools, albeit included in the official TensorFlow 1.4 and 1.5 releases, have some bugs when optimizing some more complicated models. The updated and right tool is the C++ based `transform_graph` tool, now the official tool recommended at the TensorFlow Mobile site (https://www.tensorflow.org/mobile). Run the following commands to get rid of the float incompatible with the int64 error when deployed on mobile devices:

```
bazel build tensorflow/tools/graph_transforms:transform_graph

bazel-bin/tensorflow/tools/graph_transforms/transform_graph \
--in_graph=/tmp/image2text_frozen.pb \
--out_graph=/tmp/image2text_frozen_transformed.pb \
--inputs="convert_image/Cast,input_feed,lstm/state_feed" \
--outputs="softmax,lstm/initial_state,lstm/state" \
--transforms='
  strip_unused_nodes(type=float, shape="299,299,3")
  fold_constants(ignore_errors=true, clear_output_shapes=true)
  fold_batch_norms
  fold_old_batch_norms'
```

We won't go into the details of all the `--transforms` options, which are fully documented at https://github.com/tensorflow/tensorflow/tree/master/tensorflow/tools/graph_transforms. Basically, the `--transforms` setting correctly gets rid of unused nodes such as `DecodeJpeg` for our model and also does a few other optimizations.

Now if you load the `image2text_frozen_transformed.pb` file in your iOS and Android apps, the incompatible error will be gone. Of course, we haven't written any real iOS and Android code yet, but we know the model is good and ready for us to have fun with. Good, but can be better.

Optimizing the transformed model

The truly final step, and a crucial one, especially when running a complicated frozen and transformed model such as the one we have trained on an older iOS device, is to use another tool called `convert_graphdef_memmapped_format`, located at `tensorflow/contrib/util`, to convert the frozen and transformed model to a memmapped format. A memmapped file allows modern operating systems such as iOS and Android to map the file to the main memory directly, so there's no need to allocate the memory for the file and no writing back to disk as the file data is read-only, a significant performance increase.

More importantly, the memmapped file doesn't get treated as memory usage by iOS so, when there's too much memory pressure, an app using a memmapped file even with a large size won't be killed by iOS due to its large memory use and crash. In fact, as we'll see soon in the next section, the transformed version of our model file, if not converted to the memmapped format, will crash on older mobile devices such as iPhone 6, in which case the conversion is a must-have.

The command to build and run the tool is pretty straightfoward:

```
bazel build tensorflow/contrib/util:convert_graphdef_memmapped_format

bazel-bin/tensorflow/contrib/util/convert_graphdef_memmapped_format \
--in_graph=/tmp/image2text_frozen_transformed.pb \
--out_graph=/tmp/image2text_frozen_transformed_memmapped.pb
```

We'll show you how to use the `image2text_frozen_transformed_memmapped.pb` model file in the iOS app in the next section. It can also be used in Android using native code, but due to time restrictions, we won't be able to cover it in this chapter.

We have gone some extra miles to get a complicated image captioning model finally ready for our mobile apps. It's time to appreciate the simplicity of using the model. Actually, using the model is more than just a single `session->Run` call in iOS or an `mInferenceInterface.run` call in Android as we have done in all the previous chapters; the inference from an input image to a natural language output, as you have seen when looking into how `run_inference.py` works in a previous section, involves multiple calls to the `run` method to the model. This is how LSTM models work: "keep sending me a new input (based on my previous state and output) and I'll send you back the next state and output." By simplicity, we mean we'll show you how to use as little clean code as possible to build iOS and Android apps that use the model to describe an image in natural language. That way, you can easily integrate the model and its inference code in your own apps, if desired.

Using the image captioning model in iOS

As the CNN part of the model is based on Inception v3, the same model we used in Chapter 2, *Classifying Images with Transfer Learning*, we can and will use the simpler TensorFlow pod to create our Objective-C iOS app. Follow the steps here to see how to use both the `image2text_frozen_transformed.pb` and `image2text_frozen_transformed_memmapped.pb` model files in a new iOS app:

1. Similar to the first four steps in Chapter 2, *Classifying Images with Transfer Learning*, in the *Adding TensorFlow to your Objective-C iOS app* section, create a new iOS project named `Image2Text`, add a new file named `Podfile` with the following content:

   ```
   target 'Image2Text'
       pod 'TensorFlow-experimental'
   ```

   Then run `pod install` on a Terminal and open the `Image2Text.xcworkspace` file. Drag and drop `ios_image_load.h`, `ios_image_load.mm`, `tensorflow_utils.h` and `tensorflow_utils.mm` files from the TensorFlow iOS example Camera app located at `tensorflow/examples/ios/camera` to the `Image2Text` project in Xcode. We've reused the `ios_image_load.*` files before and the `tensorflow_utils.*` files are used here mainly for loading the memmapped model file. There are two methods, `LoadModel` and `LoadMemoryMappedModel` in `tensorflow_utils.mm`: one loads a non-memmapped model in a way we used to, and the other loads a memmapped model. Take a look at how `LoadMemoryMappedModel` gets implemented there, if you're interested, and you may also find the documentation at https://www.tensorflow.org/mobile/optimizing#reducing_model_loading_time_andor_memory_footprint helpful.

Describing Images in Natural Language

2. Add the two model files we generated at the end of the last section, the `word_counts.txt` file generated in Step 2 of the subsection, *Training and testing caption generation*, as well as a few test images – we saved and use the four images at the top of the TensorFlow im2txt model page (https://github.com/tensorflow/models/tree/master/research/im2txt) so we can compare the captioning results of our models with those generated with a model supposedly trained with a lot more steps. Also rename `ViewController.m` to `.mm` and from now on we'll just work on the `ViewController.mm` file to complete the app. Now your Xcode `Image2Text` project should look like Figure 6.5:

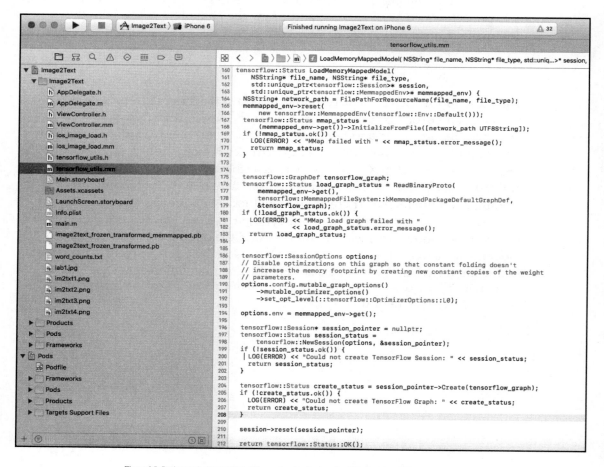

Figure 6.5: Setting up the Image2Text iOS app, also showing how LoadMemoryMappedModel is implemented

3. Open `ViewController.mm` and add a bunch of Objective-C and C++ constants as follows:

```
static NSString* MODEL_FILE = @"image2text_frozen_transformed";
static NSString* MODEL_FILE_MEMMAPPED =
@"image2text_frozen_transformed_memmapped";
static NSString* MODEL_FILE_TYPE = @"pb";
static NSString* VOCAB_FILE = @"word_counts";
static NSString* VOCAB_FILE_TYPE = @"txt";
static NSString *image_name = @"im2txt4.png";

const string INPUT_NODE1 = "convert_image/Cast";
const string OUTPUT_NODE1 = "lstm/initial_state";
const string INPUT_NODE2 = "input_feed";
const string INPUT_NODE3 = "lstm/state_feed";
const string OUTPUT_NODE2 = "softmax";
const string OUTPUT_NODE3 = "lstm/state";

const int wanted_width = 299;
const int wanted_height = 299;
const int wanted_channels = 3;

const int CAPTION_LEN = 20;
const int START_ID = 2;
const int END_ID = 3;
const int WORD_COUNT = 12000;
const int STATE_COUNT = 1024;
```

They're all self explanatory and should all look familiar if you have followed through the chapter, except probably the last five constants: `CAPTION_LEN` is the max number of words we want to generate in a caption and `START_ID` is the ID of the sentence start word, `<S>`, defined as the line number in the `word_counts.txt` file; so 2 means the word at the second line, 3 the third line. The first few lines of the `word_counts.txt` file are like this:

```
a 969108
<S> 586368
</S> 586368
. 440479
on 213612
of 202290
```

`WORD_COUNT` is the number of total words the model assumes, and for each inference call you'll see soon, the model returns a total of 12,000 probability scores, along with 1,024 state values of the LSTM model.

4. Add a few global variables and a function signature:

```
unique_ptr<tensorflow::Session> session;
unique_ptr<tensorflow::MemmappedEnv> tf_memmapped_env;

std::vector<std::string> words;

UIImageView *_iv;
UILabel *_lbl;

NSString* generateCaption(bool memmapped);
```

This simple UI-related code is similar to that of the iOS app in Chapter 2, *Classifying Images with Transfer Learning*. Basically you tap anywhere after the app starts, and pick one of the two models and the image description result will appear at the top. When the memmapped model is picked by the user in an `alert` action, the following code runs:

```
dispatch_async(dispatch_get_global_queue(0, 0), ^{
    NSString *caption = generateCaption(true);
    dispatch_async(dispatch_get_main_queue(), ^{
        _lbl.text = caption;
    });
});
```

If the non-memmapped model is selected, `generateCaption(false)` is used.

5. At the end of the `viewDidLoad` method, add the code to load the `word_counts.txt` and save the words to a vector line by line in Objective-C and C++:

```
NSString* voc_file_path = FilePathForResourceName(VOCAB_FILE,
VOCAB_FILE_TYPE);
if (!voc_file_path) {
    LOG(FATAL) << "Couldn't load vocabuary file: " <<
voc_file_path;
}
ifstream t;
t.open([voc_file_path UTF8String]);
string line;
while(t){
    getline(t, line);
    size_t pos = line.find(" ");
    words.push_back(line.substr(0, pos));
}
t.close();
```

6. All we need to do for the rest is to implement the `generateCaption` function. Inside it, first load the right model:

   ```
   tensorflow::Status load_status;
   if (memmapped)
       load_status = LoadMemoryMappedModel(MODEL_FILE_MEMMAPPED,
   MODEL_FILE_TYPE, &session, &tf_memmapped_env);
   else
       load_status = LoadModel(MODEL_FILE, MODEL_FILE_TYPE, &session);
   if (!load_status.ok()) {
       return @"Couldn't load model";
   }
   ```

7. Then, use similar image processing code to prepare the image tensor to be fed into the model:

   ```
   int image_width;
   int image_height;
   int image_channels;
   NSArray *name_ext = [image_name componentsSeparatedByString:@"."];
   NSString* image_path = FilePathForResourceName(name_ext[0],
   name_ext[1]);
   std::vector<tensorflow::uint8> image_data =
   LoadImageFromFile([image_path UTF8String], &image_width,
   &image_height, &image_channels);

   tensorflow::Tensor image_tensor(tensorflow::DT_FLOAT,
   tensorflow::TensorShape({wanted_height, wanted_width,
   wanted_channels}));
   auto image_tensor_mapped = image_tensor.tensor<float, 3>();
   tensorflow::uint8* in = image_data.data();
   float* out = image_tensor_mapped.data();
   for (int y = 0; y < wanted_height; ++y) {
       const int in_y = (y * image_height) / wanted_height;
       tensorflow::uint8* in_row = in + (in_y * image_width *
   image_channels);
       float* out_row = out + (y * wanted_width * wanted_channels);
       for (int x = 0; x < wanted_width; ++x) {
           const int in_x = (x * image_width) / wanted_width;
           tensorflow::uint8* in_pixel = in_row + (in_x *
   image_channels);
           float* out_pixel = out_row + (x * wanted_channels);
           for (int c = 0; c < wanted_channels; ++c) {
               out_pixel[c] = in_pixel[c];
           }
       }
   }
   ```

8. We can now send the image to the model and get the returned `initial_state` tensor vector, which contains 1,200 (`STATE_COUNT`) values:

   ```
   vector<tensorflow::Tensor> initial_state;

   if (session.get()) {
       tensorflow::Status run_status = session->Run({{INPUT_NODE1,
   image_tensor}}, {OUTPUT_NODE1}, {}, &initial_state);
       if (!run_status.ok()) {
           return @"Getting initial state failed";
       }
   }
   ```

9. Define the `input_feed` and `state_feed` tensors and set their values to the ID of the start word and the returned `initial_state` values, respectively:

   ```
   tensorflow::Tensor input_feed(tensorflow::DT_INT64,
   tensorflow::TensorShape({1,}));
   tensorflow::Tensor state_feed(tensorflow::DT_FLOAT,
   tensorflow::TensorShape({1, STATE_COUNT}));

   auto input_feed_map = input_feed.tensor<int64_t, 1>();
   auto state_feed_map = state_feed.tensor<float, 2>();
   input_feed_map(0) = START_ID;
   auto initial_state_map = initial_state[0].tensor<float, 2>();
   for (int i = 0; i < STATE_COUNT; i++){
       state_feed_map(0,i) = initial_state_map(0,i);
   }
   ```

10. Create a `for` loop over the `CAPTION_LEN` and, inside the loop, first create the `output_feed` and `output_states` tensor vectors, then feed the `input_feed` and `state_feed` we set previously and run the model to get back the `output` tensor vector which consists of the `softmax` tensor and the `new_state` tensor:

    ```
    vector<int> captions;
    for (int i=0; i<CAPTION_LEN; i++) {
        vector<tensorflow::Tensor> output;
        tensorflow::Status run_status = session->Run({{INPUT_NODE2,
    input_feed}, {INPUT_NODE3, state_feed}}, {OUTPUT_NODE2,
    OUTPUT_NODE3}, {}, &output);
        if (!run_status.ok()) {
            return @"Getting LSTM state failed";
        }
        else {
            tensorflow::Tensor softmax = output[0];
    ```

```
tensorflow::Tensor state = output[1];
auto softmax_map = softmax.tensor<float, 2>();
auto state_map = state.tensor<float, 2>();
```

11. Now, find the word ID with the largest probability (softmax value). If it's the ID of the end word, end the `for` loop; otherwise add the word `id` with the max softmax value to the vector `captions`. Notice here we use a greedy search, always selecting the word with the max probability, instead of a beam search with size set to 3 as in the `run_inference.py` script. At the end of the `for` loop, update the `input_feed` value with the max word `id` and the `state_feed` value with the previously returned `state` value, before feeding the two inputs again to the model for the softmax values of all the next words and the next state value:

```
float max_prob = 0.0f;
int max_word_id = 0;
for (int j = 0; j < WORD_COUNT; j++){
    if (softmax_map(0,j) > max_prob) {
        max_prob = softmax_map(0,j);
        max_word_id = j;
    }
}
if (max_word_id == END_ID) break;
captions.push_back(max_word_id);
input_feed_map(0) = max_word_id;
for (int j = 0; j < STATE_COUNT; j++){
    state_feed_map(0,j) = state_map(0,j);
}
}
}
```

We probably have never explained in detail how you get and set a TensorFlow tensor value in C++. But if you have read through the code in the book so far, you should have learned how. This is like RNN learning: if you're trained with enough code examples, you'll be able to write code that makes sense. In summary, first you define a variable with the `Tensor` type, specified with the variable's data type and shape, then you call the `Tensor` class's `tensor` method, passing in the C++ version of the data type and the dimension of the shape, to create a map variable for the tensor. After that, you can simply use the map to get or set the tensor's values.

12. Finally, just go through the `captions` vector and convert each word ID stored in the vector to a word before adding the word to a `sentence` string, ignoring the start ID and end ID, then return the sentence, hopefully in a sensible natural language:

```
NSString *sentence = @"";
for (int i=0; i<captions.size(); i++) {
    if (captions[i] == START_ID) continue;
    if (captions[i] == END_ID) break;
    sentence = [NSString stringWithFormat:@"%@ %s", sentence,
words[captions[i]].c_str()];
}

return sentence;
```

That's all it takes to run the model in an iOS app. Now run the app in an iOS simulator or device, tap, and select a model, as in Figure 6.6:

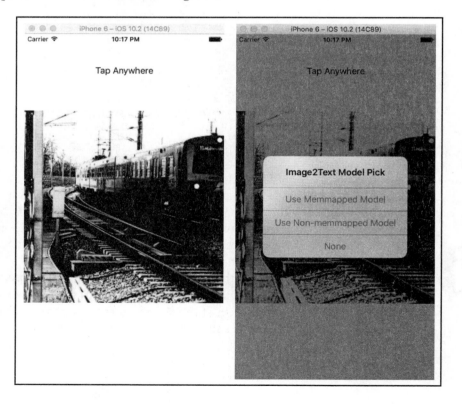

Figure 6.6: Running the Image2Text iOS app and selecting a model

On an iOS simulator, it takes over 10 seconds to run the non-memmapped model and about 5 seconds to run the memmapped model. On iPhone 6, it also takes about 5 seconds to run the memmapped model but crash as when running the non-memmapped model due to the large model file and memory pressure.

As for the results, Figure 6.7 shows the four test image results:

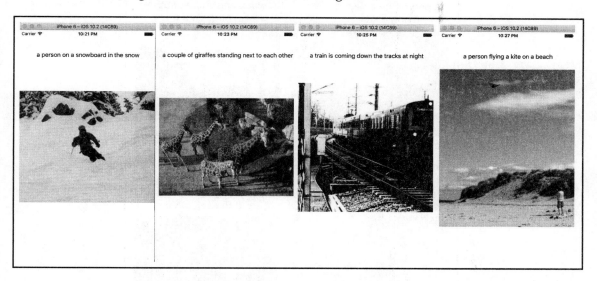

Figure 6.7: Showing the image captioning results

Figure 6.8 shows the results on the TensorFlow im2txt site and you can see our simpler greedy search results look pretty good as well; however for the giraffe picture, it looks like our model or the inference code can't count well enough. With the work done in the chapter, hopefully you'll have some fun ahead improving the training or model inference:

Figure 6.8: The example captioning results in the TensorFlow im2txt model website

It's time to give Android developers out there a nice treat before we move on to our next intelligent task.

Using the image captioning model in Android

Following the same simplicity considerations, we'll develop a new Android app with a minimal UI and focus on how to use the model in Android:

1. Create a new Android app called `Image2Text`, add `compile 'org.tensorflow:tensorflow-android:+'` to the end of your app `build.gradle` file's dependencies, create an `assets` folder, and drag and drop to it the `image2text_frozen_transformed.pb` model file, the `word_counts.txt` file, and a few test image files.

2. Add an ImageView and a button to the `activity_main.xml` file:

    ```
    <ImageView
        android:id="@+id/imageview"
        android:layout_width="match_parent"
        android:layout_height="match_parent"
        app:layout_constraintBottom_toBottomOf="parent"
        app:layout_constraintHorizontal_bias="0.0"
        app:layout_constraintLeft_toLeftOf="parent"
        app:layout_constraintRight_toRightOf="parent"
        app:layout_constraintTop_toTopOf="parent"
        app:layout_constraintVertical_bias="1.0"/>


    <Button
        android:id="@+id/button"
        android:layout_width="wrap_content"
        android:layout_height="wrap_content"
        android:text="DESCRIBE ME"
        app:layout_constraintBottom_toBottomOf="parent"
        app:layout_constraintHorizontal_bias="0.5"
        app:layout_constraintLeft_toLeftOf="parent"
        app:layout_constraintRight_toRightOf="parent"
        app:layout_constraintTop_toTopOf="parent"
        app:layout_constraintVertical_bias="1.0"/>
    ```

3. Open `MainActivity.java`, make it implement the `Runnable` interface, then add the following constants, the last five of which are explained in the previous section while others are self explanatory:

    ```
    private static final String MODEL_FILE = "file:///android_asset/image2text_frozen_transformed.pb";
    private static final String VOCAB_FILE =
    ```

Describing Images in Natural Language

```
            "file:///android_asset/word_counts.txt";
    private static final String IMAGE_NAME = "im2txt1.png";

    private static final String INPUT_NODE1 = "convert_image/Cast";
    private static final String OUTPUT_NODE1 = "lstm/initial_state";
    private static final String INPUT_NODE2 = "input_feed";
    private static final String INPUT_NODE3 = "lstm/state_feed";
    private static final String OUTPUT_NODE2 = "softmax";
    private static final String OUTPUT_NODE3 = "lstm/state";

    private static final int IMAGE_WIDTH = 299;
    private static final int IMAGE_HEIGHT = 299;
    private static final int IMAGE_CHANNEL = 3;

    private static final int CAPTION_LEN = 20;
    private static final int WORD_COUNT = 12000;
    private static final int STATE_COUNT = 1024;
    private static final int START_ID = 2;
    private static final int END_ID = 3;
```

And the following instance variables and a Handler implementation:

```
    private ImageView mImageView;
    private Button mButton;

    private TensorFlowInferenceInterface mInferenceInterface;
    private String[] mWords = new String[WORD_COUNT];
    private int[] intValues;
    private float[] floatValues;

    Handler mHandler = new Handler() {
        @Override
        public void handleMessage(Message msg) {
            mButton.setText("DESCRIBE ME");
            String text = (String)msg.obj;
            Toast.makeText(MainActivity.this, text,
    Toast.LENGTH_LONG).show();
            mButton.setEnabled(true);
        }  };
```

4. In the `onCreate` method, first add the code that shows a test image in the `ImageView` and handles the button click event:

```
    mImageView = findViewById(R.id.imageview);
    try {
        AssetManager am = getAssets();
        InputStream is = am.open(IMAGE_NAME);
        Bitmap bitmap = BitmapFactory.decodeStream(is);
```

```
        mImageView.setImageBitmap(bitmap);
    } catch (IOException e) {
        e.printStackTrace();
    }

    mButton = findViewById(R.id.button);
    mButton.setOnClickListener(new View.OnClickListener() {
        @Override
        public void onClick(View v) {
            mButton.setEnabled(false);
            mButton.setText("Processing...");
            Thread thread = new Thread(MainActivity.this);
            thread.start();
        }
    });
```

Then add the code that reads each line of word_counts.txt, and saves each word in the mWords array:

```
    String filename = VOCAB_FILE.split("file:///android_asset/")[1];
    BufferedReader br = null;
    int linenum = 0;
    try {
        br = new BufferedReader(new
InputStreamReader(getAssets().open(filename)));
        String line;
        while ((line = br.readLine()) != null) {
            String word = line.split(" ")[0];
            mWords[linenum++] = word;
        }
        br.close();
    } catch (IOException e) {
        throw new RuntimeException("Problem reading vocab file!" , e);
    }
```

5. Now, in the public void run() method, started when the DESCRIBE ME button's onClick event occurs, add the code to resize the test image, read the pixel values from the resized bitmap, then convert them to float numbers – we've seen code like this in three earlier chapters:

```
    intValues = new int[IMAGE_WIDTH * IMAGE_HEIGHT];
    floatValues = new float[IMAGE_WIDTH * IMAGE_HEIGHT *
    IMAGE_CHANNEL];

    Bitmap bitmap =
    BitmapFactory.decodeStream(getAssets().open(IMAGE_NAME));
    Bitmap croppedBitmap = Bitmap.createScaledBitmap(bitmap,
```

```
            IMAGE_WIDTH, IMAGE_HEIGHT, true);
    croppedBitmap.getPixels(intValues, 0, IMAGE_WIDTH, 0, 0,
            IMAGE_WIDTH, IMAGE_HEIGHT);
    for (int i = 0; i < intValues.length; ++i) {
        final int val = intValues[i];
        floatValues[i * IMAGE_CHANNEL + 0] = ((val >> 16) & 0xFF);
        floatValues[i * IMAGE_CHANNEL + 1] = ((val >> 8) & 0xFF);
        floatValues[i * IMAGE_CHANNEL + 2] = (val & 0xFF);
    }
```

6. Create a `TensorFlowInferenceInterface` instance that loads the model file and make the first inference using the model by feeding the image values to it and then fetching the return result in `initialState`:

```
AssetManager assetManager = getAssets();
mInferenceInterface = new
TensorFlowInferenceInterface(assetManager, MODEL_FILE);

float[] initialState = new float[STATE_COUNT];
mInferenceInterface.feed(INPUT_NODE1, floatValues, IMAGE_WIDTH,
        IMAGE_HEIGHT, 3);
mInferenceInterface.run(new String[] {OUTPUT_NODE1}, false);
mInferenceInterface.fetch(OUTPUT_NODE1, initialState);
```

7. Set the first `input_feed` value to be the start ID, and the first `state_feed` value to the returned `initialState` value:

```
long[] inputFeed = new long[] {START_ID};
float[] stateFeed = new float[STATE_COUNT * inputFeed.length];
for (int i=0; i < STATE_COUNT; i++) {
    stateFeed[i] = initialState[i];
}
```

As you can see, it's simpler in Android than iOS to get and set tensor values and make the inference, thanks to the `TensorFlowInferenceInterface` implementation in Android. Before we start using the `inputFeed` and `stateFeed` repeatedly to make inferences using the model, we create a `captions` list of a pair of integers and floats, with the integer as the word ID with the max softmax value (among all the softmax values returned by the model for each inference call) and the float as the softmax value of the word. We could just use a simple vector to hold the word with the max softmax value from each inference return, but using the list of pairs makes it easier when later we want to switch from the greedy search method to a beam search:

```
List<Pair<Integer, Float>> captions = new ArrayList<Pair<Integer,
```

```
Float>>();
```

8. In a `for` loop over the caption length, we feed `input_feed` and `state_feed` the values we set above, then fetch the returned `softmax` and `newstate` values:

```
for (int i=0; i<CAPTION_LEN; i++) {
    float[] softmax = new float[WORD_COUNT * inputFeed.length];
    float[] newstate = new float[STATE_COUNT * inputFeed.length];

    mInferenceInterface.feed(INPUT_NODE2, inputFeed, 1);
    mInferenceInterface.feed(INPUT_NODE3, stateFeed, 1, STATE_COUNT);
    mInferenceInterface.run(new String[]{OUTPUT_NODE2, OUTPUT_NODE3}, false);
    mInferenceInterface.fetch(OUTPUT_NODE2, softmax);
    mInferenceInterface.fetch(OUTPUT_NODE3, newstate);
```

9. Now, create another list of pairs of integers and floats, add the ID and softmax value of each word to the list, and sort the list in descending order:

```
    List<Pair<Integer, Float>> prob_id = new ArrayList<Pair<Integer, Float>>();
    for (int j = 0; j < WORD_COUNT; j++) {
        prob_id.add(new Pair(j, softmax[j]));
    }

    Collections.sort(prob_id, new Comparator<Pair<Integer, Float>>() {
        @Override
        public int compare(final Pair<Integer, Float> o1, final Pair<Integer, Float> o2) {
            return o1.second > o2.second ? -1 : (o1.second == o2.second ? 0 : 1);
        }
    });
```

10. If the word with the max probability is the end word, we end the loop; otherwise, add the pair to the `captions` list, and update `input_feed` with the word ID with the max softmax value and `state_feed` with the returned state values, to continue with the next inference:

```
if (prob_id.get(0).first == END_ID) break;

captions.add(new Pair(prob_id.get(0).first, prob_id.get(0).first));

inputFeed = new long[] {prob_id.get(0).first};
for (int j=0; j < STATE_COUNT; j++) {
```

```
        stateFeed[j] = newstate[j];
    }
}
```

11. Finally, go through each pair in the `captions` list and add each word, if not the start and end ones, to the `sentence` string, which is returned via the Handler to show the natural language output to the user:

```
String sentence = "";
for (int i=0; i<captions.size(); i++) {
    if (captions.get(i).first == START_ID) continue;
    if (captions.get(i).first == END_ID) break;

    sentence = sentence + " " + mWords[captions.get(i).first];
}

Message msg = new Message();
msg.obj = sentence;
mHandler.sendMessage(msg);
```

Run the app in your virtual or real Android device. It takes about 10 seconds to see the result. You can play with the four different test images shown in the previous section, and see the results in Figure 6.9:

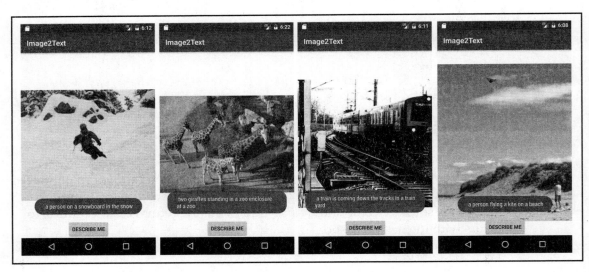

Figure 6.9: Showing image captioning results in Android

Some results are slightly different from the iOS results and the results on the TensorFlow im2txt site. But they all look pretty good. Also, running the non-memmapped version of the model in a relatively older Android device such as Nexus 5 works fine. But it'd be nice to load the memmapped model in Android to see a significant performance increase, which we may cover in a chapter later in the book.

So, this completes the step-by-step, Android app-building process using the powerful image captioning model. Whether you're working with an iOS or Android app, you should be able to easily integrate our trained model and inference code in your own apps, or go back to the training process to fine-tune the model and then prepare and optimize a better model to be used in your mobile apps.

Summary

In this chapter, we first discussed how image captioning powered by modern end-to-end deep learning works, then summarized how to train such a model using the TensorFlow im2txt model project. We discussed in detail how to find the correct input node names and output node names, and how to freeze the model and then use the latest graph transformation tool and the memmapped conversion tool to fix some nasty bugs while loading the model on mobiles. After that, we showed detailed tutorials on how to build iOS and Android apps using the model and making new sequence inferences with the LSTM RNN component of the model.

It's pretty amazing that, after training with tens of thousands of image captioning examples, and powered by modern CNN and LSTM models, we can build and use a model that can generate a sensible natural language description of a picture on our mobile devices. It's not hard to imagine what kind of useful apps can be built on top of this. Are we like Holmes yet? Definitely not. Are we on the road already? We hope so. The world of AI is so fascinating and challenging at the same time, but as long as we keep making steady progress and improving our own learning process, while avoiding the vanishing and exploding gradient problems, there's a good chance we'll be able to build a Holmes-like model and use it in mobile apps anytime, anywhere, one day.

Having gone through a long chapter with the practical use of a CNN- and LSTM-based network model, we deserve some fun. In the next chapter, you'll see how to use another CNN- and LSTM-based model to develop fun iOS and Android apps that let you draw objects and then recognize what they are. For a quick fun play of the online version of the game, go to `https://quickdraw.withgoogle.com`.

Recognizing Drawing with CNN and LSTM

In the previous chapter, we saw the power of using a deep learning model that integrates CNN with LSTM RNN to generate a natural language description of an image. If deep learning-powered AI is like the new electricity, we certainly expect to see the application of such hybrid neural network models in many different areas. What's the opposite of a serious application such as image captioning? A fun drawing app such as Quick Draw (https://quickdraw.withgoogle.com, see https://quickdraw.withgoogle.com/data for fun sample data), which uses a model trained and based on 50 million drawings in 345 categories, and classifies new drawings into those categories, sounds like a good one. And there's an official TensorFlow tutorial (https://www.tensorflow.org/tutorials/recurrent_quickdraw) on how to build such a model to help us start quickly.

It turns out that the task of using the model built with this tutorial on iOS and Android apps offers a great opportunity to:

- Strengthen our understanding of finding out the right input and output node names of a model so we can appropriately prepare the model for mobile apps
- Use additional methods to fix new model loading and inference errors in iOS
- Build, for the first time, a custom TensorFlow native library for Android to fix new model loading and prediction errors in Android
- See more examples of how to feed a TensorFlow model with input in its expected format and get and process its output in iOS and Android

In addition, in the process of dealing with all the tedious yet important details so the model can work like magic in making cool drawing classifications, you get to have some fun doodling on your iOS and Android devices.

So, in this chapter, we'll cover the following topics:

- Drawing classification – how it works
- Training and preparing the drawing classification model
- Using the drawing classification model in iOS
- Using the drawing classification model in Android

Drawing classification – how it works

The drawing classification model built into the TensorFlow tutorial (https://www.tensorflow.org/tutorials/recurrent_quickdraw) first takes the user drawing input represented as a list of points and converts the normalized input to a tensor of the deltas of consecutive points along with information about whether each point is the beginning of a new stroke. Then it passes the tensor through several convolutional layers and LSTM layers, and finally a softmax layer, as shown in Figure 7.1, to classify the user drawing:

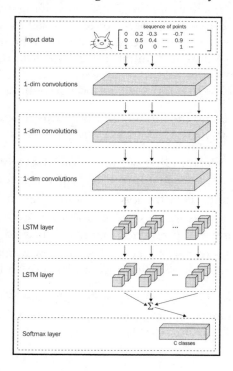

Figure 7.1: The drawing classification mode

Unlike the 2D convolution API `tf.layers.conv2d` that accepts a 2D image input, the 1D convolution API `tf.layers.conv1d` is used here for temporal convolution such as drawing. By default, in the drawing classification model, three 1D convolutional layers are used and each layer has 48, 64, and 96 filters, the lengths of which are 5, 5, and 3, respectively. After the convolutional layers, 3 LSTM layers, with 128 forward BasicLSTMCell nodes and 128 backward BasicLSTMCell nodes per layer, are created and then used to create a dynamic bidirectional recurrent neural network, whose output is sent to a final fully connected layer for the calculation of logits (unnormalized log probabilities).

If you don't have a good understanding of all these details, don't worry; to develop powerful mobile apps using a model built by others, you don't have to understand all the details, but in the next chapter we'll also discuss in greater detail how you can build a RNN model from scratch for stock prediction, and with that, you'll have a better understanding of all the RNN stuff.

The simple and elegant model and the Python implementation of building the model are described in detail in the fun tutorial mentioned earlier, with the source code at `tutorials/rnn/quickdraw` in the repo https://github.com/tensorflow/models. There's only one thing we'd like to say before we move on to the next section: the code for building, training, evaluating, and predicting with the model, unlike the code you've seen in the previous chapters, uses a high-level TensorFlow API called `Estimator` (https://www.tensorflow.org/api_docs/python/tf/estimator/Estimator), or more accurately, a custom `Estimator`. If you're interested in the model implementation details, you should take a look at this guide on creating and using the custom `Estimator` (https://www.tensorflow.org/get_started/custom_estimators) and the helpful source code for the guide at `models/samples/core/get_started/custom_estimator.py` of https://github.com/tensorflow/models. Basically, you first implement a function that defines your model, specifies the loss and accuracy measures, and sets the optimizer and `training` operation, then you create an instance of the `tf.estimator.Estimator` class and call its `train`, `evaluate`, and `predict` methods. As you'll see soon, using `Estimator` simplifies how to build, train, and infer with neural network models, but because it's a high-level API, it also makes some lower-level tasks, such as finding out input and output node names for inference on mobile devices, more difficult.

Training, predicting, and preparing the drawing classification model

It's pretty straightforward to train the model but a little tricky to prepare the model for mobile deployment. Before we can start training, first make sure you already have the TensorFlow model repo (`https://github.com/tensorflow/models`) cloned in your TensorFlow root directory, as we did in the previous two chapters. Then download the drawing classification training dataset at `http://download.tensorflow.org/data/quickdraw_tutorial_dataset_v1.tar.gz`, which is about 1.1 GB, create a new folder called `rnn_tutorial_data`, and unzip the `dataset tar.gz` file to it. You'll see 10 training TFRecord files and 10 evaluation TFRecord files, as well as two files with the `.classes` extension, which have the same content and are just plain text for the 345 categories that the dataset can be used to classify, such as "sheep", "skull", "donut", and "apple".

Training the drawing classification model

To train the model, simply open a Terminal, `cd` to `tensorflow/models/tutorials/rnn/quickdraw`, then run the following script:

```
python train_model.py \
  --training_data=rnn_tutorial_data/training.tfrecord-?????-of-????? \
  --eval_data=rnn_tutorial_data/eval.tfrecord-?????-of-????? \
  --model_dir quickdraw_model/ \
  --classes_file=rnn_tutorial_data/training.tfrecord.classes
```

By default, the training steps are 100k, and it takes about 6 hours on our GTX 1070 GPU to finish the training. After the training is completed, you'll see a familiar file listing in the model directory (omitting the other four sets of `model.ckpt*` files):

```
ls -lt quickdraw_model/
-rw-rw-r-- 1 jeff jeff 164419871 Feb 12 05:56 events.out.tfevents.1518422507.AiLabby
-rw-rw-r-- 1 jeff jeff 1365548 Feb 12 05:56 model.ckpt-100000.meta
-rw-rw-r-- 1 jeff jeff 279 Feb 12 05:56 checkpoint
-rw-rw-r-- 1 jeff jeff 13707200 Feb 12 05:56 model.ckpt-100000.data-00000-of-00001
-rw-rw-r-- 1 jeff jeff 2825 Feb 12 05:56 model.ckpt-100000.index
-rw-rw-r-- 1 jeff jeff 2493402 Feb 12 05:47 graph.pbtxt
drwxr-xr-x 2 jeff jeff 4096 Feb 12 00:11 eval
```

If you run `tensorboard --logdir quickdraw_model` then launch the TensorBoard at `http://localhost:6006` from a browser, you'll see the accuracy gets to about 0.55 and the loss to about 2.0. If you continue the training for about 200k more steps, the accuracy will improve to about 0.65 and the loss will drop to 1.3, as shown in Figure 7.2:

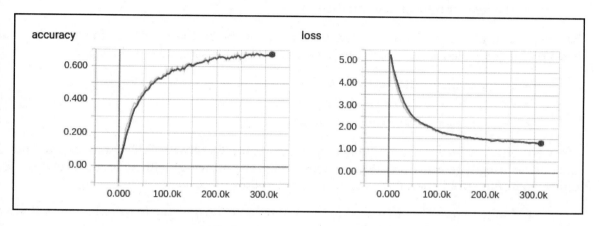

Figure 7.2: The accuracy and loss of the model after 300k training steps

Now we can run the `freeze_graph.py` tool, as we did in the previous chapter, to generate a model file for mobile devices. But before we do that, let's first see how we can use the model in Python to make some inferences, like the `run_inference.py` script in the previous chapter.

Predicting with the drawing classification model

Take a look at the `train_model.py` file in the `models/tutorial/rnn/quickdraw` folder. When it starts to run, an `Estimator` instance gets created in the `create_estimator_and_specs` function:

```
estimator = tf.estimator.Estimator(
    model_fn=model_fn,
    config=run_config,
    params=model_params)
```

The key parameter passed to the `Estimator` class is a model function called `model_fn` that defines:

- Functions to get the input tensor and create convolutional, RNN, and final layers
- Code to call those functions to build the model
- The loss, optimizer, and prediction

Before returning a `tf.estimator.EstimatorSpec` instance, the `model_fn` function also has a parameter called `mode` which can have one of the following three values:

- `tf.estimator.ModeKeys.TRAIN`
- `tf.estimator.ModeKeys.EVAL`
- `tf.estimator.ModeKeys.PREDICT`

The way `train_model.py` is implemented supports the TRAIN and EVAL modes, but you can't use it directly to make an inference (classifying the drawing) with a specific drawing input. To test the prediction with a specific input, follow these steps:

1. Make a copy of `train_model.py` and rename the new file to `predict.py` – that way you can play with the prediction more freely.
2. In `predict.py`, define an input function for prediction with the `features` set as the drawing input (the deltas of the consecutive points with a third number indicating whether the point is the beginning of a stroke) expected by the model:

```
def predict_input_fn():
    def _input_fn():
        features = {'shape': [[16, 3]], 'ink': [[
            -0.23137257, 0.31067961, 0. ,
            -0.05490196, 0.1116505 , 0. ,
            0.00784314, 0.09223297, 0. ,
            0.19215687, 0.07766992, 0. ,
            ...
            0.12156862, 0.05825245, 0. ,
            0. , -0.06310678, 1. ,
            0. , 0., 0. ,
            ...
            0. , 0., 0. ,
        ]]}
        features['shape'].append( features['shape'][0])
        features['ink'].append( features['ink'][0])
        features=dict(features)
        
        dataset = tf.data.Dataset.from_tensor_slices(features)
```

```
            dataset = dataset.batch(FLAGS.batch_size)

            return dataset.make_one_shot_iterator().get_next()

      return _input_fn
```

We're not showing all the point values but they are created using the sample cat example data shown in the TensorFlow RNN for Drawing Classification tutorial with the `parse_line` function applied (see the tutorial or `create_dataset.py` in the `models/tutorials/rnn/quickdraw` folder for details).

Also notice that we use `tf.data.Dataset`'s `make_one_shot_iterator` method to create an iterator that returns an example from the dataset (in the case here we only have one example in the dataset), the same way the model gets data during training and evaluation when working with a large dataset – this is why you'll see the `OneShotIterator` operation in the model's graph later.

3. In the main function, call the estimator's `predict` method, which yields predictions for the given feature, and then print the next prediction:

```
predictions = estimator.predict(input_fn=predict_input_fn())
print(next(predictions)['argmax'])
```

4. In the `model_fn` function, after `logits = _add_fc_layers(final_state)`, add the following code:

```
argmax = tf.argmax(logits, axis=1)

if mode == tf.estimator.ModeKeys.PREDICT:
  predictions = {
    'argmax': argmax,
    'softmax': tf.nn.softmax(logits),
    'logits': logits,
  }

  return tf.estimator.EstimatorSpec(mode, predictions=predictions)
```

Now if you run `predict.py`, you'll get the class ID with the max value returned for the input data in Step 2.

With a basic understanding of how to predict with the model built using the `Estimator` high-level API, we're now ready to freeze the model so it can be used on mobile devices, which requires us to first figure out what the output node names should be.

Preparing the drawing classification model

Let's use TensorBoard to see what we can find. In the GRAPHS section of the TensorBoard view of our model, you can see, as shown in Figure 7.3, that the **BiasAdd** node, highlighted in red is the input of an **ArgMax** operation used to calculate accuracy, as well as the input of a softmax operation. We can use the **SparseSoftmaxCrossEntropyWithLogits** (shown in Figure 7.3 only as **SparseSiftnaxCr...**) operation or just **dense/BiasAdd** as the output node name, but we'll use **ArgMax** and **dense/BiasAdd** as the two output node names for the `freeze_graph` tool so we can more easily see the output of the final dense layer as well as the **ArgMax** result:

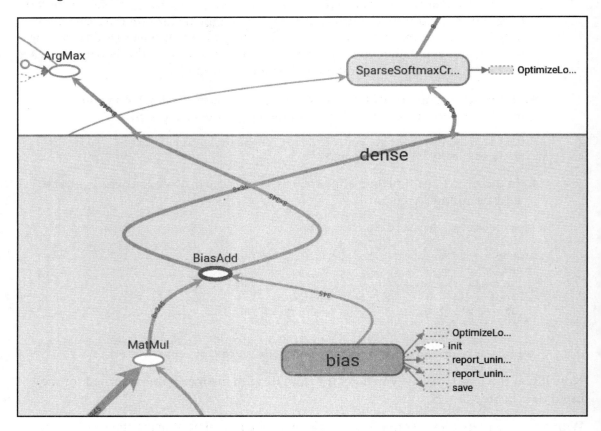

Figure 7.3: Showing possible output node names of the model

Run the following script in your TensorFlow root to get a frozen graph after replacing the `--input_graph` and `--input_checkpoint` values with the path of your `graph.pbtxt` file and the latest model checkpoint prefix:

```
python tensorflow/python/tools/freeze_graph.py --
input_graph=/tmp/graph.pbtxt --input_checkpoint=/tmp/model.ckpt-314576 --
output_graph=/tmp/quickdraw_frozen_dense_biasadd_argmax.pb --
output_node_names="dense/BiasAdd,ArgMax"
```

You'll see `quickdraw_frozen_dense_biasadd_argmax.pb` gets created successfully. But if you try to load the model in your iOS or Android app, you'll get an error message that says `Could not create TensorFlow Graph: Not found: Op type not registered 'OneShotIterator' in binary. Make sure the Op and Kernel are registered in the binary running in this process.`

We talked about what `OneShotIterator` means in the previous subsection. Back in the TensorBoard GRAPHS section, we can see, as shown in Figure 7.4, the `OneShotIterator`, highlighted with red and also shown in the right info panel, in the bottom of the graph, and a couple of levels above, there's one `Reshape` operation that serves as the input to the first convolutional layer:

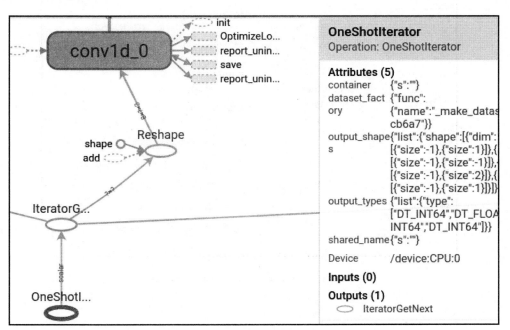

Figure 7.4: Finding out possible input node names

You may wonder why we can't fix the `Not found: Op type not registered 'OneShotIterator'` error with a technique we used before, which is to first find out which source file contains the op using the command `grep 'REGISTER.*"OneShotIterator"' tensorflow/core/ops/*.cc` (and you'll see the output as `tensorflow/core/ops/dataset_ops.cc:REGISTER_OP("OneShotIterator")`) then add `tensorflow/core/ops/dataset_ops.cc` to `tf_op_files.txt` and rebuild the TensorFlow library. Even if this were feasible, it would complicate the solution as now we need to feed the model with some data related to `OneShotIterator`, instead of the direct user drawing in points.

Furthermore, one more level above on the right side (Figure 7.5), there's another operation, `Squeeze`, which acts as the input to the `rnn_classification` subgraph:

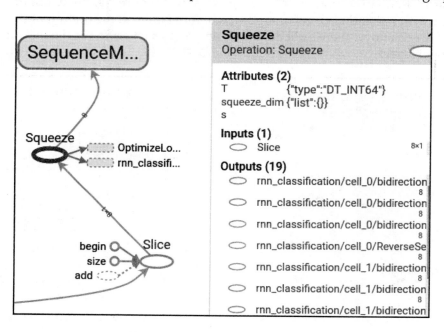

Figure 7.5: Further research to find out input node names

We don't need to worry about the operation Shape that's on the right hand of Reshape, as it's actually an output of the rnn_classification subgraph. So the intuition behind all this research is that we can use Reshape and Squeeze as the two input nodes and then with the transform_graph tool that we saw in the previous chapter, we should be able to strip out the nodes below Reshape and Squeeze, including OneShotIterator.

Now run the following commands in your TensorFlow root directory:

```
bazel-bin/tensorflow/tools/graph_transforms/transform_graph --
in_graph=/tmp/quickdraw_frozen_dense_biasadd_argmax.pb --
out_graph=/tmp/quickdraw_frozen_strip_transformed.pb --
inputs="Reshape,Squeeze" --outputs="dense/BiasAdd,ArgMax" --transforms='
strip_unused_nodes(name=Squeeze,type_for_name=int64,shape_for_name="8",name
=Reshape,type_for_name=float,shape_for_name="8,16,3")'
```

Here we use a more advanced format for strip_unused_nodes: for each input node name (Squeeze and Reshape), we specify its particular type and shape to avoid the model loading error later. For more detail on the strip_unused_nodes of the transform_graph tool, see its documentation at https://github.com/tensorflow/tensorflow/tree/master/tensorflow/tools/graph_transforms.

Now loading the model in iOS or Android, the OneShotIterator error will be gone. But as you have probably learned to expect, a new error occurs: Could not create TensorFlow Graph: Invalid argument: Input 0 of node IsVariableInitialized was passed int64 from global_step:0 incompatible with expected int64_ref.

We first need to know more about IsVariableInitialized. If we go back to the TensorBoard **GRAPHS** tab, we see, on the left side, there's an IsVariableInitialized operation, highlighted with red and shown in the info panel on the right, with global_step as its input (Figure 7.6).

Even if we don't know exactly what it's used for, we can be sure that it has nothing to do with the model inference, which simply expects some inputs (Figure 7.4 and Figure 7.5) and generates drawing classification as outputs (Figure 7.3):

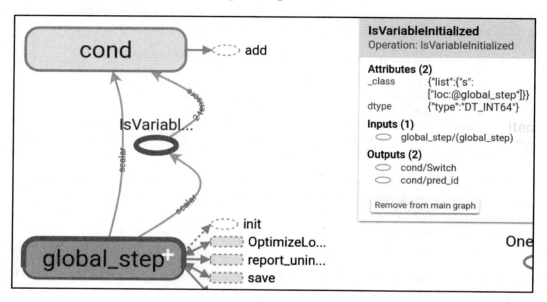

Figure 7.6: Finding out nodes causing model loading error but not related to model inference

So how can we get rid of the `global_step` as well as the other related cond nodes that, because of their isolation, won't be stripped out by the transform graph tool? Luckily, the `freeze_graph` script supports that – it's only documented in its source code at https://github.com/tensorflow/tensorflow/blob/master/tensorflow/python/tools/freeze_graph.py. We can use the `variable_names_blacklist` parameter for the script to specify the nodes that should be removed in the frozen model:

```
python tensorflow/python/tools/freeze_graph.py --
input_graph=/tmp/graph.pbtxt --input_checkpoint=/tmp/model.ckpt-314576 --
output_graph=/tmp/quickdraw_frozen_long_blacklist.pb --
output_node_names="dense/BiasAdd,ArgMax" --
variable_names_blacklist="IsVariableInitialized,global_step,global_step/Ini
tializer/zeros,cond/pred_id,cond/read/Switch,cond/read,cond/Switch_1,cond/M
erge"
```

Here we just list the nodes inside the `global_step` and cond scopes. Now run the `transform_graph` tool again:

```
bazel-bin/tensorflow/tools/graph_transforms/transform_graph --
in_graph=/tmp/quickdraw_frozen_long_blacklist.pb --
out_graph=/tmp/quickdraw_frozen_long_blacklist_strip_transformed.pb --
inputs="Reshape,Squeeze" --outputs="dense/BiasAdd,ArgMax" --transforms='
strip_unused_nodes(name=Squeeze,type_for_name=int64,shape_for_name="8",name
=Reshape,type_for_name=float,shape_for_name="8,16,3")'
```

Load the resulting model file `quickdraw_frozen_long_blacklist_strip_transformed.pb` in iOS or Android and you won't see the `IsVariableInitialized` error anymore. Of course, there's one more error you'll see – on both iOS and Android. Loading the preceding model will result in this error:

```
Couldn't load model: Invalid argument: No OpKernel was registered to
support Op 'RefSwitch' with these attrs. Registered devices: [CPU],
Registered kernels:
  device='GPU'; T in [DT_FLOAT]
  device='GPU'; T in [DT_INT32]
  device='GPU'; T in [DT_BOOL]
  device='GPU'; T in [DT_STRING]
  device='CPU'; T in [DT_INT32]
  device='CPU'; T in [DT_FLOAT]
  device='CPU'; T in [DT_BOOL]

[[Node: cond/read/Switch = RefSwitch[T=DT_INT64,
_class=["loc:@global_step"], _output_shapes=[[], []]](global_step,
cond/pred_id)]]
```

To fix this error, we have to build custom TensorFlow libraries for iOS and Android, in their own different ways. Before we discuss how to do that in the following iOS and Android sections, let's do one more thing first: convert the model to the memmapped version so it loads faster and uses less memory in iOS:

```
bazel-bin/tensorflow/contrib/util/convert_graphdef_memmapped_format \
--in_graph=/tmp/quickdraw_frozen_long_blacklist_strip_transformed.pb \
--
out_graph=/tmp/quickdraw_frozen_long_blacklist_strip_transformed_memmapped.
pb
```

Using the drawing classification model in iOS

To fix the previous RefSwitch error, which will occur no matter whether you use the TensorFlow Pod as we did in Chapter 2, *Classifying Images with Transfer Learning*, and Chapter 6, *Describing Images in Natural Language*, or the manually built TensorFlow library, as in the other chapters, we have to use some new trick. The error occurs because the INT64 data type is required for a RefSwitch operation, but it's not one of the registered data types built into the TensorFlow library because, by default, to make the library as small as possible, only common data types for each operation are included. We may fix this from the model building end in Python, but here we'll just show you how to fix this from the iOS end, which can be useful when you don't have access to the source code to build a model.

Building custom TensorFlow library for iOS

Open the Makefile from tensorflow/contrib/makefile/Makefile then, if you use TensorFlow 1.4, search for IOS_ARCH. For each of the architectures (5 in total: ARMV7, ARMV7S, ARM64, I386, X86_64), change -D__ANDROID_TYPES_SLIM__ to -D__ANDROID_TYPES_FULL__. The Makefile in TensorFlow 1.5 (or 1.6/1.7) is a little different, although it's still in the same folder; for 1.5/1.6/1.7, search for ANDROID_TYPES_SLIM and change it to ANDROID_TYPES_FULL. Now rebuild the TensorFlow library by running tensorflow/contrib/makefile/build_all_ios.sh. After this, the RefSwitch error will be gone when loading the model file. The app size built with the TensorFlow library with full data type support will be about 70 MB versus 37 MB, the app size when built with the default slim data types.

As if enough were not enough, yet another model loading error occurs:

```
Could not create TensorFlow Graph: Invalid argument: No OpKernel was
registered to support Op 'RandomUniform' with these attrs. Registered
devices: [CPU], Registered kernels: <no registered kernels>.
```

Luckily, you should be pretty familiar with how to fix this kind of error if you have gone through the previous chapters. Here's a quick recap: first find out which op and kernel files define and implement the operation, then check to see if the op or kernel file is included in the `tf_op_files.txt` file, and there should be at least one missing, causing the error; now just add the op or kernel file to `tf_op_files.txt` and rebuild the library. In our case, run the following commands:

```
grep RandomUniform tensorflow/core/ops/*.cc
grep RandomUniform tensorflow/core/kernels/*.cc
```

And you'll see these files as output:

```
tensorflow/core/ops/random_grad.cc
tensorflow/core/ops/random_ops.cc:
tensorflow/core/kernels/random_op.cc
```

The `tensorflow/contrib/makefile/tf_op_files.txt` file only has the first two files, so just add the last one, `tensorflow/core/kernels/random_op.cc`, to the end of `tf_op_files.txt`, and run `tensorflow/contrib/makefile/build_all_ios.sh` again.

Finally, all errors are gone when loading the model, and we can start having some real fun by implementing the app logic to handle user drawing, converting the points to a format expected by the model, and getting the classification results back.

Developing an iOS app to use the model

Let's create a new Xcode project using Objective-C, then drag and drop the `tensorflow_util.h` and `tensorflow_util.mm` files from the `Image2Text` iOS project created in the previous chapter. Also, drag and drop the two model files, `quickdraw_frozen_long_blacklist_strip_transformed.pb` and `quickdraw_frozen_long_blacklist_strip_transformed_memmapped.pb`, and the `training.tfrecord.classes` file from `models/tutorials/rnn/quickdraw/rnn_tutorial_data` to the QuickDraw project, and rename `training.tfrecord.classes` to `classes.txt`.

Recognizing Drawing with CNN and LSTM

Also rename `ViewController.m` to `ViewController.mm`, and comment the `GetTopN` function definition in `tensorflow_util.h` and its implementation in `tensorflow_util.mm`, as we'll implement a modified version in `ViewController.mm`. Your project should now look like Figure 7.7:

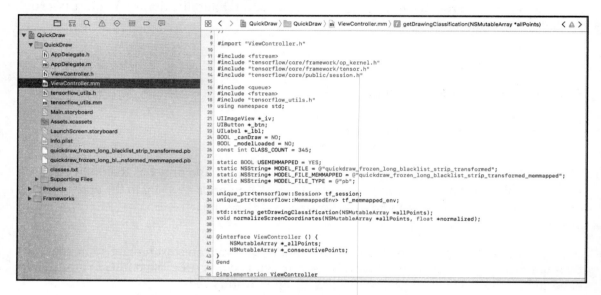

Figure 7.7: Showing the QuickDraw Xcode project with initial content of ViewController.mm

We are now ready to work on `ViewController.mm` alone to finish our mission.

1. After setting basic constants and variables and two function prototypes as in Figure 7.6, in the `viewDidLoad` of `ViewController`, instantiate a `UIButton`, `UILabel`, and a `UIImageView`. Each UI control has been set with several `NSLayoutConstraint` (see the source code repo for the complete code listing). The `UIImageView` related code is as follows:

```
_iv = [[UIImageView alloc] init];
_iv.contentMode = UIViewContentModeScaleAspectFit;
[_iv setTranslatesAutoresizingMaskIntoConstraints:NO];
[self.view addSubview:_iv];
```

The `UIImageView` will be used to display user drawing implemented with `UIBezierPath`. Also, initialize two arrays used to hold every consecutive point and all the points the user has drawn:

```
_allPoints = [NSMutableArray array];
_consecutivePoints = [NSMutableArray array];
```

2. After the button, which has an initial title "Start," is tapped, the user can start drawing; the button title is changed to "Restart" and a few other resets are done:

```
- (IBAction)btnTapped:(id)sender {
    _canDraw = YES;
    [_btn setTitle:@"Restart" forState:UIControlStateNormal];
    [_lbl setText:@""];
    _iv.image = [UIImage imageNamed:@""];
    [_allPoints removeAllObjects];
}
```

3. To handle user drawing, we first implement a `touchesBegan` method:

```
- (void)touchesBegan:(NSSet *)touches withEvent:(UIEvent *)event {
    if (!_canDraw) return;
    [_consecutivePoints removeAllObjects];
    UITouch *touch = [touches anyObject];
    CGPoint point = [touch locationInView:self.view];
    [_consecutivePoints addObject:[NSValue valueWithCGPoint:point]];
    _iv.image = [self createDrawingImageInRect:_iv.frame];
}
```

Then a `touchesMoved` method:

```
- (void)touchesMoved:(NSSet *)touches withEvent:(UIEvent *)event {
    if (!_canDraw) return;
    UITouch *touch = [touches anyObject];
    CGPoint point = [touch locationInView:self.view];
    [_consecutivePoints addObject:[NSValue valueWithCGPoint:point]];
    _iv.image = [self createDrawingImageInRect:_iv.frame];
}
```

And finally a `touchesEnd` method:

```
- (void)touchesEnded:(NSSet *)touches withEvent:(UIEvent *)event {
    if (!_canDraw) return;
    UITouch *touch = [touches anyObject];
    CGPoint point = [touch locationInView:self.view];
```

```
        [_consecutivePoints addObject:[NSValue
valueWithCGPoint:point]];
        [_allPoints addObject:[NSArray
arrayWithArray:_consecutivePoints]];
        [_consecutivePoints removeAllObjects];
        _iv.image = [self createDrawingImageInRect:_iv.frame];
        dispatch_async(dispatch_get_global_queue(0, 0), ^{
            std::string classes = getDrawingClassification(_allPoints);
            dispatch_async(dispatch_get_main_queue(), ^{
                NSString *c = [NSString
stringWithCString:classes.c_str() encoding:[NSString
defaultCStringEncoding]];
                [_lbl setText:c];
            });
        });
    }
```

The code here is pretty self explanatory, except for two methods, `createDrawingImageInRect` and `getDrawingClassification`, which we'll cover next.

4. The method `createDrawingImageInRect` uses `UIBezierPath`'s `moveToPoint` and `addLineToPoint` methods to show user drawing. It first prepares all the completed strokes with all the points stored in the `_allPoints` array by the touch events:

```
- (UIImage *)createDrawingImageInRect:(CGRect)rect
{
UIGraphicsBeginImageContextWithOptions(CGSizeMake(rect.size.width,
rect.size.height), NO, 0.0);
    UIBezierPath *path = [UIBezierPath bezierPath];
    for (NSArray *cp in _allPoints) {
        bool firstPoint = TRUE;
        for (NSValue *pointVal in cp) {
            CGPoint point = pointVal.CGPointValue;
            if (firstPoint) {
                [path moveToPoint:point];
                firstPoint = FALSE;
            }
            else
                [path addLineToPoint:point];
        }
    }
```

Then it prepares all the points in the current ongoing stroke, stored in `_consecutivePoints`:

```
bool firstPoint = TRUE;
for (NSValue *pointVal in _consecutivePoints) {
    CGPoint point = pointVal.CGPointValue;
    if (firstPoint) {
        [path moveToPoint:point];
        firstPoint = FALSE;
    }
    else
        [path addLineToPoint:point];
}
```

Finally, it does the actual drawing and returns the drawing as an `UIImage` to be shown in the `UIImageView`:

```
path.lineWidth = 6.0;
[[UIColor blackColor] setStroke];
[path stroke];
UIImage *image = UIGraphicsGetImageFromCurrentImageContext();
UIGraphicsEndImageContext();
return image;
}
```

5. The `getDrawingClassification` first uses the same code as we did in the last chapter to load a model or its memmapped version:

```
std::string getDrawingClassification(NSMutableArray *allPoints) {
    if (!_modelLoaded) {
        tensorflow::Status load_status;
        if (USEMEMMAPPED) {
            load_status =
LoadMemoryMappedModel(MODEL_FILE_MEMMAPPED, MODEL_FILE_TYPE,
&tf_session, &tf_memmapped_env);
        }
        else {
            load_status = LoadModel(MODEL_FILE, MODEL_FILE_TYPE,
&tf_session);
        }
        if (!load_status.ok()) {
            LOG(FATAL) << "Couldn't load model: " << load_status;
            return "";
        }
        _modelLoaded = YES;
    }
```

Then it gets the number of total points and allocates an array of floating numbers, before calling another function `normalizeScreenCoordinates`, which we'll cover shortly, to convert the points to the format expected by the model:

```
if ([allPoints count] == 0) return "";
int total_points = 0;
for (NSArray *cp in allPoints) {
    total_points += cp.count;
}
float *normalized_points = new float[total_points * 3];
normalizeScreenCoordinates(allPoints, normalized_points);
```

Next, we define input and output node names and create a tensor that holds the number of total points:

```
        std::string input_name1 = "Reshape";
        std::string input_name2 = "Squeeze";
        std::string output_name1 = "dense/BiasAdd";
        std::string output_name2 = "ArgMax";
        const int BATCH_SIZE = 8;

        tensorflow::Tensor seqlen_tensor(tensorflow::DT_INT64,
    tensorflow::TensorShape({BATCH_SIZE}));
        auto seqlen_mapped = seqlen_tensor.tensor<int64_t, 1>();
        int64_t* seqlen_mapped_data = seqlen_mapped.data();
        for (int i=0; i<BATCH_SIZE; i++) {
            seqlen_mapped_data[i] = total_points;
        }
```

Note that we have to use the same `BATCH_SIZE` as the `BATCH_SIZE` when used running `train_model.py` to train the model, which is 8 by default.

The other tensor that holds all the converted point values is created here:

```
        tensorflow::Tensor points_tensor(tensorflow::DT_FLOAT,
    tensorflow::TensorShape({8, total_points, 3}));
        auto points_tensor_mapped = points_tensor.tensor<float, 3>();
        float* out = points_tensor_mapped.data();
        for (int i=0; i<BATCH_SIZE; i++) {
            for (int j=0; j<total_points*3; j++)
                out[i*total_points*3+j] = normalized_points[j];
        }
```

6. Now we run the model and get the expected outputs:

   ```
   std::vector<tensorflow::Tensor> outputs;
   tensorflow::Status run_status = tf_session->Run({{input_name1,
   points_tensor}, {input_name2, seqlen_tensor}}, {output_name1,
   output_name2}, {}, &outputs);
       if (!run_status.ok()) {
           LOG(ERROR) << "Getting model failed:" << run_status;
           return "";
       }

       tensorflow::string status_string = run_status.ToString();
       tensorflow::Tensor* logits_tensor = &outputs[0];
   ```

7. Use a modified version of `GetTopN` and parse the `logits` for top results:

   ```
   const int kNumResults = 5;
   const float kThreshold = 0.1f;
   std::vector<std::pair<float, int> > top_results;
   const Eigen::TensorMap<Eigen::Tensor<float, 1,
   Eigen::RowMajor>, Eigen::Aligned>& logits =
   logits_tensor->flat<float>();

       GetTopN(logits, kNumResults, kThreshold, &top_results);
       string result = "";
       for (int i=0; i<top_results.size(); i++) {
           std::pair<float, int> r = top_results[i];
           if (result == "")
               result = classes[r.second];
           else result += ", " + classes[r.second];
       }
   ```

8. Change `GetTopN` by converting the logits values to the softmax values and return the top softmax values with their positions:

```
float sum = 0.0;
for (int i = 0; i < CLASS_COUNT; ++i) {
    sum += expf(prediction(i));
}
for (int i = 0; i < CLASS_COUNT; ++i) {
    const float value = expf(prediction(i)) / sum;
    if (value < threshold) {
        continue;
    }
    top_result_pq.push(std::pair<float, int>(value, i));
    if (top_result_pq.size() > num_results) {
        top_result_pq.pop();
```

        }
    }

9. Finally, the `normalizeScreenCoordinates` function converts all the points from its screen coordinates, captured in the touch events, to the delta differences – this is pretty much a port of the Python method `parse_line` in https://github.com/tensorflow/models/blob/master/tutorials/rnn/quickdraw/create_dataset.py:

```
void normalizeScreenCoordinates(NSMutableArray *allPoints, float *normalized) {
    float lowerx=MAXFLOAT, lowery=MAXFLOAT, upperx=-MAXFLOAT, uppery=-MAXFLOAT;
    for (NSArray *cp in allPoints) {
        for (NSValue *pointVal in cp) {
            CGPoint point = pointVal.CGPointValue;
            if (point.x < lowerx) lowerx = point.x;
            if (point.y < lowery) lowery = point.y;
            if (point.x > upperx) upperx = point.x;
            if (point.y > uppery) uppery = point.y;
        }
    }
    float scalex = upperx - lowerx;
    float scaley = uppery - lowery;
    int n = 0;
    for (NSArray *cp in allPoints) {
        int m=0;
        for (NSValue *pointVal in cp) {
            CGPoint point = pointVal.CGPointValue;
            normalized[n*3] = (point.x - lowerx) / scalex;
            normalized[n*3+1] = (point.y - lowery) / scaley;
            normalized[n*3+2] = (m ==cp.count-1 ? 1 : 0);
            n++; m++;
        }
    }
    for (int i=0; i<n-1; i++) {
        normalized[i*3] = normalized[(i+1)*3] - normalized[i*3];
        normalized[i*3+1] = normalized[(i+1)*3+1] - normalized[i*3+1];
        normalized[i*3+2] = normalized[(i+1)*3+2];
    }
}
```

Chapter 7

Now you can run the app in an iOS simulator or device, start drawing, and see what the model thinks you're drawing. Figure 7.8 shows a few drawings and the classification results – not the best drawings, but the whole process works!:

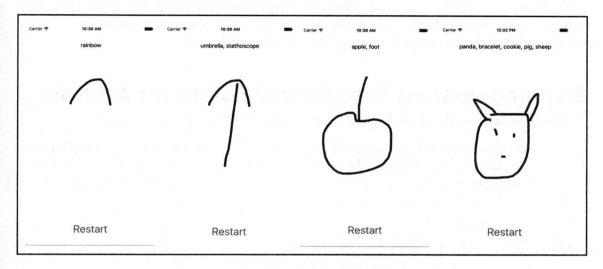

Figure 7.8: Showing drawing and classification results on iOS

Using the drawing classification model in Android

It's time to see how we can load and use the model in Android. In the previous chapters, we added TensorFlow support just by using the Android apps' `build.gradle` file and adding a line `compile 'org.tensorflow:tensorflow-android:+'`. Compared to iOS, where we have to build a custom TensorFlow library to fix different model loading or running errors (for example, in `Chapter 3`, *Detecting Objects and Their Locations*, `Chapter 4`, *Transforming Pictures with Amazing Art Styles*, and `Chapter 5`, *Understanding Simple Speech Commands*), the default TensorFlow library for Android has better support for registered operations and data types, which could be because Android is Google's first-class citizen while iOS is second, even a close second.

The fact is that, when we deal with all kinds of amazing models out there, it's just a matter of time before we have to face the inevitable: we have to build the TensorFlow library for Android manually to fix some errors that the default TensorFlow library simply cannot deal with. The `No OpKernel was registered to support Op 'RefSwitch' with these attrs`. error is one such error. For optimistic developers, this just means another opportunity to add some new tricks to your skill set.

Building custom TensorFlow library for Android

Follow these steps to manually build a custom TensorFlow library for Android:

1. In your TensorFlow root directory, there's a file named WORKSPACE. Edit it and make the `android_sdk_repository` and `android_ndk_repository` look like the following settings (replace `build_tools_version` and the SDK and NDK paths with your own settings):

   ```
   android_sdk_repository(
       name = "androidsdk",
       api_level = 23,
       build_tools_version = "26.0.1",
       path = "$HOME/Library/Android/sdk",
   )

   android_ndk_repository(
       name="androidndk",
       path="$HOME/Downloads/android-ndk-r15c",
       api_level=14)
   ```

2. If you have also worked with iOS apps in the book and have changed the `tensorflow/core/platform/default/mutex.h` from `#include "nsync_cv.h"` and `#include "nsync_mu.h"` to `#include "nsync/public/nsync_cv.h"` and `#include "nsync/public/nsync_mu.h"`, as shown in Chapter 3, *Detecting Objects and Their Locations*, you need to change them back to successfully build the TensorFlow Android library (later when you work on Xcode and iOS apps using the manually built TensorFlow library, you need to add `nsync/public` before the two headers.

   Changing `tensorflow/core/platform/default/mutex.h` back and forth certainly is not an ideal solution. It's supposed to be just as a workaround. As it only needs to be changed when you start using a manually built TensorFlow iOS library or when you build a custom TensorFlow library, we can live with it for now.

3. Run the following command to build the native TensorFlow library if you have a virtual emulator or Android device that supports x86 CPU:

```
bazel build -c opt --copt="-D__ANDROID_TYPES_FULL__"
//tensorflow/contrib/android:libtensorflow_inference.so \
   --crosstool_top=//external:android/crosstool \
   --host_crosstool_top=@bazel_tools//tools/cpp:toolchain \
   --cpu=x86_64
```

If your Android device supports armeabi-v7a, as most Android devices do, run this:

```
bazel build -c opt --copt="-D__ANDROID_TYPES_FULL__"
//tensorflow/contrib/android:libtensorflow_inference.so \
   --crosstool_top=//external:android/crosstool \
   --host_crosstool_top=@bazel_tools//tools/cpp:toolchain \
   --cpu=armeabi-v7a
```

When using a manually built native library in an Android app, you need to let the app know which CPU instruction sets, also known as **Application Binary Interface (ABI)**, the library is built for. There are two main categories of ABIs supported by Android: ARM and X86, and armeabi-v7a is the most popular ABI on Android. To find out which ABI your device or emulator uses, run `adb -s <device_id> shell getprop ro.product.cpu.abi`. For example, this command returns `armeabi-v7a` for my Nexus 7 tablet and `x86_64` for my emulator.

You may want to build both if you have a virtual emulator with x86_64 support for a quick test during development and on a device for the final performance test.

After the build completes, you'll see the TensorFlow native library file `libtensorflow_inference.so` generated in the `bazel-bin/tensorflow/contrib/android` folder. Drag it to your `app` folder at `android/app/src/main/jniLibs/armeabi-v7a` or `android/app/src/main/jniLibs/x86_64`, as shown in Figure 7.9:

Figure 7.9: Showing the TensorFlow native library file

Recognizing Drawing with CNN and LSTM

4. Build the Java interface to the TensorFlow native library by running:

   ```
   bazel build
   //tensorflow/contrib/android:android_tensorflow_inference_java
   ```

This will generate the file `libandroid_tensorflow_inference_java.jar` at `bazel-bin/tensorflow/contrib/android`. Move the file to the `android/app/lib` folder, as shown in Figure 7.10:

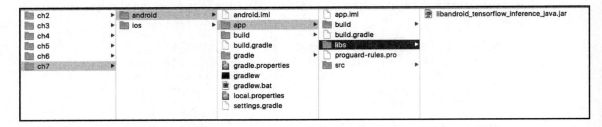

Figure 7.10: Showing the Java interface file to the TensorFlow library

We're now ready to code and test the model in Android.

Developing an Android app to use the model

Follow these steps to create a new Android app using the TensorFlow library and the model we built previously:

1. In Android Studio, create a new Android app named QuickDraw, accepting all the defaults. Then in the app's `build.gradle`, add `compile files('libs/libandroid_tensorflow_inference_java.jar')` to the end of dependencies. Create a new `assets` folder as we did before and drag and drop `quickdraw_frozen_long_blacklist_strip_transformed.pb` and `classes.txt` to it.

2. Create a new Java class called `QuickDrawView` that extends `View`, and set the fields and its constructor as follows:

   ```
   public class QuickDrawView extends View {
       private Path mPath;
       private Paint mPaint, mCanvasPaint;
       private Canvas mCanvas;
       private Bitmap mBitmap;
       private MainActivity mActivity;
       private List<List<Pair<Float, Float>>> mAllPoints = new
   ```

[218]

```
                ArrayList<List<Pair<Float, Float>>>();
        private List<Pair<Float, Float>> mConsecutivePoints = new
ArrayList<Pair<Float, Float>>();

        public QuickDrawView(Context context, AttributeSet attrs) {
            super(context, attrs);
            mActivity = (MainActivity) context;
            setPathPaint();
        }
```

`mAllPoints` is used to save a list of `mConsecutivePoints`. The `QuickDrawView` is used in the main activity's layout to show the user's drawing.

3. Define the `setPathPaint` method as follows:

```
private void setPathPaint() {
    mPath = new Path();
    mPaint = new Paint();
    mPaint.setColor(0xFF000000);
    mPaint.setAntiAlias(true);
    mPaint.setStrokeWidth(18);
    mPaint.setStyle(Paint.Style.STROKE);
    mPaint.setStrokeJoin(Paint.Join.ROUND);
    mCanvasPaint = new Paint(Paint.DITHER_FLAG);
}
```

Add two overriding methods that instantiate a `Bitmap` and `Canvas` object and show the user drawing on the canvas:

```
        @Override
        protected void onSizeChanged(int w, int h, int oldw, int oldh) {
            super.onSizeChanged(w, h, oldw, oldh);
            mBitmap = Bitmap.createBitmap(w, h, Bitmap.Config.ARGB_8888);
            mCanvas = new Canvas(mBitmap);
        }

        @Override
        protected void onDraw(Canvas canvas) {
            canvas.drawBitmap(mBitmap, 0, 0, mCanvasPaint);
            canvas.drawPath(mPath, mPaint);
        }
```

4. The overriding method `onTouchEvent` is used to populate `mConsecutivePoints` and `mAllPoints`, call the canvas's `drawPath` method, invalidate the drawing (to call the `onDraw` method), and (every time a stroke is completed with `MotionEvent.ACTION_UP`), to launch a new thread to use the model to classify the drawing:

```
@Override
public boolean onTouchEvent(MotionEvent event) {
    if (!mActivity.canDraw()) return true;
    float x = event.getX();
    float y = event.getY();
    switch (event.getAction()) {
        case MotionEvent.ACTION_DOWN:
            mConsecutivePoints.clear();
            mConsecutivePoints.add(new Pair(x, y));
            mPath.moveTo(x, y);
            break;
        case MotionEvent.ACTION_MOVE:
            mConsecutivePoints.add(new Pair(x, y));
            mPath.lineTo(x, y);
            break;
        case MotionEvent.ACTION_UP:
            mConsecutivePoints.add(new Pair(x, y));
            mAllPoints.add(new ArrayList<Pair<Float, Float>>
            (mConsecutivePoints));
            mCanvas.drawPath(mPath, mPaint);
            mPath.reset();
            Thread thread = new Thread(mActivity);
            thread.start();
            break;
        default:
            return false;
    }
    invalidate();
    return true;
}
```

5. Define two public methods that will be called by `MainActivity` to get all the points and reset the drawing after the user taps the **Restart** button:

```
public List<List<Pair<Float, Float>>> getAllPoints() {
    return mAllPoints;
}

public void clearAllPointsAndRedraw() {
    mBitmap = Bitmap.createBitmap(mBitmap.getWidth(),
```

```
            mBitmap.getHeight(), Bitmap.Config.ARGB_8888);
        mCanvas = new Canvas(mBitmap);
        mCanvasPaint = new Paint(Paint.DITHER_FLAG);
        mCanvas.drawBitmap(mBitmap, 0, 0, mCanvasPaint);
        setPathPaint();
        invalidate();
        mAllPoints.clear();
    }
```

6. Now open `MainActivity`, and make it implement `Runnable` and its fields as follows:

```
public class MainActivity extends AppCompatActivity implements Runnable {

    private static final String MODEL_FILE = "file:///android_asset/quickdraw_frozen_long_blacklist_strip_transformed.pb";
    private static final String CLASSES_FILE = "file:///android_asset/classes.txt";

    private static final String INPUT_NODE1 = "Reshape";
    private static final String INPUT_NODE2 = "Squeeze";
    private static final String OUTPUT_NODE1 = "dense/BiasAdd";
    private static final String OUTPUT_NODE2 = "ArgMax";

    private static final int CLASSES_COUNT = 345;
    private static final int BATCH_SIZE = 8;
    private String[] mClasses = new String[CLASSES_COUNT];
    private QuickDrawView mDrawView;
    private Button mButton;
    private TextView mTextView;
    private String mResult = "";
    private boolean mCanDraw = false;

    private TensorFlowInferenceInterface mInferenceInterface;
```

7. In the main layout file `activity_main.xml`, create a `QuickDrawView` element, in addition to a `TextView` and a `Button` as we did before:

```
<com.ailabby.quickdraw.QuickDrawView
    android:id="@+id/drawview"
    android:layout_width="fill_parent"
    android:layout_height="fill_parent"
    app:layout_constraintBottom_toBottomOf="parent"
    app:layout_constraintLeft_toLeftOf="parent"
    app:layout_constraintRight_toRightOf="parent"
```

```
                app:layout_constraintTop_toTopOf="parent"/>
```

8. Back to `MainActivity`; in its `onCreate` method, bind the UI element IDs with the fields, set up a click listener for the **Start/Restart** button, and then read the `classes.txt` file into a string array:

```
@Override
protected void onCreate(Bundle savedInstanceState) {
    super.onCreate(savedInstanceState);
    setContentView(R.layout.activity_main);

    mDrawView = findViewById(R.id.drawview);
    mButton = findViewById(R.id.button);
    mTextView = findViewById(R.id.textview);
    mButton.setOnClickListener(new View.OnClickListener() {
        @Override
        public void onClick(View v) {
            mCanDraw = true;
            mButton.setText("Restart");
            mTextView.setText("");
            mDrawView.clearAllPointsAndRedraw();
        }
    });

    String classesFilename =
CLASSES_FILE.split("file:///android_asset/")[1];
    BufferedReader br = null;
    int linenum = 0;
    try {
        br = new BufferedReader(new
InputStreamReader(getAssets().open(classesFilename)));
        String line;
        while ((line = br.readLine()) != null) {
            mClasses[linenum++] = line;
        }
        br.close();
    } catch (IOException e) {
        throw new RuntimeException("Problem reading classes file!"
, e);
    }
}
```

9. Then call a synchronized method `classifyDrawing` from the Thread's `run` method:

```
public void run() {
    classifyDrawing();
}

private synchronized void classifyDrawing() {
    try {
        double normalized_points[] = normalizeScreenCoordinates();
        long total_points = normalized_points.length / 3;
        float[] floatValues = new float[normalized_points.length*BATCH_SIZE];

        for (int i=0; i<normalized_points.length; i++) {
            for (int j=0; j<BATCH_SIZE; j++)
                floatValues[j*normalized_points.length + i] = (float)normalized_points[i];
        }

        long[] seqlen = new long[BATCH_SIZE];
        for (int i=0; i<BATCH_SIZE; i++)
            seqlen[i] = total_points;
```

The `normalizeScreenCoordinates` method, which will be implemented soon, converts user drawing points to the format expected by the model. `floatValues` and `seqlen` will be fed to the model as input. Note that we have to use `float` for `floatValues` and `long` for `seqlen` here as the model expects those exact data types (float and int64), or a runtime error will occur when using the model.

10. Create a Java interface to the TensorFlow library to load the model, feed the model with inputs, and fetch the outputs:

```
AssetManager assetManager = getAssets();
mInferenceInterface = new TensorFlowInferenceInterface(assetManager, MODEL_FILE);

mInferenceInterface.feed(INPUT_NODE1, floatValues, BATCH_SIZE, total_points, 3);
mInferenceInterface.feed(INPUT_NODE2, seqlen, BATCH_SIZE);

float[] logits = new float[CLASSES_COUNT * BATCH_SIZE];
float[] argmax = new float[CLASSES_COUNT * BATCH_SIZE];

mInferenceInterface.run(new String[] {OUTPUT_NODE1, OUTPUT_NODE2}, false);
```

```
mInferenceInterface.fetch(OUTPUT_NODE1, logits);
mInferenceInterface.fetch(OUTPUT_NODE1, argmax);
```

11. Normalize the fetched `logits` probabilities and sort them in decreasing order:

```
double sum = 0.0;
for (int i=0; i<CLASSES_COUNT; i++)
    sum += Math.exp(logits[i]);

List<Pair<Integer, Float>> prob_idx = new ArrayList<Pair<Integer, Float>>();
for (int j = 0; j < CLASSES_COUNT; j++) {
    prob_idx.add(new Pair(j, (float)(Math.exp(logits[j]) / sum) ));
}

Collections.sort(prob_idx, new Comparator<Pair<Integer, Float>>() {
    @Override
    public int compare(final Pair<Integer, Float> o1, final Pair<Integer, Float> o2) {
        return o1.second > o2.second ? -1 : (o1.second == o2.second ? 0 : 1);
    }
});
```

Get the top five results and show them in the `TextView`:

```
mResult = "";
for (int i=0; i<5; i++) {
    if (prob_idx.get(i).second > 0.1) {
        if (mResult == "") mResult = "" + mClasses[prob_idx.get(i).first];
        else mResult = mResult + ", " + mClasses[prob_idx.get(i).first];
    }
}

runOnUiThread(
    new Runnable() {
        @Override
        public void run() {
            mTextView.setText(mResult);
        }
});
```

12. Finally, implement the `normalizeScreenCoordinates` method, which is an easy port of the iOS implementation:

```java
private double[] normalizeScreenCoordinates() {
    List<List<Pair<Float, Float>>> allPoints =
mDrawView.getAllPoints();
    int total_points = 0;
    for (List<Pair<Float, Float>> cp : allPoints) {
        total_points += cp.size();
    }

    double[] normalized = new double[total_points * 3];
    float lowerx=Float.MAX_VALUE, lowery=Float.MAX_VALUE, upperx=-Float.MAX_VALUE, uppery=-Float.MAX_VALUE;
    for (List<Pair<Float, Float>> cp : allPoints) {
        for (Pair<Float, Float> p : cp) {
            if (p.first < lowerx) lowerx = p.first;
            if (p.second < lowery) lowery = p.second;
            if (p.first > upperx) upperx = p.first;
            if (p.second > uppery) uppery = p.second;
        }
    }
    float scalex = upperx - lowerx;
    float scaley = uppery - lowery;

    int n = 0;
    for (List<Pair<Float, Float>> cp : allPoints) {
        int m = 0;
        for (Pair<Float, Float> p : cp) {
            normalized[n*3] = (p.first - lowerx) / scalex;
            normalized[n*3+1] = (p.second - lowery) / scaley;
            normalized[n*3+2] = (m ==cp.size()-1 ? 1 : 0);
            n++; m++;
        }
    }

    for (int i=0; i<n-1; i++) {
        normalized[i*3] = normalized[(i+1)*3] - normalized[i*3];
        normalized[i*3+1] = normalized[(i+1)*3+1] -
                                        normalized[i*3+1];
        normalized[i*3+2] = normalized[(i+1)*3+2];
    }
    return normalized;
}
```

Run the app in your Android emulator or device, and enjoy some doodling with the classification results. You should see something like Figure 7.11:

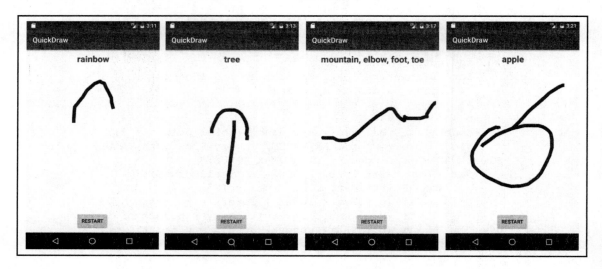

Figure 7.11: Showing drawing and classification results on Android

Now that you've seen the whole process of training a Quick Draw model and used it in your iOS and Android apps, you can certainly fine-tune the training to make it more accurate and also improve mobile apps to have more fun.

One final tip before we have to end the fun ride in this chapter is that, if you build a TensorFlow native library for Android with an incorrect ABI, you'll still be able to build and run the app from Android Studio but you'll get a runtime error `java.lang.RuntimeException: Native TF methods not found; check that the correct native libraries are present in the APK`. This means you don't have the correct TensorFlow native library in your app's `jniLibs` folder (Figure 7.9). To find out if the file is missing in the specific ABI folder inside `jniLibs`, you can open `Device File Explorer` from **Android Studio** | **View** | **Tool Windows**, then select the device's **data** | **app** | **package** | **lib** to take a look, as shown in Figure 7.12. If you prefer the command line, you can also use the `adb` tool to find this out.

Chapter 7

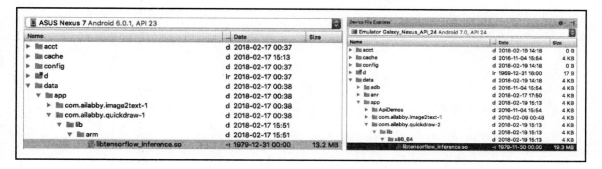

Figure 7.12: Checking out the TensorFlow native library file with Device File Explorer

Summary

In this chapter, we first described how a drawing classification model works, then covered how to train such a model using the high-level TensorFlow Estimator API. We looked at how to write Python code to make predictions with a trained model, then discussed in great detail how to find the right input and output node names and how to freeze and transform the model in the right way so mobile apps can use it. We also offered a new method to build a new TensorFlow custom iOS library, and a step-by-step tutorial on building a TensorFlow custom library for Android, to fix runtime errors when using the model. Finally, we showed the iOS and Android code that captures and shows user drawings, converts them to the data expected by the model, and processes and presents the classification results returned by the model. Hopefully, you have learned as much as you have had fun over the long journey.

So far, other than a couple of models from other open source projects, all the models, pretrained or trained by ourselves, we have used in our iOS and Android apps are from the TensorFlow open source project, which certainly offers a great list of powerful models, some of which were trained for weeks on powerful GPUs. But if you're interested in building your own model from scratch, and if you're also a little confused about the powerful RNN model and concepts used and applied in this chapter, the next chapter is what you need: we'll discuss how to build your own RNN model from scratch and use it in your mobile apps, to have another kind of fun – make money from the stock market - or at least we'll try our best to do so. Of course, nobody can guarantee you'll always make money from each stock trade, but let's at least see how our RNN model can help us improve the chances of doing so.

8
Predicting Stock Price with RNN

If you had fun doodling and building (and running the model to recognize doodling) on your mobile devices in the last chapter, you'll possibly also have fun when you make money in the stock market, or get serious when you don't. On the one hand, stock prices are time-series data, a sequence of discrete-time data, and the best deep learning method to deal with time series data is RNN, which we have used in the last two chapters. Aurélien Géron, in his best seller, *Hands-On Machine Learning with Scikit-Learn and TensorFlow*, suggested using RNN to "analyze time series data such as stock prices, and tell you when to buy or sell." On the other hand, others think the past performance of a stock cannot predict its future returns, and a randomly selected portfolio would do just as well as one carefully selected by experts. In fact, François Chollet, the author of Keras, a very popular high-level deep learning library running on top of TensorFlow and several other libraries, stated in his best seller, *Deep Learning with Python*, that using RNN with only public data to beat markets is "a very difficult endeavor, and you are likely to waste your time and resources with nothing to show for it."

So at the risk of "likely" wasting our time and resources, but with the certainty that we'll at least learn more about RNN and why it is possible or not to predict stock prices better than a random 50% strategy, we'll first give an overview of how to use RNN for stock price predictions, then discuss how to build an RNN model with the TensorFlow API to predict stock prices, and how to build an RNN LSTM model with the easier-to-use Keras API for stock price predictions. We'll test to see whether such models can beat a random buy-or-sell strategy. If we feel good about our models for improving our chances of beating the market, or just for the sake of know-how, we'll see how we can freeze and prepare the TensorFlow and Keras models to run on iOS and Android apps. If the model can improve our chances, then our mobile apps powered with the models may help us wherever and whenever we want to make a buy-or-sell decision. Feeling a little uncertain and excited? Welcome to the market.

In summary, we'll cover the following topics in this chapter:

- RNN and stock price prediction: what and how
- Using the TensorFlow RNN API for stock price prediction
- Using the Keras RNN LSTM API for stock price prediction
- Running TensorFlow and Keras models on iOS
- Running TensorFlow and Keras models on Android

RNN and stock price prediction – what and how

Feedforward networks, such as densely connected networks, have no memory and treat each input as a whole. For example, an image input represented as a vector of pixels gets processed by a feedforward network in one single step. But time series data, such as stock prices for the last 10 or 20 days, are better processed with a network with memory; assume the prices of the past 10 days are $X_1, X_2, ..., X_{10}$, with X_1 being the oldest and X_{10} the latest, then all 10-day prices can be treated as one sequence input, and when RNN processes such an input, the following steps occur:

1. A specific RNN cell, connected to the first element, X_1, in the sequence, processes X_1 and gets its output, y_1
2. Another RNN cell, connected to the next element, X_2, in the sequence input, uses X_2, as well as the previous output, y_1, to get the next output, y_2
3. The process gets repeated: when using an RNN cell to process the X_i element in the input sequence at time step i, the previous output, y_{i-1}, at time step i-1 is used together with X_i to generate the new output, y_i, at time step i

So each y_i output at time step i has information about all the elements in the input sequence up to and including time step i: $X_1, X_2, ... X_{i-1}$, and X_i. During RNN training, the predicted prices, $y_1, y_2, ..., y_9$, and y_{10}, at each time step are compared with the true target prices at each time step, that is, $X_2, X_3, ..., X_{10}$, and X_{11}, and a loss function is thus defined and used for optimization to update the network parameters. After the training is done, during prediction, X_{11} is used as the prediction for the input sequence, $X_1, X_2, ..., X_{10}$.

This is why we say RNN has memory. And RNN seems to make sense for dealing with stock price data, because the intuition is that a stock's price today (and tomorrow and the day after tomorrow, and so on) is likely to be affected by its previous N days' prices.

LSTM is just one type of RNN that solves RNN's known vanishing gradient problem, which we introduced in Chapter 6, *Describing Images in Natural Language*. Basically, during the training of an RNN model, if the time steps of an input sequence to RNN are too long, then updating the network weights of an earlier time step using back propagation may get gradient values of 0, causing no learning to occur. For example, when we use 50 days of prices as the input and if using 50, or even 40, time steps turns out to be too long, then regular RNN will be untrainable. LSTM solves this problem by adding a longterm state that decides what information can be discarded and what information needs to be stored and carried over many time steps.

The other type of RNN that solves the vanishing gradient problem nicely is called **Gated Recurrent Unit (GRU)**, which simplifies standard LSTM models a bit and is becoming more popular. Both TensorFlow and Keras APIs support basic RNN and LSTM/GRU models. In the next two sections, you'll see concrete TensorFlow and Keras APIs for using RNN and standard LSTM, and you can simply replace "LSTM" with "GRU" in the code to use and compare the results of using the GRU model with RNN and standard LSTM models.

Three common techniques are used to make LSTM models perform better:

- Stacking LSTM layers and increasing the number of neurons in the layers: This normally would lead to more powerful and accurate network models if no overfitting is produced. If you haven't, you should definitely play with the TensorFlow Playground (http://playground.tensorflow.org) to get a feel for on this.
- Using dropout to deal with overfitting. Dropout means randomly dropping out hidden and input units in a layer.
- Using bidirectional RNNs that process each input sequence in two directions (the regular direction and the reverse one), hoping to detect patterns that may be overlooked with the regular one-directional RNN.

All these techniques have been implemented and can be easily accessed in both the TensorFlow and Keras APIs.

So how do we test stock price prediction with RNN and LSTM? We'll collect daily stock price data for a specific stock symbol using a free API at `https://www.alphavantage.co`, parse the data into a training set and a test set, feed it every time a batch of training inputs (each with 20 time steps, that is, prices of 20 consecutive days) to a RNN/LSTM model, train the model, and test to see how accurately the model can be on the testing dataset. We'll test with both the TensorFlow and Keras APIs, and compare the differences between regular RNN and LSTM models. We'll also test with three slightly different sequence input and output to see which one is the best:

- Predict one day's price based on the past N days
- Predict M day's price based on the past N days
- Predict based on shifting the past N days by 1 and using the last output of the predicted sequence as the predicted price for the following day

Let's now dive into the TensorFlow RNN API and coding for training a model to predict stock prices to see how accurate it can be.

Using the TensorFlow RNN API for stock price prediction

First, you need to claim your free API key at `https://www.alphavantage.co` so you can get stock price data for any stock symbol. After you get your API key, open a Terminal and run the following command (after replacing `<your_api_key>` with your own key) to get the daily stock data for Amazon (amzn) and Google (goog) or replace them with any symbol of your interest:

```
curl -o daily_amzn.csv
"https://www.alphavantage.co/query?function=TIME_SERIES_DAILY&symbol=amzn&a
pikey=<your_api_key>&datatype=csv&outputsize=full"

curl -o daily_goog.csv
"https://www.alphavantage.co/query?function=TIME_SERIES_DAILY&symbol=goog&a
pikey=<your_api_key>&datatype=csv&outputsize=full"
```

This will generate a `daily_amzn.csv` or `daily_goog.csv` csv file with the top line as "timestamp, open, high, low, close, volume" and the rest of the lines as daily stock info. We only care about the closing prices so run the following command to get all the closing prices:

```
cut -d ',' -f 5 daily_amzn.csv | tail -n +2 > amzn.txt
```

```
cut -d ',' -f 5 daily_goog.csv | tail -n +2 > goog.txt
```

As of February 26, 2018, the number of lines in `amzn.txt` or `goog.txt` is 4,566 or 987, which is the number of days Amazon or Google has been traded. Now let's look at the complete Python code that uses the TensorFlow RNN API to train and predict with a model.

Training an RNN model in TensorFlow

1. Import the needed Python packages and define a few constants:

    ```
    import numpy as np
    import tensorflow as tf
    from tensorflow.contrib.rnn import *
    import matplotlib.pyplot as plt

    num_neurons = 100
    num_inputs = 1
    num_outputs = 1
    symbol = 'goog' # amzn
    epochs = 500
    seq_len = 20
    learning_rate = 0.001
    ```

    numpy (http://www.numpy.org) is the most popular Python library for n-dimensional array operation, and Matplotlib (https://matplotlib.org) is the leading Python 2D plotting library. We'll use numpy to process the dataset and Matplotlib to visualize the stock prices and predictions. `num_neurons` is the number of neurons the RNN, or more accurately, an RNN cell, has at each time step - each neuron receives both the input element of the input sequence at the time step, and the output from the previous time step. `num_inputs` and `num_outputs` specify the number of inputs and outputs at each time step - we'll feed one stock price from a 20-day input sequence at each time step to an RNN cell with `num_neurons` neurons and expect one predicted stock output at each step. `seq_len` is the number of time steps. So we'll use 20-day stock prices of Google as an input sequence and send such input to an RNN cell with 100 neurons.

2. Open and read the text file with all the prices, parse the prices into a list of `float` numbers, reverse the list order so the oldest price starts first, then add `seq_len+1` values every time (the first `seq_len` values will be an input sequence to RNN, the last `seq_len` values will be the target output sequence), starting from the first one in the list and shifting 1 every time till the end of the list, to a numpy `result` array:

```
f = open(symbol + '.txt', 'r').read()
data = f.split('\n')[:-1] # get rid of the last '' so float(n) works
data.reverse()
d = [float(n) for n in data]

result = []
for i in range(len(d) - seq_len - 1):
    result.append(d[i: i + seq_len + 1])

result = np.array(result)
```

3. The `result` array contains the whole dataset for our model now, but we need to further process it into the format expected by the RNN API. Let's first split it into a training set (90% of the whole dataset) and a testing set (10%):

```
row = int(round(0.9 * result.shape[0]))
train = result[:row, :]
test = result[row:, :]
```

Then shuffle the training set randomly, as a standard practice in machine learning model training:

```
np.random.shuffle(train)
```

Formulate the input sequences of the training set and testing set, `X_train` and `X_test`, as well as the target output sequences of the training set and testing set, `y_train` and `y_test`. Notice the upper case X and the lower case y are common naming conventions used in machine learning to represent the input and target output, respectively:

```
X_train = train[:, :-1] # all rows with all columns except the last one
X_test = test[:, :-1] # each row contains seq_len + 1 columns

y_train = train[:, 1:]
y_test = test[:, 1:]
```

Finally, reshape the four arrays to 3-D (batch size, number of time steps, and number of input or output) to complete the preparation of the training and testing datasets:

```
X_train = np.reshape(X_train, (X_train.shape[0], X_train.shape[1],
num_inputs))
X_test = np.reshape(X_test, (X_test.shape[0], X_test.shape[1],
num_inputs))
y_train = np.reshape(y_train, (y_train.shape[0], y_train.shape[1],
num_outputs))
y_test = np.reshape(y_test, (y_test.shape[0], y_test.shape[1],
num_outputs))
```

Notice that `X_train.shape[1]`, `X_test.shape[1]`, `y_train.shape[1]`, and `y_test.shape[1]` are all the same as `seq_len`.

4. We're ready to build the model. Create two placeholders to be fed with `X_train` and `y_train` during training, and `X_test` during testing:

```
X = tf.placeholder(tf.float32, [None, seq_len, num_inputs])
y = tf.placeholder(tf.float32, [None, seq_len, num_outputs])
```

Use `BasicRNNCell` to create an RNN cell, each with `num_neurons` of neurons, for each time step:

```
cell = tf.contrib.rnn.OutputProjectionWrapper(
    tf.contrib.rnn.BasicRNNCell(num_units=num_neurons,
activation=tf.nn.relu), output_size=num_outputs)
outputs, _ = tf.nn.dynamic_rnn(cell, X, dtype=tf.float32)
```

`OutputProjectionWrapper` is used to add a fully connected layer on top of the output in each cell, so at each time step the RNN cell's output, which would be a sequence of `num_neurons` values, gets reduced to a single value. That's how the RNN outputs one value for each value in an input sequence at each time step, or outputs a total of `seq_len` number of values for each input sequence of the `seq_len` number of values per each instance.

`dynamic_rnn` is used to loop over RNN cells over all time steps, with the total being `seq_len` (defined in the X shape), and it returns two values: a list of the output at each time step, and the final state of the network. We'll use the reshaped value of the first `outputs` return value to define the loss function next.

5. Complete the model definition by specifying the prediction tensor, loss, optimizer, and training op in the standard way:

   ```
   preds = tf.reshape(outputs, [1, seq_len], name="preds")
   loss = tf.reduce_mean(tf.square(outputs - y))
   optimizer = tf.train.AdamOptimizer(learning_rate=learning_rate)
   training_op = optimizer.minimize(loss)
   ```

   Notice that "preds" will be used as the output node name when we use the freeze_graph tool to prepare the model to be deployed on mobile it'll also be used in iOS and Android to run the model for prediction. As you can see, it's definitely nice to know that piece of information before we even start training the model, and this benefit comes with a model we built from scratch.

6. Start the training process. For each epoch, we feed the X_train and y_train data to run training_op to minimize loss, then we save the model checkpoint files, and print the loss value every 10 epochs:

   ```
   init = tf.global_variables_initializer()
   saver = tf.train.Saver()

   with tf.Session() as sess:
       init.run()

       count = 0
       for _ in range(epochs):
           n=0
           sess.run(training_op, feed_dict={X: X_train, y: y_train})
           count += 1
           if count % 10 == 0:
               saver.save(sess, "/tmp/" + symbol + "_model.ckpt")
               loss_val = loss.eval(feed_dict={X: X_train, y: y_train})
               print(count, "loss:", loss_val)
   ```

   If you run the preceding code above, you'll see output like this:

   ```
   (10, 'loss:', 243802.61)
   (20, 'loss:', 80629.57)
   (30, 'loss:', 40018.996)
   (40, 'loss:', 28197.496)
   (50, 'loss:', 24306.758)
   ...
   (460, 'loss:', 93.095985)
   (470, 'loss:', 92.864082)
   (480, 'loss:', 92.33461)
   ```

```
(490, 'loss:', 92.09893)
(500, 'loss:', 91.966286)
```

 You can replace BasicRNNCell in step 4 with BasicLSTMCell and run the training code, but the training using BasicLSTMCell is much slower and the loss value is still pretty big after 500 epochs. We won't experiment with BasicLSTMCell further in this section, but for comparison, you'll see detailed use of stacked LSTM layers, dropout, and bidirectional RNNs all in the next section of using Keras.

Testing the TensorFlow RNN model

To see whether the loss value after 500 epochs is good enough, let's add the following code using the testing dataset to calculate the number of correct predictions among the total testing examples (by correct, we mean the predicted price goes up or down in the same direction as the target price, relative to the price of the previous day):

```
correct = 0
y_pred = sess.run(outputs, feed_dict={X: X_test})
targets = []
predictions = []
for i in range(y_pred.shape[0]):
    input = X_test[i]
    target = y_test[i]
    prediction = y_pred[i]

    targets.append(target[-1][0])
    predictions.append(prediction[-1][0])

    if target[-1][0] >= input[-1][0] and prediction[-1][0] >=
input[-1][0]:
        correct += 1
    elif target[-1][0] < input[-1][0] and prediction[-1][0] <
input[-1][0]:
        correct += 1
```

Now we can visualize the correct prediction ratio using the `plot` method:

```
total = len(X_test)
xs = [i for i, _ in enumerate(y_test)]
plt.plot(xs, predictions, 'r-', label='prediction')
plt.plot(xs, targets, 'b-', label='true')
plt.legend(loc=0)
plt.title("%s - %d/%d=%.2f%%" %(symbol, correct, total,
          100*float(correct)/total))
plt.show()
```

Running the code now will show something like Figure 8.1, with the ratio of correct prediction as 56.25%:

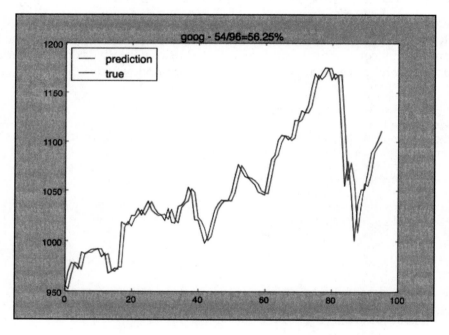

Figure 8.1 Showing the stock price prediction correctness trained with TensorFlow RNN

Notice you'll likely get a somewhat different ratio every time you run this training and testing code. By fine-tuning the hyper-parameters of the model, you may possibly get the ratio over 60%, which seems better than random prediction. If you're optimistic, you'd think at least we have something better than 50% to show (56.25%) and would probably love to see the model running on-mobile. But let's first see whether we can use the cool Keras library to build a better model - before we do that, let's freeze the trained TensorFlow model by simply running:

```
python tensorflow/python/tools/freeze_graph.py --
input_meta_graph=/tmp/amzn_model.ckpt.meta --
input_checkpoint=/tmp/amzn_model.ckpt --output_graph=/tmp/amzn_tf_frozen.pb
--output_node_names="preds" --input_binary=true
```

Using the Keras RNN LSTM API for stock price prediction

Keras is a very easy-to-use high-level deep learning Python library running on top of other popular deep learning libraries, including TensorFlow, Theano, and CNTK. As you'll see soon, Keras makes building and playing with models a lot easier. To install and use Keras, along with TensorFlow as Keras' backend, it's best to set up a virtualenv first:

```
sudo pip install virtualenv
```

Then run the following commands, if you have TensorFlow 1.4 source on your machine and your iOS and Android apps; use the TensorFlow 1.4 custom library:

```
cd
mkdir ~/tf14_keras
virtualenv --system-site-packages ~/tf14_keras/
cd ~/tf14_keras/
source ./bin/activate
easy_install -U pip
pip install --upgrade
https://storage.googleapis.com/tensorflow/mac/cpu/tensorflow-1.4.0-py2-none
-any.whl
pip install keras
```

If you have TensorFlow 1.5 source on your machine, you should install TensorFlow 1.5 with Keras because the model created using Keras needs to have the same TensorFlow version as that used by TensorFlow mobile apps or an error occurs when you try to load the model:

```
cd
mkdir ~/tf15_keras
virtualenv --system-site-packages ~/tf15_keras/
cd ~/tf15_keras/
source ./bin/activate
easy_install -U pip
pip install --upgrade
https://storage.googleapis.com/tensorflow/mac/cpu/tensorflow-1.5.0-py2-none-any.whl
pip install keras
```

If your OS is not Mac or if your computer has GPU, you need to replace the TensorFlow Python package URL with the right one, which you can find at https://www.tensorflow.org/install.

Training an RNN model in Keras

Let's now see what it takes to build and train an LSTM model for stock price prediction in Keras. First, some imports and constant settings:

```
import keras
from keras import backend as K
from keras.layers.core import Dense, Activation, Dropout
from keras.layers.recurrent import LSTM
from keras.layers import Bidirectional
from keras.models import Sequential
import matplotlib.pyplot as plt

import tensorflow as tf
import numpy as np

symbol = 'amzn'
epochs = 10
num_neurons = 100
seq_len = 20
pred_len = 1
shift_pred = False
```

shift_pred is used to indicate whether we want to predict an output sequence of prices versus just a single output price. If it's True, we'll predict $X_2, X_3, ..., X_{n+1}$ from the $X_1, X_2, X_3, ..., X_n$ input, as we did in the last section of using the TensorFlow API. If shift_pred is False, we'll predict pred_len of output based on the $X_1, X_2, ..., X_n$ input. For example, if pred_len is 1, we'll predict X_{n+1}, and if pred_len is 3, we'll predict $X_{n+1}, X_{n+2},$ and X_{n+3}, which may make sense as we'd love to know whether the price will continue to go up 3 days in a row or just up 1 day then down 2 days.

Now let's create a method, modified based on the data loading code in the last section, that prepares the appropriate training and testing datasets based on the pred_len and shift_pred settings:

```
def load_data(filename, seq_len, pred_len, shift_pred):
    f = open(filename, 'r').read()
    data = f.split('\n')[:-1] # get rid of the last '' so float(n) works
    data.reverse()
    d = [float(n) for n in data]
    lower = np.min(d)
    upper = np.max(d)
    scale = upper-lower
    normalized_d = [(x-lower)/scale for x in d]

    result = []
    if shift_pred:
        pred_len = 1
    for i in range((len(normalized_d) - seq_len - pred_len)/pred_len):
        result.append(normalized_d[i*pred_len: i*pred_len + seq_len + pred_len])
    result = np.array(result)
    row = int(round(0.9 * result.shape[0]))
    train = result[:row, :]
    test = result[row:, :]

    np.random.shuffle(train)

    X_train = train[:, :-pred_len]
    X_test = test[:, :-pred_len]

    if shift_pred:
        y_train = train[:, 1:]
        y_test = test[:, 1:]
    else:
        y_train = train[:, -pred_len:]
        y_test = test[:, -pred_len:]
    X_train = np.reshape(X_train, (X_train.shape[0], X_train.shape[1], 1))
```

Predicting Stock Price with RNN

```
            X_test = np.reshape(X_test, (X_test.shape[0], X_test.shape[1], 1))

        return [X_train, y_train, X_test, y_test, lower, scale]
```

Notice that we also apply normalization here, using the same kind of normalization as we did in the previous chapter, to see whether it improves our model. We also return the `lower` and `scale` values as they're needed for denormalization when predicting with the trained model.

Now we can call `load_data` to get the training and testing datasets, as well as the `lower` and `scale` values:

```
X_train, y_train, X_test, y_test, lower, scale = load_data(symbol + '.txt',
seq_len, pred_len, shift_pred)
```

The complete model-building code is as follows:

```
model = Sequential()
model.add(Bidirectional(LSTM(num_neurons, return_sequences=True,
input_shape=(None, 1)), input_shape=(seq_len, 1)))
model.add(Dropout(0.2))

model.add(LSTM(num_neurons, return_sequences=True))
model.add(Dropout(0.2))

model.add(LSTM(num_neurons, return_sequences=False))
model.add(Dropout(0.2))

if shift_pred:
    model.add(Dense(units=seq_len))
else:
    model.add(Dense(units=pred_len))

model.add(Activation('linear'))
model.compile(loss='mse', optimizer='rmsprop')

model.fit(
    X_train,
    y_train,
    batch_size=512,
    epochs=epochs,
    validation_split=0.05)

print(model.output.op.name)
print(model.input.op.name)
```

This code is pretty much self-explanatory and simpler than the model-building code in TensorFlow, even with the newly added `Bidirectional`, `Dropout`, `validation_split`, and stacking LSTM layers. Notice that the `return_sequences` parameter in the LSTM call needs to be True, so the output of the LSTM cell will be the full output sequence instead of just the last output in the output sequence, unless it's the last stacked layer. The last two `print` statements will print the input node name (**bidirectional_1_input**) and output node name (**activation_1/Identity**), needed when we freeze the model and run the model on mobile.

Now if you run the preceding code, you'll see output like this:

```
824/824 [==============================] - 7s 9ms/step - loss: 0.0833 - val_loss: 0.3831
Epoch 2/10
824/824 [==============================] - 2s 3ms/step - loss: 0.2546 - val_loss: 0.0308
Epoch 3/10
824/824 [==============================] - 2s 2ms/step - loss: 0.0258 - val_loss: 0.0098
Epoch 4/10
824/824 [==============================] - 2s 2ms/step - loss: 0.0085 - val_loss: 0.0035
Epoch 5/10
824/824 [==============================] - 2s 2ms/step - loss: 0.0044 - val_loss: 0.0026
Epoch 6/10
824/824 [==============================] - 2s 2ms/step - loss: 0.0038 - val_loss: 0.0022
Epoch 7/10
824/824 [==============================] - 2s 2ms/step - loss: 0.0033 - val_loss: 0.0019
Epoch 8/10
824/824 [==============================] - 2s 2ms/step - loss: 0.0030 - val_loss: 0.0019
Epoch 9/10
824/824 [==============================] - 2s 2ms/step - loss: 0.0028 - val_loss: 0.0017
Epoch 10/10
824/824 [==============================] - 2s 3ms/step - loss: 0.0027 - val_loss: 0.0019
```

Both the training loss and validation loss are printed with a simple call of `model.fit`.

Testing the Keras RNN model

It's time to save the model checkpoint and use the testing dataset to calculate the number of correct predictions in the sense we explained in the last section:

```
saver = tf.train.Saver()
saver.save(K.get_session(), '/tmp/keras_' + symbol + '.ckpt')

predictions = []
correct = 0
total = pred_len*len(X_test)
for i in range(len(X_test)):
    input = X_test[i]
    y_pred = model.predict(input.reshape(1, seq_len, 1))
    predictions.append(scale * y_pred[0][-1] + lower)
    if shift_pred:
        if y_test[i][-1] >= input[-1][0] and y_pred[0][-1] >= input[-1][0]:
            correct += 1
        elif y_test[i][-1] < input[-1][0] and y_pred[0][-1] < input[-1][0]:
            correct += 1
    else:
        for j in range(len(y_test[i])):
            if y_test[i][j] >= input[-1][0] and y_pred[0][j] >= input[-1][0]:
                correct += 1
            elif y_test[i][j] < input[-1][0] and y_pred[0][j] < input[-1][0]:
                correct += 1
```

We mainly call `model.predict` to get a prediction for each instance in X_test, and use it with the true value and the previous day's price to see whether it's a correct prediction in terms of direction. Finally, let's plot the true prices from the testing dataset and the predictions:

```
y_test = scale * y_test + lower
y_test = y_test[:, -1]
xs = [i for i, _ in enumerate(y_test)]
plt.plot(xs, y_test, 'g-', label='true')
plt.plot(xs, predictions, 'r-', label='prediction')
plt.legend(loc=0)
if shift_pred:
    plt.title("%s - epochs=%d, shift_pred=True, seq_len=%d: %d/%d=%.2f%%" 
        %(symbol, epochs, seq_len, correct, total, 100*float(correct)/total))
else:
    plt.title("%s - epochs=%d, lens=%d,%d: %d/%d=%.2f%%" %(symbol, epochs, 
        seq_len, pred_len, correct, total, 100*float(correct)/total))
```

```
plt.show()
```

And you'll see something like Figure 8.2:

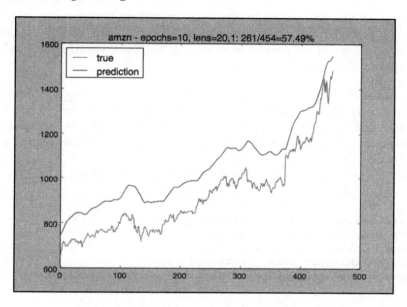

Figure 8.2 Making stock price predictions with the Keras bidirectional and stacked LSTM layers

It's easy to add more LSTM layers to the stack, or play with the hyper-parameters, such as learning rate and dropout rate, and many of the constant settings. But we haven't found significant difference in the correctness ratio for different settings of using `pred_len` and `shift_pred`. Maybe we should just be realistically satisfied with the close-to-60% correctness ratio for now, and see how we can use the TensorFlow- and Keras-trained models on iOS and Android - we can try to keep improving the model later, but it'd be valuable to know whether we'd have any issues using our RNN models trained with TensorFlow and Keras.

As François Chollet points out, "deep learning is more an art than a science... every problem is unique and you will have to try and evaluate different strategies empirically. There is currently no theory that will tell you in advance precisely what you should do to optimally solve a problem. You must try and iterate." Hopefully we have provided a good starting point for you to improve the stock price prediction models using TensorFlow and Keras APIs.

The last thing we need to do in this section is to freeze the Keras model from the checkpoint - because we installed TensorFlow and Keras in our virtual environment and TensorFlow is the only installed and supported deep learning library in the virtualenv, Keras uses the TensorFlow backend and generates the checkpoint in the TensorFlow format with the `saver.save(K.get_session(), '/tmp/keras_' + symbol + '.ckpt')` call. Now run the following command to freeze the checkpoint (recall that we get `output_node_name` from `print(model.input.op.name)` during training):

```
python tensorflow/python/tools/freeze_graph.py --
input_meta_graph=/tmp/keras_amzn.ckpt.meta --
input_checkpoint=/tmp/keras_amzn.ckpt --
output_graph=/tmp/amzn_keras_frozen.pb --
output_node_names="activation_1/Identity" --input_binary=true
```

Because our model is pretty simple and straightforward, we will try the two frozen models directly on-mobile without using the `transform_graph` tool as we did in the previous two chapters.

Running the TensorFlow and Keras models on iOS

We won't bore you by repeating the project setup step - just follow what we did before to create a new Objective-C project named StockPrice that will use the manually built TensorFlow library (see the iOS section of Chapter 7, *Recognizing Drawing with CNN and LSTM*, if you need detailed info). Then add the two `amzn_tf_frozen.pb` and `amzn_keras_frozen.pb` model files to the project and you should have your StockPrice project in Xcode, as in Figure 8.3:

Chapter 8

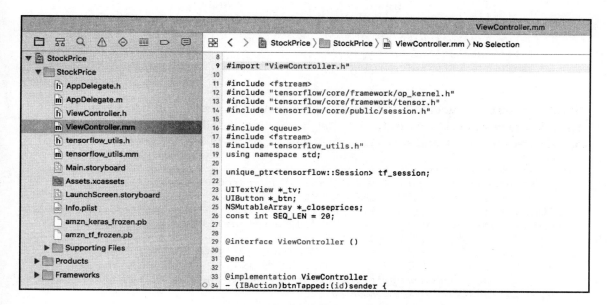

Figure 8.3 iOS app using the TensorFlow- and Keras-trained models in Xcode

In **ViewController.mm**, we'll first declare some variables and one constant:

```
unique_ptr<tensorflow::Session> tf_session;
UITextView *_tv;
UIButton *_btn;
NSMutableArray *_closeprices;
const int SEQ_LEN = 20;
```

Then create a button-tap handler to let the user choose either the TensorFlow or Keras model (the button is created in the `viewDidLoad` method as before):

```
- (IBAction)btnTapped:(id)sender {
    UIAlertAction* tf = [UIAlertAction actionWithTitle:@"Use TensorFlow
Model" style:UIAlertActionStyleDefault handler:^(UIAlertAction * action) {
        [self getLatestData:NO];
    }];
    UIAlertAction* keras = [UIAlertAction actionWithTitle:@"Use Keras
Model" style:UIAlertActionStyleDefault handler:^(UIAlertAction * action) {
        [self getLatestData:YES];
    }];
    UIAlertAction* none = [UIAlertAction actionWithTitle:@"None"
style:UIAlertActionStyleDefault handler:^(UIAlertAction * action) {}];
    UIAlertController* alert = [UIAlertController
alertControllerWithTitle:@"RNN Model Pick" message:nil
preferredStyle:UIAlertControllerStyleAlert];
```

```objc
        [alert addAction:tf];
        [alert addAction:keras];
        [alert addAction:none];
        [self presentViewController:alert animated:YES completion:nil];
}
```

The `getLatestData` method first makes a URL request to get the compact version of the Alpha Vantage API that returns the last 100 data points of daily stock data for Amazon, then it parses the result and saves the last 20 closing prices in the _closeprices array:

```objc
-(void)getLatestData:(BOOL)useKerasModel {
    NSURLSession *session = [NSURLSession sharedSession];
    [[session dataTaskWithURL:[NSURL URLWithString:@"https://www.alphavantage.co/query?function=TIME_SERIES_DAILY&symbol=amzn&apikey=<your_api_key>&datatype=csv&outputsize=compact"]
            completionHandler:^(NSData *data,
                                NSURLResponse *response,
                                NSError *error) {
                NSString *stockinfo = [[NSString alloc] initWithData:data encoding:NSASCIIStringEncoding];
                NSArray *lines = [stockinfo componentsSeparatedByString:@"\n"];
                _closeprices = [NSMutableArray array];
                for (int i=0; i<SEQ_LEN; i++) {
                    NSArray *items = [lines[i+1] componentsSeparatedByString:@","];
                    [_closeprices addObject:items[4]];
                }
                if (useKerasModel)
                    [self runKerasModel];
                else
                    [self runTFModel];
    }] resume];
}
```

The `runTFModel` method is defined as follows:

```objc
- (void) runTFModel {
    tensorflow::Status load_status;
    load_status = LoadModel(@"amzn_tf_frozen", @"pb", &tf_session);
    tensorflow::Tensor prices(tensorflow::DT_FLOAT,
             tensorflow::TensorShape({1, SEQ_LEN, 1}));
    auto prices_map = prices.tensor<float, 3>();
    NSString *txt = @"Last 20 Days:\n";
    
    for (int i = 0; i < SEQ_LEN; i++){
        prices_map(0,i,0) = [_closeprices[SEQ_LEN-i-1] floatValue];
```

```
            txt = [NSString stringWithFormat:@"%@%@\n", txt,
                                    _closeprices[SEQ_LEN-i-1]];
    }
    std::vector<tensorflow::Tensor> output;
    tensorflow::Status run_status = tf_session->Run({{"Placeholder",
                                    prices}}, {"preds"}, {}, &output);
    if (!run_status.ok()) {
        LOG(ERROR) << "Running model failed:" << run_status;
    }
    else {
        tensorflow::Tensor preds = output[0];
        auto preds_map = preds.tensor<float, 2>();
        txt = [NSString stringWithFormat:@"%@\nPrediction with TF RNN
                    model:\n%f", txt, preds_map(0,SEQ_LEN-1)];
        dispatch_async(dispatch_get_main_queue(), ^{
            [_tv setText:txt];
            [_tv sizeToFit];
        });
    }
}
```

`preds_map(0,SEQ_LEN-1)` is the predicted price for the next day, based on the last 20 days; `Placeholder` is the input node name defined in X = `tf.placeholder(tf.float32, [None, seq_len, num_inputs])` in step 4 of the *Training an RNN model in TensorFlow* subsection. After the prediction is generated by the model, we display it along with the last 20 days' prices in a TextView.

The `runKeras` method is defined similarly, but with denormalization and different input and output node names. Because our Keras model is trained to output just one prediction price, instead of a sequence of `seq_len` prices, we use `preds_map(0,0)` to get the prediction:

```
- (void) runKerasModel {
    tensorflow::Status load_status;
    load_status = LoadModel(@"amzn_keras_frozen", @"pb", &tf_session);
    if (!load_status.ok()) return;
    tensorflow::Tensor prices(tensorflow::DT_FLOAT,
    tensorflow::TensorShape({1, SEQ_LEN, 1}));
    auto prices_map = prices.tensor<float, 3>();
    float lower = 5.97;
    float scale = 1479.37;
    NSString *txt = @"Last 20 Days:\n";
    for (int i = 0; i < SEQ_LEN; i++){
        prices_map(0,i,0) = ([_closeprices[SEQ_LEN-i-1] floatValue] -
                                                    lower)/scale;
        txt = [NSString stringWithFormat:@"%@%@\n", txt,
```

Predicting Stock Price with RNN

```
                                    _closeprices[SEQ_LEN-i-1]];
        }
        std::vector<tensorflow::Tensor> output;
        tensorflow::Status run_status =
tf_session->Run({{"bidirectional_1_input", prices}},
{"activation_1/Identity"},
                                                    {}, &output);
        if (!run_status.ok()) {
            LOG(ERROR) << "Running model failed:" << run_status;
        }
        else {
            tensorflow::Tensor preds = output[0];
            auto preds_map = preds.tensor<float, 2>();
            txt = [NSString stringWithFormat:@"%@\nPrediction with Keras
                RNN model:\n%f", txt, scale * preds_map(0,0) + lower];
            dispatch_async(dispatch_get_main_queue(), ^{
                [_tv setText:txt];
                [_tv sizeToFit];
            });
        }
    }
}
```

If you run the app now and tap the **Predict** button, you'll see the model selection message (Figure 8.4):

Figure 8.4 Selecting the TensorFlow or Keras RNN model

[250]

If you select the TensorFlow model, you may get an error:

```
Could not create TensorFlow Graph: Invalid argument: No OpKernel was
registered to support Op 'Less' with these attrs. Registered devices:
[CPU], Registered kernels:
 device='CPU'; T in [DT_FLOAT]
[[Node: rnn/while/Less = Less[T=DT_INT32,
_output_shapes=[[]]](rnn/while/Merge, rnn/while/Less/Enter)]]
```

If you pick the Keras model, a slightly different error may occur:

```
Could not create TensorFlow Graph: Invalid argument: No OpKernel was
registered to support Op 'Less' with these attrs. Registered devices:
[CPU], Registered kernels:
 device='CPU'; T in [DT_FLOAT]
[[Node: bidirectional_1/while_1/Less = Less[T=DT_INT32,
_output_shapes=[[]]](bidirectional_1/while_1/Merge,
bidirectional_1/while_1/Less/Enter)]]
```

We've seen a similar error with the `RefSwitch` op in the previous chapter and know a fix for this kind of error is to build the TensorFlow library with –D__ANDROID_TYPES_FULL__ enabled. If you don't see these errors, it means you've built such a library when going through the iOS app in the previous chapter; otherwise, follow the instructions in the beginning of *Building a custom TensorFlow library for iOS* in the *Using the drawing classification model in iOS* section of the previous chapter to build the new TensorFlow library, then run the app again.

Now pick the TensorFlow model and you'll see results as in Figure 8.5:

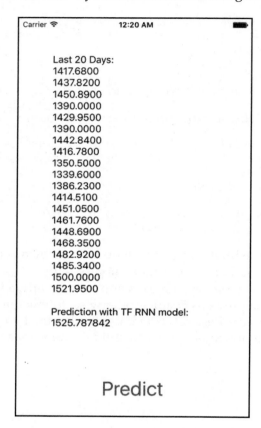

Figure 8.5 Predicting with the TensorFlow RNN model

Using the Keras model outputs a different prediction, as shown in Figure 8.6:

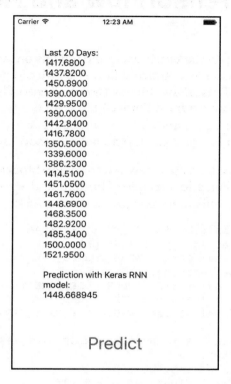

Figure 8.6 Predicting with the Keras RNN model

We can't be sure which model works better without further research, but what we can be sure of is that both our RNN models trained from scratch using the TensorFlow and Keras APIs, with accuracy close to 60%, run fine on iOS, which is probably worth our effort as we're trying to build a model that many experts think will achieve about the same performance as a random pick, and during the process, we have learned a few new cool things - using TensorFlow and Keras to build RNN models and running them on iOS. There's only one thing left before we continue our journey in the next chapter: how about using the models on Android? Will we have any new obstacles?

Running the TensorFlow and Keras models on Android

It turns out it's like walking on the beach using the models on Android - we don't even need to use the custom TensorFlow Android library as we did in the last chapter, although we have to use the custom TensorFlow library (not the TensorFlow pod as of February 2018) on iOS. The TensorFlow Android library built with `compile 'org.tensorflow:tensorflow-android:+'` in the `build.gradle` file must have more complete data type support for the `Less` Op than the TensorFlow pod for iOS does.

To test the models in Android, create a new Android app StockPrice and add the two model files to its `assets` folder. Then add a couple of buttons and a `TextView` in the layout and define some fields and constants in the `MainActivity.java`:

```
    private static final String TF_MODEL_FILENAME =
"file:///android_asset/amzn_tf_frozen.pb";
    private static final String KERAS_MODEL_FILENAME =
"file:///android_asset/amzn_keras_frozen.pb";
    private static final String INPUT_NODE_NAME_TF = "Placeholder";
    private static final String OUTPUT_NODE_NAME_TF = "preds";
    private static final String INPUT_NODE_NAME_KERAS =
"bidirectional_1_input";
    private static final String OUTPUT_NODE_NAME_KERAS =
"activation_1/Identity";
    private static final int SEQ_LEN = 20;
    private static final float LOWER = 5.97f;
    private static final float SCALE = 1479.37f;

    private TensorFlowInferenceInterface mInferenceInterface;

    private Button mButtonTF;
    private Button mButtonKeras;
    private TextView mTextView;
    private boolean mUseTFModel;
    private String mResult;
```

Make `onCreate` as follows:

```
    protected void onCreate(Bundle savedInstanceState) {
        super.onCreate(savedInstanceState);
        setContentView(R.layout.activity_main);

        mButtonTF = findViewById(R.id.tfbutton);
        mButtonKeras = findViewById(R.id.kerasbutton);
```

```
        mTextView = findViewById(R.id.textview);
        mTextView.setMovementMethod(new ScrollingMovementMethod());
        mButtonTF.setOnClickListener(new View.OnClickListener() {
            @Override
            public void onClick(View v) {
                mUseTFModel = true;
                Thread thread = new Thread(MainActivity.this);
                thread.start();
            }
        });
        mButtonKeras.setOnClickListener(new View.OnClickListener() {
            @Override
            public void onClick(View v) {
                mUseTFModel = false;
                Thread thread = new Thread(MainActivity.this);
                thread.start();
            }
        });

}
```

The rest of the code is all in the `run` method, started in a worker thread when either the TF PREDICTION or KERAS PREDICTION button is tapped, which needs little explanation other than that using the Keras model needs normalization and denormalization before and after running the model:

```
public void run() {
    runOnUiThread(
            new Runnable() {
                @Override
                public void run() {
                    mTextView.setText("Getting data...");
                }
            });

    float[] floatValues = new float[SEQ_LEN];

    try {
        URL url = new URL("https://www.alphavantage.co/query?function=TIME_SERIES_DAILY&symbol=amzn&apikey=4SOSJM2XCRIB5IUS&datatype=csv&outputsize=compact");
        HttpURLConnection urlConnection = (HttpURLConnection) url.openConnection();
        InputStream in = new BufferedInputStream(urlConnection.getInputStream());
        Scanner s = new Scanner(in).useDelimiter("\\n");
        mResult = "Last 20 Days:\n";
```

```java
            if (s.hasNext()) s.next(); // get rid of the first title line
            List<String> priceList = new ArrayList<>();
            while (s.hasNext()) {
                String line = s.next();
                String[] items = line.split(",");
                priceList.add(items[4]);
            }

            for (int i=0; i<SEQ_LEN; i++)
                mResult += priceList.get(SEQ_LEN-i-1) + "\n";

            for (int i=0; i<SEQ_LEN; i++) {
                if (mUseTFModel)
                    floatValues[i] = Float.parseFloat(priceList.get(SEQ_LEN-i-1));
                else
                    floatValues[i] = (Float.parseFloat(priceList.get(SEQ_LEN-i-1)) - LOWER) / SCALE;
            }

            AssetManager assetManager = getAssets();
            mInferenceInterface = new TensorFlowInferenceInterface(assetManager, mUseTFModel ? TF_MODEL_FILENAME : KERAS_MODEL_FILENAME);

            mInferenceInterface.feed(mUseTFModel ? INPUT_NODE_NAME_TF : INPUT_NODE_NAME_KERAS, floatValues, 1, SEQ_LEN, 1);

            float[] predictions = new float[mUseTFModel ? SEQ_LEN : 1];

            mInferenceInterface.run(new String[] {mUseTFModel ? OUTPUT_NODE_NAME_TF : OUTPUT_NODE_NAME_KERAS}, false);
            mInferenceInterface.fetch(mUseTFModel ? OUTPUT_NODE_NAME_TF : OUTPUT_NODE_NAME_KERAS, predictions);
            if (mUseTFModel) {
                mResult += "\nPrediction with TF RNN model:\n" + predictions[SEQ_LEN - 1];
            }
            else {
                mResult += "\nPrediction with Keras RNN model:\n" + (predictions[0] * SCALE + LOWER);
            }

            runOnUiThread(
                    new Runnable() {
                        @Override
                        public void run() {
                            mTextView.setText(mResult);
```

```
                    }
                });

        } catch (Exception e) {
            e.printStackTrace();
        }
    }
}
```

Now run the app and tap the **TF PREDICTION** button, and you'll see the results in Figure 8.7:

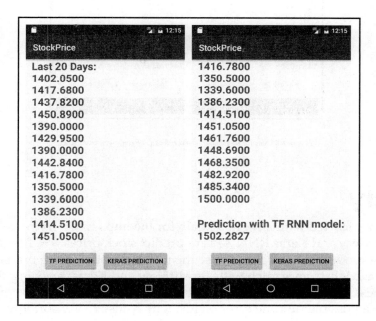

Figure 8.7 Using the TensorFlow model to make stock price predictions on Amazon

Selecting the **KERAS PREDICTION** will give you the results in Figure 8.8:

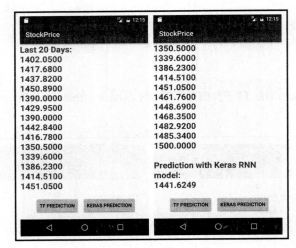

Figure 8.8 Using the Keras model to make stock price predictions on Amazon

Summary

In this chapter, we started with a slight disdain for the impossible, trying to beat the market by using TensorFlow and Keras RNN APIs to predict stock prices. We first discussed what RNN and LSTM models are and how to use them to make stock price predictions. Then we built two RNN models from scratch with TensorFlow and Keras, reaching close to 60% of testing correctness. Finally, we covered how to freeze the models and use them on iOS and Android, fixing a possible runtime error on iOS with a custom TensorFlow library.

If you're a little bit disappointed that we haven't built a model with an 80% or 90% correct prediction ratio, you may want to continue the "try and iterate" process to see whether predicting stock prices with that correct ratio is possible at all. But the skills you have learned from RNN model building, training, and testing using TensorFlow and Keras APIs and running on iOS and Android are yours to take away for sure.

If you're interested in and excited about beating the market using deep learning technologies, let's work in the next chapter on GANs (Generative Adversarial Networks), a model that tries to beat an adversary who can tell the difference between real and fake data, and gets better and better at generating data that looks like real data, fooling the adversary. GANs are actually hailed by some top researchers in deep learning as the most interesting and exciting idea in deep learning in the last 10 years.

9
Generating and Enhancing Images with GAN

Since deep learning took off in 2012, some people think no new idea has been more interesting or promising than **Generative Adversarial Network (GAN)**, introduced by Ian Goodfellow in 2014 in the paper, *Generative Adversarial Networks* (https://arxiv.org/abs/1406.2661). In fact, Yann LeCun, the Facebook AI research director and one of the pioneering deep learning researchers, referred to GAN and adversarial training as *"the most interesting idea in the last 10 years in machine learning."* Because of this, how can we not cover it here, to understand why GAN is so exciting and how to build GAN models and run them on iOS and Android?

In this chapter, we'll first give an overview of what a GAN is, how it works, and why it has such great potential. Then we'll go through two GAN models: one basic GAN model that can be used to generate human-like handwritten digits, the other more advanced GAN model that can enhance low-resolution images to high-resolution ones. We'll show you how to build and train such models in Python and TensorFlow and how to prepare the models for mobile deployment. Then we'll present iOS and Android apps, with the complete source code, that use the models to generate handwritten digits and enhance images. By the end of the chapter, you should be ready to further explore all kinds of GAN-based models out there or start building your own ones and understand how to run them in your mobile apps.

In summary, we'll cover the following topics in this chapter:

- GAN – what and why
- Building and training GAN models with TensorFlow
- Using the GAN models in iOS
- Using the GAN models in Android

GAN – what and why

GANs are neural networks that learn to generate data similar to real data, or the data in the training set. The key idea of a GAN is to have a generator network and a discriminator network playing against each other: the generator tries to generate data that looks like real data, while the discriminator tries to tell whether the generated data is real (from the known real data) or fake (generated by the generator). The generator and the discriminator are trained together, and during the training process, the generator learns to generate data that looks more and more like real data, while the discriminator learns to distinguish real data from fake data. The generator learns by trying to make the discriminator's probability of output being real data, when fed with the generator's output as the discriminator's input, as close to 1.0 as possible, while the discriminator learns by trying to accomplish two goals:

- Make its probability of output being real, when fed with the generator's output as input, as close to 0.0 as possible, which happens to be the exact opposite goal of the generator
- Make its probability of output being real, when fed with the real data as input, as close to 1.0 as possible

In the next section, you'll see a detailed code snippet that matches the given description of the generator and the discriminator networks and their training process. If you feel like understanding more about GANs, in addition to our summary overview here, you can search for *"Introduction to GANs"* on YouTube and watch Ian Goodfellow's introduction and tutorial videos on GAN at NIPS (Neural Information Processing Systems) Conference 2016 and ICCV (International Conference on Computer Vision) 2017. In fact, there are 7 NIPS 2016 Workshop on Adversarial Training videos and 12 ICCV 2017 GAN Tutorial videos on YouTube that you can immerse yourself in.

With the competing goals of the two players, the generator and the discriminator, a GAN is a system that seeks the equilibrium between two opponents. If both players have unlimited capabilities and can train optimally, then the Nash Equilibrium (after John Nash, the Nobel Prize winner in Economic Sciences in 1994 and subject of the movie, *A Beautiful Mind*), which is a stable state where no players can gain by changing only its own strategy, corresponds to the state of the generator generating data that looks just like real data and the discriminator unable to tell real data from fake data.

If you're interested in knowing more about the Nash Equilibrium, Google *"khan academy nash equilibrium"* and watch the two fun videos on it by Sal Khan. The Wikipedia page on Nash Equilibrium and the article, *"What is the Nash equilibrium and why does it matter?"* in The Economist explaining economics (https://www.economist.com/blogs/economist-explains/2016/09/economist-explains-economics) are also a good read. Understanding the basic intuition and idea behind GANs will help you appreciate more why it has great potential.

The potential of the generator being able to generate data that looks like real data means all kinds of cool applications can be developed with GANs, such as:

- Generating high-quality images from low-quality images
- Image inpainting (fixing lost or corrupted images)
- Translating images (for example, from edge sketches to photos, or adding or removing objects such as glasses on human faces)
- Generating images from text (the opposite of Text2Image we saw in `Chapter 6`, *Describing Images in Natural Language*)
- Writing news articles that look like real ones
- Generating audio waveforms that are similar to the audio in the training set

Basically, GANs have the potential to generate realistic images, texts or audio data from random input; if you have a training set of source data and target data, GANs can also generate data similar to the target data from input similar to the source data. It is this universal feature of how the generator and the discriminator in a GAN model works in a dynamic way to allow GANs to generate any kind of realistic output that makes GANs very exciting.

But, because of the dynamics or competing goals of the generator and the discriminator, training GANs to reach the Nash Equilibrium state is a tricky and difficult problem. In fact, it's still an open research problem – Ian Goodfellow in his *Heroes of Deep Learning* interview with Andrew Ng in August 2017 (search *ian goodfellow andrew ng* on YouTube) says that if we can make GANs become as reliable as deep learning has become, we'll see GANs with much greater success, otherwise we'll eventually replace them with other forms of generative models.

Despite the challenges in the training of GANs, there have already been lots of effective known hacks that you can apply during the training (https://github.com/soumith/ganhacks) – we won't cover them here, but they can be useful if you're interested in tweaking the models we'll describe in this chapter, or many other GAN models (https://github.com/eriklindernoren/Keras-GAN), or building your own GAN models.

Building and training GAN models with TensorFlow

In general, a GAN model has two neural networks: G for the generator and D for the discriminator. x is some real data input from the training set, and z is random input noise. During the training, D(x) is the probability of x being real and D tries to make D(x) close to 1; G(z) is the generated output with the random input z, and D tries to make D(G(z)) close to 0, but at the same time, G tries to make D(G(z)) close to 1. Now, let's first see how we can build a basic GAN model in TensorFlow and Python that can write or generate handwritten digits.

Basic GAN model of generating handwritten digits

The training model for handwritten digits is based on the repo https://github.com/jeffxtang/generative-adversarial-networks, which is a fork of https://github.com/jonbruner/generative-adversarial-networks, with added script that shows generated digits and saves the TensorFlow trained model with an input placeholder, so our iOS and Android apps can use the model. You should check out the blog https://www.oreilly.com/learning/generative-adversarial-networks-for-beginners for the original repo, if you need a basic understanding of the GAN model with code before moving on.

Before we look into the core code snippet that defines the generator and discriminator networks and does the GAN training, let's first run the scripts to train and test the model after cloning the repo and going to the repo directory:

```
git clone https://github.com/jeffxtang/generative-adversarial-networks
cd generative-adversarial-networks
```

This fork added checkpoint saving code to the `gan-script-fast.py` script and also added a new script, `gan-script-test.py` to test and save new checkpoints with a random input placeholder – so the model frozen with the new checkpoint can be used in the iOS and Android apps.

Run the command `python gan-script-fast.py` to train the model, which takes less than one hour on our GTX-1070 GPU on Ubuntu. After the training completes, the checkpoint files will be saved in the model directory. Now run `python gan-script-test.py` to see some generated handwritten digits. The script also reads the checkpoint files from the model directory, saved when running `gan-script-fast.py`, and re-saves the updated checkpoint files, along with a random input placeholder, in the `newmodel` directory:

```
ls -lt newmodel
-rw-r--r--  1 jeffmbair  staff  266311 Mar  5 16:43 ckpt.meta
-rw-r--r--  1 jeffmbair  staff      65 Mar  5 16:42 checkpoint
-rw-r--r--  1 jeffmbair  staff  69252168 Mar  5 16:42 ckpt.data-00000-of-00001
-rw-r--r--  1 jeffmbair  staff    2660 Mar  5 16:42 ckpt.index
```

The next code snippet in `gan-script-test.py` shows the input node name (`z_placeholder`) and the output node name (`Sigmoid_1`), as printed by `print(generated_images)`:

```
z_placeholder = tf.placeholder(tf.float32, [None, z_dimensions],
name='z_placeholder')
...
saver.restore(sess, 'model/ckpt')
generated_images = generator(z_placeholder, 5, z_dimensions)
print(generated_images)
images = sess.run(generated_images, {z_placeholder: z_batch})
saver.save(sess, "newmodel/ckpt")
```

In the `gan-script-fast.py` script, the method `def discriminator(images, reuse_variables=None)` defines the discriminator network, which takes a real handwritten image input or one generated by the generator, goes through a typical small CNN network with two `conv2d` layers, each of which is followed by a `relu` activation and an average pooling layer, and two fully connected layers to output a scalar value that holds the probability of the input image being real or fake. The other method `def generator(batch_size, z_dim)` defines the generator network, which takes a random input image vector and converts it to a 28 x 28 image with 3 `conv2d` layers.

The two methods can now be used to define three outputs:

- Gz, the generator output of a random image input: `Gz = generator(batch_size, z_dimensions)`
- Dx, the discriminator output of a real image input: `Dx = discriminator(x_placeholder)`
- Dg, the discriminator output of Gz: `Dg = discriminator(Gz, reuse_variables=True)`

and three loss functions:

- `d_loss_real`, the difference between Dx and 1: `d_loss_real = tf.reduce_mean(tf.nn.sigmoid_cross_entropy_with_logits(logits = Dx, labels = tf.ones_like(Dx)))`
- `d_loss_fake`, the difference between Dg and 0: `d_loss_fake = tf.reduce_mean(tf.nn.sigmoid_cross_entropy_with_logits(logits = Dg, labels = tf.zeros_like(Dg)))`
- `g_loss`, the difference between Dg and 1: `g_loss = tf.reduce_mean(tf.nn.sigmoid_cross_entropy_with_logits(logits = Dg, labels = tf.ones_like(Dg)))`

Notice that the discriminator tries to minimize `d_loss_fake`, while the generator tries to minimize `g_loss`, the difference between `Dg` in both cases and 0 and 1, respectively.

Finally, three optimizers for the three loss functions can be set now: `d_trainer_fake`, `d_trainer_real`, and `g_trainer`, all defined with the `tf.train.AdamOptimizer`'s `minimize` method.

Now the script just creates a TensorFlow session, trains the generator and the discriminator for 100,000 steps by running the three optimizers, with random image inputs fed into the generator and both real and fake image inputs fed into the discriminator.

After you run both `gan-script-fast.py` and `gan-script-test.py`, copy the checkpoint files from the `newmodel` directory to `/tmp`, then go to the TensorFlow source root directory and run:

```
python tensorflow/python/tools/freeze_graph.py \
--input_meta_graph=/tmp/ckpt.meta \
--input_checkpoint=/tmp/ckpt \
--output_graph=/tmp/gan_mnist.pb \
--output_node_names="Sigmoid_1" \
--input_binary=true
```

This creates the frozen model `gan_mnist.pb` that we can use on mobile apps. But before we do that, let's take a look at a more advanced GAN model that can enhance low-resolution images.

Advanced GAN model of enhancing image resolution

The model we'll use to enhance low-resolution blurry images is based on the paper *Image-to-Image Translation with Conditional Adversarial Networks* (https://arxiv.org/abs/1611.07004) and its TensorFlow implementation, pix2pix (https://affinelayer.com/pix2pix/). In our fork of the repo (https://github.com/jeffxtang/pix2pix-tensorflow), we added two scripts:

- `tools/convert.py` creates blurry images from ordinary images
- `pix2pix_runinference.py` adds a placeholder for a low-resolution image input and an operation to return the enhanced image, and saves the new checkpoint files, which we'll freeze to generate the model file used on mobile devices.

Basically, pix2pix uses GAN to map an input image to an output image. You can use different types of input images and output images to create many interesting image translations:

- Map to aerial
- Day to night
- Edges to photo
- Black/white images to color images
- Corrupted images to original images
- Low-resolution images to high-resolution images

In all the cases, the generator transforms an input image to an output image, trying to make the output look like the real target image, and the discriminator takes as input a sample from the training set or an output from the generator, and tries to tell whether it's a real image or one produced by the generator. Naturally, the generator and discriminator networks in pix2pix are built in a more complicated way than the model to generate handwritten digits, and the training also applies some tricks to make the process stable – for details, you can read the paper or the TensorFlow implementation link supplied earlier. We'll just show you here how to set up a training set and train the pix2pix model to enhance low-resolution images.

1. Clone the repo by running on your Terminal:

    ```
    git clone https://github.com/jeffxtang/pix2pix-tensorflow
    cd pix2pix-tensorflow
    ```

2. Create a new directory `photos/original` and copy some image files to it – for example, we copied all the Labrador Retriever photos from the Stanford Dog Dataset (http://vision.stanford.edu/aditya86/ImageNetDogs) used in Chapter 2, *Classifying Images with Transfer Learning*, to the `photos/original` directory
3. Run the script `python tools/process.py --input_dir photos/original --operation resize --output_dir photos/resized` to resize the images in the `photo/original` directory and save the resized ones to the `photos/resized` directory
4. Run `mkdir photos/blurry` then `python tools/convert.py` to convert the resized images to blurry ones using the popular ImageMagick's `convert` command. The code for `convert.py` is as follows:

```
import os
file_names = os.listdir("photos/resized/")
for f in file_names:
    if f.find(".png") != -1:
        os.system("convert photos/resized/" + f + " -blur 0x3 photos/blurry/" + f)
```

5. Combine each file in `photos/resized` and `photos/blurry` to a pair and save all the paired images (one resized image, and the other the blurry version) to the `photos/resized_blurry` directory:

```
python tools/process.py    --input_dir photos/resized    --b_dir photos/blurry    --operation combine    --output_dir photos/resized_blurry
```

6. Run the split tool `python tools/split.py --dir photos/resized_blurry` to convert the files to a `train` directory and a `val` directory
7. Train the `pix2pix` model by running:

```
python pix2pix.py \
   --mode train \
   --output_dir photos/resized_blurry/ckpt_1000 \
   --max_epochs 1000 \
   --input_dir photos/resized_blurry/train \
   --which_direction BtoA
```

The direction BtoA means translating from a blurry image to the original one. The training takes about four hours on GTX-1070 GPU and the resulting checkpoint files in the photos/resized_blurry/ckpt_1000 directory look like this:

```
-rw-rw-r-- 1 jeff jeff 1721531 Mar  2 18:37 model-136000.meta
-rw-rw-r-- 1 jeff jeff      81 Mar  2 18:37 checkpoint
-rw-rw-r-- 1 jeff jeff 686331732 Mar  2 18:37 model-136000.data-00000-of-00001
-rw-rw-r-- 1 jeff jeff   10424 Mar  2 18:37 model-136000.index
-rw-rw-r-- 1 jeff jeff 3807975 Mar  2 14:19 graph.pbtxt
-rw-rw-r-- 1 jeff jeff     682 Mar  2 14:19 options.json
```

8. Optionally, you can run the script in the test mode and then check the image translation results in the directory specified by --output_dir:

```
python pix2pix.py \
   --mode test \
   --output_dir photos/resized_blurry/output_1000 \
   --input_dir photos/resized_blurry/val \
   --checkpoint photos/resized_blurry/ckpt_1000
```

9. Run the pix2pix_runinference.py script to restore the checkpoint saved in Step 7, create a new placeholder for the image input, feed it with a test image ww.png, output the translation as result.png, and finally save the new checkpoint files in the newckpt directory:

```
python pix2pix_runinference.py \
--mode test \
--output_dir photos/blurry_output \
--input_dir photos/blurry_test \
--checkpoint photos/resized_blurry/ckpt_1000
```

The code snippet in the following pix2pix_runinference.py sets and prints the input and output nodes:

```
    image_feed = tf.placeholder(dtype=tf.float32, shape=(1, 256, 256, 3), name="image_feed")
    print(image_feed) # Tensor("image_feed:0", shape=(1, 256, 256, 3), dtype=float32)
    with tf.variable_scope("generator", reuse=True):
        output_image = deprocess(create_generator(image_feed, 3))
        print(output_image)
#Tensor("generator_1/deprocess/truediv:0", shape=(1, 256, 256, 3), dtype=float32)
```

The line with `tf.variable_scope("generator", reuse=True):` is very important as the `generator` variable needs to be shared so all the trained parameter values can be used. Otherwise, you'll see weird translation results.

The following code shows how to feed the placeholder, run the GAN model and save the generator's output, as well as the checkpoint files, in the `newckpt` directory:

```
if a.mode == "test":
    from scipy import misc
    image = misc.imread("ww.png").reshape(1, 256, 256, 3)
    image = (image / 255.0) * 2 - 1
    result = sess.run(output_image, feed_dict={image_feed:image})
    misc.imsave("result.png", result.reshape(256, 256, 3))
    saver.save(sess, "newckpt/pix2pix")
```

Figure 9.1 shows the original test image, its blurry version, and our trained GAN model's generator output. The result is not ideal, but the GAN model does have a better resolution without the blurry effect:

Figure 9.1: The original, the blurry, and the generated

10. Now, copy the `newckpt` directory to `/tmp` and we can freeze the model as follows:

```
python tensorflow/python/tools/freeze_graph.py \
--input_meta_graph=/tmp/newckpt/pix2pix.meta \
--input_checkpoint=/tmp/newckpt/pix2pix \
--output_graph=/tmp/newckpt/pix2pix.pb \
--output_node_names="generator_1/deprocess/truediv" \
--input_binary=true
```

11. The generated `pix2pix.pb` model file is pretty big, about 217 MB, which would crash or cause an **Out of Memory** (**OOM**) error when loading it on an iOS or Android device. We have to transform and convert it to the memmapped format for iOS as we did with the complicated im2txt model in Chapter 6, *Describing Images in Natural Language*:

    ```
    bazel-bin/tensorflow/tools/graph_transforms/transform_graph \
    --in_graph=/tmp/newckpt/pix2pix.pb \
    --out_graph=/tmp/newckpt/pix2pix_transformed.pb \
    --inputs="image_feed" \
    --outputs="generator_1/deprocess/truediv" \
    --transforms='strip_unused_nodes(type=float, shape="1,256,256,3")
        fold_constants(ignore_errors=true, clear_output_shapes=true)
        fold_batch_norms
        fold_old_batch_norms'

    bazel-bin/tensorflow/contrib/util/convert_graphdef_memmapped_format \
    --in_graph=/tmp/newckpt/pix2pix_transformed.pb \
    --out_graph=/tmp/newckpt/pix2pix_transformed_memmapped.pb
    ```

    The `pix2pix_transformed_memmapped.pb` model file can now be used in iOS.

12. To build the model for Android, we need to quantize the frozen model to reduce the model size from 217 MB to about 54 MB:

    ```
    bazel-bin/tensorflow/tools/graph_transforms/transform_graph \
    --in_graph=/tmp/newckpt/pix2pix.pb \
    --out_graph=/tmp/newckpt/pix2pix_transformed_quantized.pb --
    inputs="image_feed" \
    --outputs="generator_1/deprocess/truediv" \
    --transforms='quantize_weights'
    ```

Now let's see how we can use the two GAN models in mobile apps.

Using the GAN models in iOS

If you try to use the TensorFlow pod in your iOS app and load the `gan_mnist.pb` file, you'll get an error:

```
Could not create TensorFlow Graph: Invalid argument: No OpKernel was
registered to support Op 'RandomStandardNormal' with these attrs.
Registered devices: [CPU], Registered kernels:
  <no registered kernels>
```

```
[[Node: z_1/RandomStandardNormal = RandomStandardNormal[T=DT_INT32,
_output_shapes=[[50,100]], dtype=DT_FLOAT, seed=0, seed2=0](z_1/shape)]]
```

Make sure your `tensorflow/contrib/makefile/tf_op_files.txt` file has `tensorflow/core/kernels/random_op.cc`, which implements the `RandomStandardNormal` operation, and the `libtensorflow-core.a` is built with `tensorflow/contrib/makefile/build_all_ios.sh` after the line is added to `tf_op_files.txt`.

Furthermore, if you try to load the `pix2pix_transformed_memmapped.pb` even in the custom TensorFlow library built with TensorFlow 1.4, you'll get the following error:

```
No OpKernel was registered to support Op 'FIFOQueueV2' with these attrs.
Registered devices: [CPU], Registered kernels:
  <no registered kernels>
  [[Node: batch/fifo_queue = FIFOQueueV2[_output_shapes=[[]], capacity=32,
component_types=[DT_STRING, DT_FLOAT, DT_FLOAT], container="", shapes=[[],
[256,256,1], [256,256,2]], shared_name=""]()]]
```

You'll need to add `tensorflow/core/kernels/fifo_queue_op.cc` to the `tf_op_files.txt` and rebuild the iOS library. But if you use TensorFlow 1.5 or 1.6, the `tensorflow/core/kernels/fifo_queue_op.cc` file has already been added to the `tf_op_files.txt` file. With each new version of TensorFlow, more and more kernels are added to `tf_op_files.txt` by default.

With the TensorFlow iOS library built for our models, let's create a new project named GAN in Xcode and set up TensorFlow in the project as we did in Chapter 8, *Predicting Stock Price with RNN*, and other chapters not using the TensorFlow pod. Then drag and drop the two model files `gan_mnist.pb` and `pix2pix_transformed_memmapped.pb` and one test image to the project. Also, copy the `tensorflow_utils.h, tensorflow_utils.mm, ios_image_load.h` and `ios_image_load.mm` files from the iOS project in Chapter 6, *Describing Images in Natural Language*, to the GAN project. Rename `ViewController.m` to `ViewController.mm`.

Now your Xcode should look like Figure 9.2:

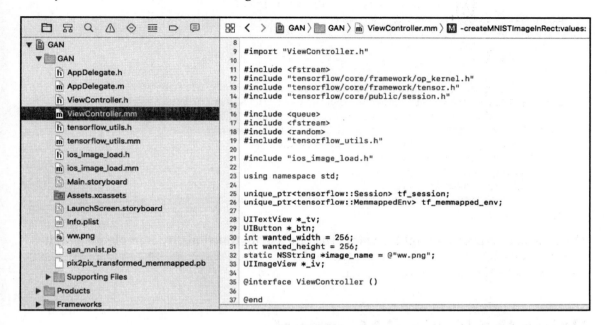

Figure 9.2: Showing the GAN app in Xcode

We'll create a button that when tapped, prompts the user to pick a model to generate digits or enhance an image:

```
- (IBAction)btnTapped:(id)sender {
    UIAlertAction* mnist = [UIAlertAction actionWithTitle:@"Generate
Digits" style:UIAlertActionStyleDefault handler:^(UIAlertAction * action) {
        _iv.image = NULL;
        dispatch_async(dispatch_get_global_queue(0, 0), ^{
            NSArray *arrayGreyscaleValues = [self runMNISTModel];
            dispatch_async(dispatch_get_main_queue(), ^{
                UIImage *imgDigit = [self createMNISTImageInRect:_iv.frame
values:arrayGreyscaleValues];
                _iv.image = imgDigit;
            });
        });
    }];
    UIAlertAction* pix2pix = [UIAlertAction actionWithTitle:@"Enhance
Image" style:UIAlertActionStyleDefault handler:^(UIAlertAction * action) {
        _iv.image = [UIImage imageNamed:image_name];
        dispatch_async(dispatch_get_global_queue(0, 0), ^{
            NSArray *arrayRGBValues = [self runPix2PixBlurryModel];
```

```
                dispatch_async(dispatch_get_main_queue(), ^{
                    UIImage *imgTranslated = [self
createTranslatedImageInRect:_iv.frame values:arrayRGBValues];
                    _iv.image = imgTranslated;
                });
            });
        }];
        UIAlertAction* none = [UIAlertAction actionWithTitle:@"None"
style:UIAlertActionStyleDefault handler:^(UIAlertAction * action) {}];
        UIAlertController* alert = [UIAlertController
alertControllerWithTitle:@"Use GAN to" message:nil
preferredStyle:UIAlertControllerStyleAlert];
        [alert addAction:mnist];
        [alert addAction:pix2pix];
        [alert addAction:none];
        [self presentViewController:alert animated:YES completion:nil];
}
```

The code here is pretty straightforward. The app's main functionality is implemented in four methods: `runMNISTModel`, `runPix2PixBlurryModel`, `createMNISTImageInRect`, and `createTranslatedImageInRect`.

Using the basic GAN model

In `runMNISTModel`, we call the helper method `LoadModel` to load the GAN model and then set the input tensor to be 6 batches of 100 random numbers with normal distribution (mean 0.0 and std 1.0). The random input with normal distribution is what the model expects. You can change 6 to any other number and get back that number of generated digits:

```
- (NSArray*) runMNISTModel {
    tensorflow::Status load_status;
    load_status = LoadModel(@"gan_mnist", @"pb", &tf_session);
    if (!load_status.ok()) return NULL;
    std::string input_layer = "z_placeholder";
    std::string output_layer = "Sigmoid_1";

    tensorflow::Tensor input_tensor(tensorflow::DT_FLOAT,
tensorflow::TensorShape({6, 100}));
    auto input_map = input_tensor.tensor<float, 2>();
    unsigned seed =
(unsigned)std::chrono::system_clock::now().time_since_epoch().count();
    std::default_random_engine generator (seed);
    std::normal_distribution<double> distribution(0.0, 1.0);
    for (int i = 0; i < 6; i++){
```

```
        for (int j = 0; j < 100; j++) {
            double number = distribution(generator);
            input_map(i,j) = number;
        }
    }
```

The rest of the code in the `runMNISTModel` method runs the model, gets the output of 6*28*28 floating numbers, representing the grayscale value at each pixel for each batch of image sized 28*28, and calls the method `createMNISTImageInRect` to render the numbers in an image context using `UIBezierPath` before converting the image context to `UIImage`, which gets returned and displayed in `UIImageView`:

```
    std::vector<tensorflow::Tensor> outputs;
    tensorflow::Status run_status = tf_session->Run({{input_layer, input_tensor}},
                                                    {output_layer}, {},
    &outputs);
    if (!run_status.ok()) {
        LOG(ERROR) << "Running model failed: " << run_status;
        return NULL;
    }
    tensorflow::string status_string = run_status.ToString();
    tensorflow::Tensor* output_tensor = &outputs[0];
    const Eigen::TensorMap<Eigen::Tensor<float, 1, Eigen::RowMajor>, Eigen::Aligned>& output = output_tensor->flat<float>();
    const long count = output.size();
    NSMutableArray *arrayGreyscaleValues = [NSMutableArray array];
    for (int i = 0; i < count; ++i) {
        const float value = output(i);
        [arrayGreyscaleValues addObject:[NSNumber numberWithFloat:value]];
    }
    return arrayGreyscaleValues;
}
```

The `createMNISTImageInRect` is defined as follows – we used the similar technique in Chapter 7, *Recognizing Drawing with CNN and LSTM*:

```
- (UIImage *)createMNISTImageInRect:(CGRect)rect values:(NSArray*)greyscaleValues
{
    UIGraphicsBeginImageContextWithOptions(CGSizeMake(rect.size.width, rect.size.height), NO, 0.0);
    int i=0;
    const int size = 3;
    for (NSNumber *val in greyscaleValues) {
        float c = [val floatValue];
        int x = i%28;
```

```
            int y = i/28;
            i++;

            CGRect rect = CGRectMake(145+size*x, 50+y*size, size, size);
            UIBezierPath *path = [UIBezierPath bezierPathWithRect:rect];
            UIColor *color = [UIColor colorWithRed:c green:c blue:c alpha:1.0];
            [color setFill];
            [path fill];
        }
        UIImage *image = UIGraphicsGetImageFromCurrentImageContext();
        UIGraphicsEndImageContext();
        return image;
    }
```

For each pixel, we draw a small rectangle of width and height both as 3, with the grayscale value returned for the pixel.

Using the advanced GAN model

In the `runPix2PixBlurryModel` method, we use the `LoadMemoryMappedModel` method to load the `pix2pix_transformed_memmapped.pb` model file, and load the test image and set the input tensor in a similar way to what we did in Chapter 4, *Transforming Pictures with Amazing Art Styles*:

```
- (NSArray*) runPix2PixBlurryModel {
    tensorflow::Status load_status;
    load_status = LoadMemoryMappedModel(@"pix2pix_transformed_memmapped", @"pb", &tf_session, &tf_memmapped_env);
    if (!load_status.ok()) return NULL;
    std::string input_layer = "image_feed";
    std::string output_layer = "generator_1/deprocess/truediv";

    NSString* image_path = FilePathForResourceName(@"ww", @"png");
    int image_width;
    int image_height;
    int image_channels;
    std::vector<tensorflow::uint8> image_data = LoadImageFromFile([image_path UTF8String], &image_width, &image_height, &image_channels);
```

Then we run the model, get the output of 256*256*3 (the image size is 256*256, with 3 values for RGB) floating numbers, and call `createTranslatedImageInRect` to convert the numbers to `UIImage`:

```
    std::vector<tensorflow::Tensor> outputs;
    tensorflow::Status run_status = tf_session->Run({{input_layer, image_tensor}},
                                                    {output_layer}, {},
&outputs);
    if (!run_status.ok()) {
        LOG(ERROR) << "Running model failed: " << run_status;
        return NULL;
    }
    tensorflow::string status_string = run_status.ToString();
    tensorflow::Tensor* output_tensor = &outputs[0];
    const Eigen::TensorMap<Eigen::Tensor<float, 1, Eigen::RowMajor>, Eigen::Aligned>& output = output_tensor->flat<float>();

    const long count = output.size(); // 256*256*3
    NSMutableArray *arrayRGBValues = [NSMutableArray array];
    for (int i = 0; i < count; ++i) {
        const float value = output(i);
        [arrayRGBValues addObject:[NSNumber numberWithFloat:value]];
    }
    return arrayRGBValues;
```

The final method `createTranslatedImageInRect` is defined as follows, all pretty self explanatory:

```
- (UIImage *)createTranslatedImageInRect:(CGRect)rect
values:(NSArray*)rgbValues
{
    UIGraphicsBeginImageContextWithOptions(CGSizeMake(wanted_width, wanted_height), NO, 0.0);
    for (int i=0; i<256*256; i++) {
        float R = [rgbValues[i*3] floatValue];
        float G = [rgbValues[i*3+1] floatValue];
        float B = [rgbValues[i*3+2] floatValue];
        const int size = 1;
        int x = i%256;
        int y = i/256;
        CGRect rect = CGRectMake(size*x, y*size, size, size);
        UIBezierPath *path = [UIBezierPath bezierPathWithRect:rect];
        UIColor *color = [UIColor colorWithRed:R green:G blue:B alpha:1.0];
        [color setFill];
        [path fill];
    }
```

```
    UIImage *image = UIGraphicsGetImageFromCurrentImageContext();
    UIGraphicsEndImageContext();
    return image;
}
```

Now, run the app in the iOS simulator or device, tap the **GAN** button and select **Generate Digits** and you'll see the results of GAN-generated handwritten digits as shown Figure 9.3:

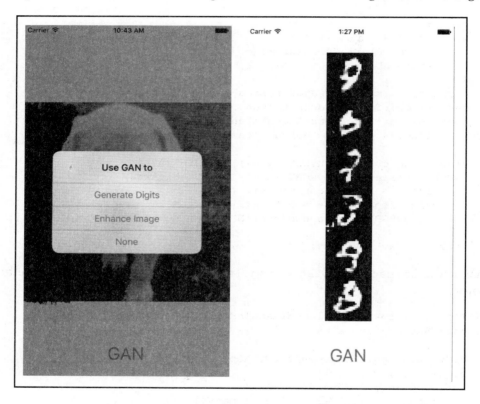

Figure 9.3: Showing GAN model selection and results of generated handwritten digits

The digits look a lot like real human handwritten digits, all done after training the basic GAN model. If you go back and review the code that does the training, and pause and think for a moment about how GAN works in general, how the generator and the discriminator play against each other and try to reach the stable Nash Equilibrium state where the generator can produce real-like fake data that the discriminator can't tell whether it's real or fake, you'll probably appreciate more how amazing GAN is or can be.

Now, let's select the **Enhance Image** option, you'll see the result in Figure 9.4, the same result as generated by the Python test code in Figure 9.1:

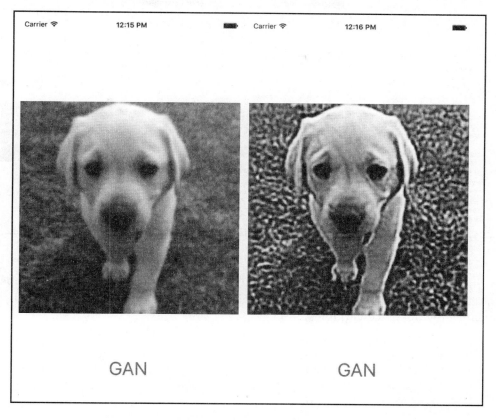

Figure 9.4: The original blurry and enhanced images on iOS

You know the drill. It's time to give our love to Android.

Using the GAN models in Android

It turns out we don't need to use the custom TensorFlow Android library, as we did in Chapter 7, *Recognizing Drawing with CNN and LSTM*, to run the GAN models in Android. Simply create a new Android Studio app called GAN with all the defaults, add `compile 'org.tensorflow:tensorflow-android:+'` to the app's `build.gradle` file, create a new assets folder and copy the two GAN model files and a test blurry image there.

Your project in Android Studio should now look like Figure 9.5:

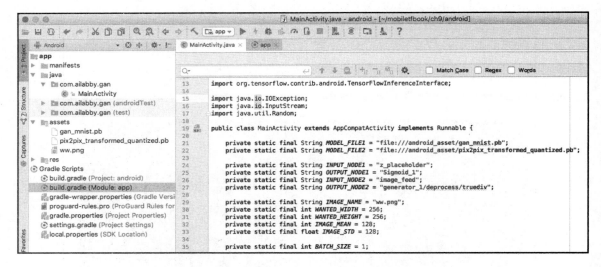

Figure 9.5: Android Studio GAN app overview, showing constant definitions

Notice that for simplicity, we set `BATCH_SIZE` to 1. You can easily set it to any number and get that many outputs back as we did in iOS. Other than the constants defined in Figure 9.5, we'll create a few instance variables:

```
private Button mButtonMNIST;
private Button mButtonPix2Pix;
private ImageView mImageView;
private Bitmap mGeneratedBitmap;
private boolean mMNISTModel;

private TensorFlowInferenceInterface mInferenceInterface;
```

The app layout consists of one ImageView and two buttons, as we've done before, and they get instantiated in the `onCreate` method:

```
protected void onCreate(Bundle savedInstanceState) {
    super.onCreate(savedInstanceState);
    setContentView(R.layout.activity_main);

    mButtonMNIST = findViewById(R.id.mnistbutton);
    mButtonPix2Pix = findViewById(R.id.pix2pixbutton);
    mImageView = findViewById(R.id.imageview);
    try {
        AssetManager am = getAssets();
        InputStream is = am.open(IMAGE_NAME);
```

```
        Bitmap bitmap = BitmapFactory.decodeStream(is);
        mImageView.setImageBitmap(bitmap);
    } catch (IOException e) {
        e.printStackTrace();
    }
```

Then, set up two click listeners for the two buttons:

```
    mButtonMNIST.setOnClickListener(new View.OnClickListener() {
        @Override
        public void onClick(View v) {
            mMNISTModel = true;
            Thread thread = new Thread(MainActivity.this);
            thread.start();
        }
    });
    mButtonPix2Pix.setOnClickListener(new View.OnClickListener() {
        @Override
        public void onClick(View v) {
            try {
                AssetManager am = getAssets();
                InputStream is = am.open(IMAGE_NAME);
                Bitmap bitmap = BitmapFactory.decodeStream(is);
                mImageView.setImageBitmap(bitmap);
                mMNISTModel = false;
                Thread thread = new Thread(MainActivity.this);
                thread.start();
            } catch (IOException e) {
                e.printStackTrace();
            }
        }
    });
}
```

When a button is tapped, the `run` method runs in a worker thread:

```
public void run() {
    if (mMNISTModel)
        runMNISTModel();
    else
        runPix2PixBlurryModel();
}
```

Using the basic GAN model

In the `runMNISTModel` method, we first prepare a random input to the model:

```
void runMNISTModel() {
    float[] floatValues = new float[BATCH_SIZE*100];

    Random r = new Random();
    for (int i=0; i<BATCH_SIZE; i++) {
        for (int j=0; i<100; i++) {
            double sample = r.nextGaussian();
            floatValues[i] = (float)sample;
        }
    }
```

Then feed the input to the model, run the model and gets the output values, which are scaled grayscale values between 0.0 and 1.0, and convert them to integers in the range of 0 and 255:

```
    float[] outputValues = new float[BATCH_SIZE * 28 * 28];
    AssetManager assetManager = getAssets();
    mInferenceInterface = new TensorFlowInferenceInterface(assetManager, MODEL_FILE1);

    mInferenceInterface.feed(INPUT_NODE1, floatValues, BATCH_SIZE, 100);
    mInferenceInterface.run(new String[] {OUTPUT_NODE1}, false);
    mInferenceInterface.fetch(OUTPUT_NODE1, outputValues);

    int[] intValues = new int[BATCH_SIZE * 28 * 28];
    for (int i = 0; i < intValues.length; i++) {
        intValues[i] = (int) (outputValues[i] * 255);
    }
```

After that, we use the returned and converted grayscale values for each pixel that gets set when creating a bitmap:

```
    try {
        Bitmap bitmap = Bitmap.createBitmap(28, 28, Bitmap.Config.ARGB_8888);
        for (int y=0; y<28; y++) {
            for (int x=0; x<28; x++) {
                int c = intValues[y*28 + x];
                int color = (255 & 0xff) << 24 | (c & 0xff) << 16 | (c & 0xff) << 8 | (c & 0xff);
                bitmap.setPixel(x, y, color);
            }
        }
```

```
            mGeneratedBitmap = Bitmap.createBitmap(bitmap);
    }
    catch (Exception e) {
        e.printStackTrace();
    }
```

Finally, we show the bitmap in the main UI thread's ImageView:

```
    runOnUiThread(
        new Runnable() {
            @Override
            public void run() {
                mImageView.setImageBitmap(mGeneratedBitmap);
            }
        });
}
```

If you run the app now, with a blank implementation of void `runPix2PixBlurryModel() {}` to avoid the build error, you'll see the initial screen and the result after you tap **GENERATE DIGITS** as in Figure 9.6:

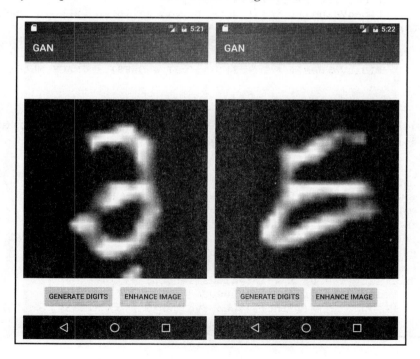

Figure 9.6: Showing the generated digits

Using the advanced GAN model

The `runPix2PixBlurryModel` method is similar to the code in the previous chapters where we used an image input to feed into our models. We first get the RGB values from the image bitmap and save them to a float array:

```
void runPix2PixBlurryModel() {
    int[] intValues = new int[WANTED_WIDTH * WANTED_HEIGHT];
    float[] floatValues = new float[WANTED_WIDTH * WANTED_HEIGHT * 3];
    float[] outputValues = new float[WANTED_WIDTH * WANTED_HEIGHT * 3];

    try {
        Bitmap bitmap = BitmapFactory.decodeStream(getAssets().open(IMAGE_NAME));
        Bitmap scaledBitmap = Bitmap.createScaledBitmap(bitmap, WANTED_WIDTH, WANTED_HEIGHT, true);
        scaledBitmap.getPixels(intValues, 0, scaledBitmap.getWidth(), 0, 0, scaledBitmap.getWidth(), scaledBitmap.getHeight());
        for (int i = 0; i < intValues.length; ++i) {
            final int val = intValues[i];
            floatValues[i * 3 + 0] = (((val >> 16) & 0xFF) - IMAGE_MEAN) / IMAGE_STD;
            floatValues[i * 3 + 1] = (((val >> 8) & 0xFF) - IMAGE_MEAN) / IMAGE_STD;
            floatValues[i * 3 + 2] = ((val & 0xFF) - IMAGE_MEAN) / IMAGE_STD;
        }
```

Then, we run the model with the input and get and convert the output values to an integer array which is then used to set a new bitmap's pixels:

```
        AssetManager assetManager = getAssets();
        mInferenceInterface = new TensorFlowInferenceInterface(assetManager, MODEL_FILE2);
        mInferenceInterface.feed(INPUT_NODE2, floatValues, 1, WANTED_HEIGHT, WANTED_WIDTH, 3);
        mInferenceInterface.run(new String[] {OUTPUT_NODE2}, false);
        mInferenceInterface.fetch(OUTPUT_NODE2, outputValues);

        for (int i = 0; i < intValues.length; ++i) {
            intValues[i] = 0xFF000000
                    | (((int) (outputValues[i * 3] * 255)) << 16)
                    | (((int) (outputValues[i * 3 + 1] * 255)) << 8)
                    | ((int) (outputValues[i * 3 + 2] * 255));
        }

        Bitmap outputBitmap = scaledBitmap.copy( scaledBitmap.getConfig() ,
```

```
true);
        outputBitmap.setPixels(intValues, 0, outputBitmap.getWidth(), 0, 0,
outputBitmap.getWidth(), outputBitmap.getHeight());
        mGeneratedBitmap = Bitmap.createScaledBitmap(outputBitmap,
bitmap.getWidth(), bitmap.getHeight(), true);

    }
    catch (Exception e) {
        e.printStackTrace();
    }
```

Finally, we show the bitmap in the main UI's ImageView:

```
    runOnUiThread(
        new Runnable() {
            @Override
            public void run() {
                mImageView.setImageBitmap(mGeneratedBitmap);
            }
        });
}
```

Run the app again and tap the **ENHANCE IMAGE** button now, you'll see in a few seconds the enhanced image in Figure 9.7:

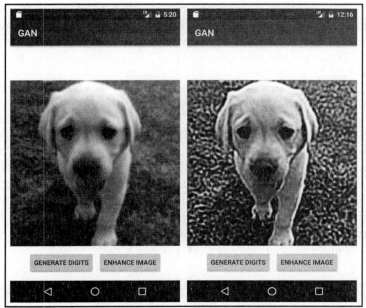

Figure 9.7: The blurry and enhanced images on Android

This completes our Android app using the two GAN models.

Summary

In this chapter, we gave the wonderful world of GANs a quick tour. We covered what GANs are and why they're so interesting – the way the generator and the discriminator play against each other and try to beat each other probably sounds appealing to most people. We then went through the detailed steps of how to train a basic GAN model and a more advanced image resolution enhancement model, and how to prepare them for mobile devices. Finally, we showed you how to build iOS and Android apps using the models. If you're excited about the whole process and the results, you'll definitely want to further explore GANs, a rapidly developing field where new types of GANs have been developed quickly to overcome the shortcomings of the previous models; for example, the same researchers who developed the pix2pix model which requires paired images for training, as we saw in the *Advanced GAN model of enhancing image resolution* subsection, have come up with a new type of GAN called CycleGAN (`https://junyanz.github.io/CycleGAN`) that removes the requirement for image paring. If you're unsatisfied with the quality of our generated digits or enhanced image, you probably should also explore GANs further to see how you can improve the GAN models. As we mentioned before, GANs are still very young, researchers are still working hard to stabilize the training, and much greater success will come, if it can be stabilized. At least by now you've gained experience of how to quickly deploy a GAN model in your mobile apps. It's up to you to decide whether to keep an eye on the latest and greatest GANs and use them on mobile devices or put your mobile developer's hat aside for a while and get your hands dirty on building new or improving existing GAN models.

If GANs have caused great excitement in the deep learning community, the accomplishment of AlphaGo in 2016 and 2017 of defeating the very best human GO players has definitely amazed everyone not living in a cave. Furthermore, in October 2017, AlphaGo Zero, a new algorithm solely based on self-study reinforcement learning without any human knowledge, was introduced unbelievably as defeating AlphaGo 100-0; in December 2017, AlphaZero, an algorithm that can achieve *"superhuman performance in many challenging domains"*, unlike AlphaGo and AlphaGo Zero targeted only in the game of GO, was published. In the next chapter, we'll see how we can use the latest and coolest AlphaZero to build and train a model for playing a simpler fun game and how to run the model on mobile devices.

10
Building an AlphaZero-like Mobile Game App

Although the ever-increasing popularity of modern **Artificial Intelligence (AI)** was essentially caused by the breakthrough of deep learning in 2012, the historic events of Google DeepMind's AlphaGo beating Lee Sedol, the 18-time world champion of GO, 4-1 in March 2016 and then beating Ke Jie, the current #1-ranked GO player, 3-0 in May 2017, contributed in large part to making AI a household acronym. Due to the complexity of the GO game, it was wildly considered an impossible mission, or impossible for at least one more decade, that a computer program would beat top GO players.

After the match between AlphaGo and Ke Jie in May 2017, Google retired AlphaGo; DeepMind, the startup Google acquired for its pioneering deep reinforcement learning technologies and the developer of AlphaGo, decided to focus their AI research on other areas. Then, interestingly, in October 2017, DeepMind published another paper on the game, *GO: Mastering the Game of GO without Human Knowledge* (https://deepmind.com/research/publications/mastering-game-go-without-human-knowledge), which describes an improved algorithm, called AlphaGo Zero, that learns how to play GO solely by self-play reinforcement learning, with no reliance on any human expert knowledge, such as a large number of professional GO games played, which AlphaGo uses to train its model. What's amazing is that AlphaGo Zero completely defeated AlphaGo, which just humbled the world's best human GO player a few months ago, with 100-0!

It turns out this is just one step toward Google's more ambitious goal of applying and improving the AI techniques behind AlphaGo to other domains. In December 2017, DeepMind published yet another paper, Mastering Chess and Shogi by Self-Play with a General Reinforcement Learning Algorithm (https://arxiv.org/pdf/1712.01815.pdf), that generalized the AlphaGo Zero program to a single algorithm, called AlphaZero, and used the algorithm to quickly learn how to play the games of Chess and Shogi from scratch, starting with random play with no domain knowledge except the game rules, and in 24 hours, achieved the superhuman level and beat world champions.

Building an AlphaZero-like Mobile Game App

In this chapter, we'll take you on a tour of the latest and coolest in AlphaZero, showing you how to build and train an AlphaZero-like model to play a simple but fun game called Connect 4 (https://en.wikipedia.org/wiki/Connect_Four) in TensorFlow and Keras, the popular high-level deep learning library we used in Chapter 8, *Predicting Stock Price with RNN*. We'll also cover how to use the trained AlphaZero-like model to get a trained expert policy to guide the gameplay on mobile, with the source code of complete iOS and Android apps that play the Connect 4 game using the model.

In summary, we'll cover the following topics in this chapter:

- AlphaZero – how does it work?
- Building and training an AlphaZero-like model for Connect 4
- Using the model in iOS to play Connect 4
- Using the model in Android to play Connect 4

AlphaZero – how does it work?

The AlphaZero algorithm consists of three main components:

- A deep convolutional neural network, which takes the board position (or state) as input and outputs a value as the predicted game result from the position and a policy that is a list of move probabilities for each possible action from the input board state.
- A general-purpose reinforcement learning algorithm, which learns via self-play from scratch with no specific domain knowledge except the game rules. The deep neural network's parameters are learned by self-play reinforcement learning to minimize the loss between the predicted value and the actual self-play game result, and maximize the similarity between the predicted policy and the search probabilities, which come from the following algorithm.
- A general-purpose (domain-independent) **Monte-Carlo Tree Search** (**MCTS**) algorithm, which simulates games of self-play from start to end, selecting each move during the simulation by considering the predicted value and policy probability values returned from the deep neural network, as well as how frequently a node has been visited—occasionally selecting a node with a low visit count is called exploration in reinforcement learning (versus taking the move with a high predicted value and policy, which is called exploitation). A nice balance between exploration and exploitation can lead to better results.

Reinforcement learning has a long history, dating back to the 1960s when the term was first used in engineering literature. But the breakthrough came in 2013 when DeepMind combined reinforcement learning with deep learning and developed deep reinforcement learning apps that learned to play Atari games from scratch, with raw pixels as input, and were able to beat humans afterward. Unlike supervised learning, which requires labelled data for training, as we have seen in many models that we built or used in the previous chapters, reinforcement learning uses a trial-and-error method to get better: an agent interacts with an environment and receives rewards (positive or negative) for every action it takes on every state. In the example of AlphaZero playing chess, the reward only comes after the game is over, with the result of winning as +1, losing as -1, and a draw as 0. The reinforcement learning algorithm in AlphaZero uses gradient descent on the loss we mentioned before to update the parameters of the deep neural network, which acts like a universal function approximation to learn and encode the gameplay expertise.

The result of the learning or training process can be a policy generated by the deep neural network that says what action should be taken on any state, or a value function that maps each state and each possible action from that state to a long-term reward.

If the policy learned by the deep neural network using self-play reinforcement learning is ideal, we may not need to let the program perform any MCTS during gameplay—the program can simply always choose the move with the maximum probability. But in complicated games such as Chess or GO, a perfect policy can't be generated so MCTS is required to work together with the trained deep network to guide the search for the best possible action for each game state.

If you're not familiar with reinforcement learning or MCTS, there's lots of information about them on the internet. Consider checking out *Richard Sutton* and *Andrew Barto's* classic book, *Reinforcement Learning: An Introduction*, which is publicly available at `http://incompleteideas.net/book/the-book-2nd.html`. You can also watch the reinforcement learning course videos by *David Silver*, the technical lead for *AlphaGo* at *DeepMind*, on YouTube (search "reinforcement learning David Silver"). A fun and useful toolkit for reinforcement learning is OpenAI Gym (`https://gym.openai.com`). In the last chapter of the book, we'll go deeper into reinforcement learning and OpenAI Gym. For MCTS, check out its Wiki page, `https://en.wikipedia.org/wiki/Monte_Carlo_tree_search`, as well as this blog: `http://tim.hibal.org/blog/alpha-zero-how-and-why-it-works`.

In the next section, we'll take a look at a Keras implementation, with TensorFlow as the backend, of the AlphaZero algorithm, with the goal of building and training a model using the algorithm to play Connect 4. You'll see what the model architecture looks like, and the key Keras code to build the model.

Training and testing an AlphaZero-like model for Connect 4

If you've never played Connect 4, you can play it for free at http://www.connectfour.org. It's a quick and fun game. Basically, two players take turns dropping different-colored discs into a grid of six rows by seven columns from the top of a column. The newly dropped disc either sits at the bottom of a column, if no discs have been dropped in that column, or on top of the last dropped disc in that column. Whoever first has four consecutive discs of their own color in any of the three possible directions (horizontally, vertically, diagonally) wins the game.

The AlphaZero model for Connect 4 was based on the https://github.com/jeffxtang/DeepReinforcementLearning repo, a fork of https://github.com/AppliedDataSciencePartners/DeepReinforcementLearning, with a nice blog on *How to build your own AlphaZero AI using Python and Keras* (https://applied-data.science/blog/how-to-build-your-own-alphazero-ai-using-python-and-keras), which you should probably read before moving on so the following steps make more sense.

Training the model

Before we take a look at some core code snippets, let's first see how to train the model. First, get the repo by running the following on your Terminal:

```
git clone https://github.com/jeffxtang/DeepReinforcementLearning
```

Then, set up a Keras and TensorFlow virtualenv if you hadn't already done so in Chapter 8, *Predicting Stock Price with RNN*:

```
cd
mkdir ~/tf_keras
virtualenv --system-site-packages ~/tf_keras/
cd ~/tf_keras/
source ./bin/activate
easy_install -U pip
```

```
#On Mac:
pip install --upgrade
https://storage.googleapis.com/tensorflow/mac/cpu/tensorflow-1.4.0-py2-none
-any.whl

#On Ubuntu:
pip install --upgrade
https://storage.googleapis.com/tensorflow/linux/gpu/tensorflow_gpu-1.4.0-cp
27-none-linux_x86_64.whl

easy_install ipython
pip install keras
```

You can also try the TensorFlow 1.5-1.8 download URL in the preceding `pip install` command.

Now, open `run.ipynb` by `cd DeepReinforcementLearning` first, then `jupyter notebook`—depending on your environment, you'll need to install the missing Python packages if you see any errors. On a browser, open `http://localhost:8888/notebooks/run.ipynb`, and run the first code block in the notebook to load all the necessary core libraries, and the second code block to start the training—the code was written to train forever, so you may want to cancel the `jupyter notebook` command after hours of training. It takes about an hour on an older Mac to see the first versions of models created in the following directory (a newer version, such as `version0004.h5`, includes weights that are more fine-tuned than the weights in an older version, such as `version0001.h5`):

```
(tf_keras) MacBook-Air:DeepReinforcementLearning jeffmbair$ ls -lt
run/models

-rw-r--r-- 1 jeffmbair staff 3781664 Mar 8 15:23 version0004.h5
-rw-r--r-- 1 jeffmbair staff 3781664 Mar 8 14:59 version0003.h5
-rw-r--r-- 1 jeffmbair staff 3781664 Mar 8 14:36 version0002.h5
-rw-r--r-- 1 jeffmbair staff 3781664 Mar 8 14:12 version0001.h5
-rw-r--r--  1 jeffmbair  staff    656600 Mar  8 12:29 model.png
```

Those files with the `.h5` extension are the HDF5-formatted Keras model files and each of them mainly contains the model architecture definition, the trained weights, and training configurations. You'll see later how to use the Keras model files to generate TensorFlow checkpoint files, which can then be frozen into model files that run on mobile.

The model.png file contains a detailed view of the deep neural network architecture. It's pretty deep with many residual blocks of convolutional layers followed by batch normalization and ReLU layers to stabilize the training. The top part of the model looks like the following diagram (we won't show the middle part as it's pretty big and you're encouraged to open the model.png file for reference):

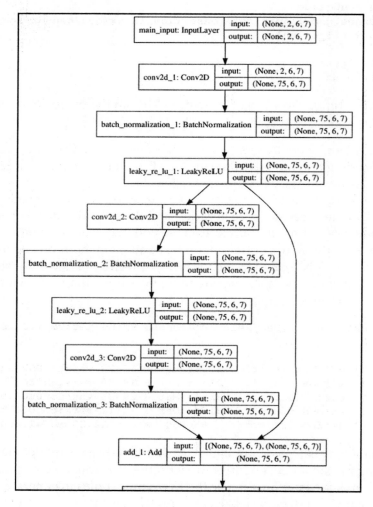

Figure 10.1 The first layers of the deep residual network

Chapter 10

It's worth noting that the neural network is called the residual network (ResNet), which was introduced by Microsoft in 2015 in their winning entries of the ImageNet and COCO 2015 competitions. In ResNet, identity mapping (the arrow on the right-hand side in Figure 10.1) is used to avoid the higher training error when the networks goes deeper. For more information on ResNet, you can check out the original paper, called *Deep Residual Learning for Image Recognition* (https://arxiv.org/pdf/1512.03385v1.pdf), and the blog *Understanding Deep Residual Networks*—a simple, modular learning framework that has redefined what constitutes state-of-the-art (https://blog.waya.ai/deep-residual-learning-9610bb62c355).

The last layers of the deep network are shown in Figure 10.2, and you can see that after a final residual block and convolutional layers with batch normalization and ReLU layers, dense fully-connected layers are applied to output the `value_head` and `policy_head` values:

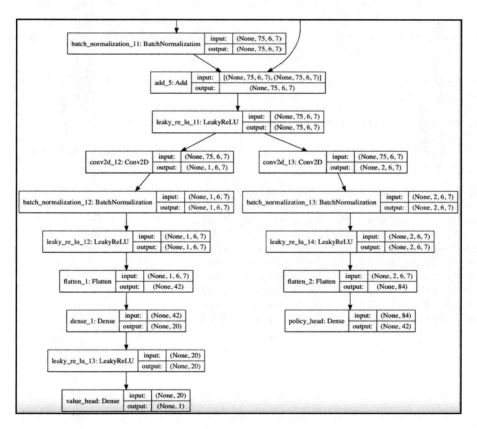

Figure 10.2 The last layers of the deep resnet

In the last part of this section, you'll see some Python code snippets that use the Keras API, which has nice support for ResNet, to build such a network. Let's now see how good those models are by letting them play against each other first, and play with us after that.

Testing the model

To let version 4, for example, of the model play against version 1, first create a new directory path by running `mkdir -p run_archive/connect4/run0001/models` and copy the `*.h5` files from `run/models` to the `run0001/models` directory. Then change your `play.py` in the `DeepReinforcementLearning` directory to be like this:

```
playMatchesBetweenVersions(env, 1, 1, 4, 10, lg.logger_tourney, 0)
```

The first value of the 1,1,4,10 parameters means the run version, so 1 means the models are in `run0001/models` of `run_archive/connect4`. The second and third values are the model versions of the two players, so 1 and 4 means version 1 of the model will play against version 4. 10 is the number of times, or episodes, to play.

After running the `python play.py` script to play games as specified, you can use the following command to find out the result:

```
grep WINS run/logs/logger_tourney.log |tail -10
```

For version 4 playing against version 1, you may see results similar to the following, meaning they're at about the same level:

```
2018-03-14 23:55:21,001 INFO player2 WINS!
2018-03-14 23:55:58,828 INFO player1 WINS!
2018-03-14 23:56:43,778 INFO player2 WINS!
2018-03-14 23:56:51,981 INFO player1 WINS!
2018-03-14 23:57:00,985 INFO player1 WINS!
2018-03-14 23:57:30,389 INFO player2 WINS!
2018-03-14 23:57:39,742 INFO player1 WINS!
2018-03-14 23:58:19,498 INFO player2 WINS!
2018-03-14 23:58:27,554 INFO player1 WINS!
2018-03-14 23:58:36,490 INFO player1 WINS!
```

There's one setting, `MCTS_SIMS = 50`, the number of simulations of MCTS, in `config.py` that has a strong impact on the play time. At every state, MCTS makes MCTS_SIMS times of simulations and, together with the trained network, comes up with the best move. So setting MCTS_SIMS to 50 makes the `play.py` script run longer, but doesn't necessarily make the player stronger, if the trained model is not good enough. You can change it to different values when playing with a specific version of the model to see how it impacts its strength level. To play against one specific version manually, change `play.py` to:

```
playMatchesBetweenVersions(env, 1, 4, -1, 10, lg.logger_tourney, 0)
```

Here, -1 means human player. So the preceding line will ask you (player 2) to play against player 1, version 4 of the model. After you run `python play.py` now, you'll see an input prompt, `Enter your chosen action:`; open another Terminal, go to the DeepReinforcementLearning directory, then type the `tail -f run/logs/logger_tourney.log` command, and you'll see the board grid printed out like this:

```
2018-03-15 00:03:43,907 INFO ====================
2018-03-15 00:03:43,907 INFO EPISODE 1 OF 10
2018-03-15 00:03:43,907 INFO ====================
2018-03-15 00:03:43,908 INFO player2 plays as X
2018-03-15 00:03:43,908 INFO --------------
2018-03-15 00:03:43,908 INFO ['-', '-', '-', '-', '-', '-', '-']
2018-03-15 00:03:43,908 INFO ['-', '-', '-', '-', '-', '-', '-']
2018-03-15 00:03:43,908 INFO ['-', '-', '-', '-', '-', '-', '-']
2018-03-15 00:03:43,909 INFO ['-', '-', '-', '-', '-', '-', '-']
2018-03-15 00:03:43,909 INFO ['-', '-', '-', '-', '-', '-', '-']
2018-03-15 00:03:43,909 INFO ['-', '-', '-', '-', '-', '-', '-']
```

Notice the last 6 lines represent the board grid of 6 rows by 7 columns: the first row corresponds to the 7 action numbers, 0, 1, 2, 3, 4, 5, 6, the second row to 7, 8, 9, 10, 11, 12, 13, and so on—so the last row maps to the 35, 36, 37, 38, 39, 40, 41 action numbers.

Now enter number 38 in the first Terminal of running `play.py`, and player 1 of version 4 of the model, playing as O, will make its move, showing new board grids, as follows:

```
2018-03-15 00:06:13,360 INFO action: 38
2018-03-15 00:06:13,364 INFO ['-', '-', '-', '-', '-', '-', '-']
2018-03-15 00:06:13,365 INFO ['-', '-', '-', '-', '-', '-', '-']
2018-03-15 00:06:13,365 INFO ['-', '-', '-', '-', '-', '-', '-']
2018-03-15 00:06:13,365 INFO ['-', '-', '-', '-', '-', '-', '-']
2018-03-15 00:06:13,365 INFO ['-', '-', '-', '-', '-', '-', '-']
2018-03-15 00:06:13,365 INFO ['-', '-', '-', 'X', '-', '-', '-']
2018-03-15 00:06:13,366 INFO --------------
2018-03-15 00:06:15,155 INFO action: 31
```

```
2018-03-15 00:06:15,155 INFO ['-', '-', '-', '-', '-', '-', '-']
2018-03-15 00:06:15,156 INFO ['-', '-', '-', '-', '-', '-', '-']
2018-03-15 00:06:15,156 INFO ['-', '-', '-', '-', '-', '-', '-']
2018-03-15 00:06:15,156 INFO ['-', '-', '-', '-', '-', '-', '-']
2018-03-15 00:06:15,156 INFO ['-', '-', '-', 'O', '-', '-', '-']
2018-03-15 00:06:15,156 INFO ['-', '-', '-', 'X', '-', '-', '-']
```

Keep entering your new action after player 1 makes a move till the end of the game, and a possible new game starts:

```
2018-03-15 00:16:03,205 INFO action: 23
2018-03-15 00:16:03,206 INFO ['-', '-', '-', '-', '-', '-', '-']
2018-03-15 00:16:03,206 INFO ['-', '-', '-', 'O', '-', '-', '-']
2018-03-15 00:16:03,206 INFO ['-', '-', '-', 'O', 'O', 'O', '-']
2018-03-15 00:16:03,207 INFO ['-', '-', 'O', 'X', 'X', 'X', '-']
2018-03-15 00:16:03,207 INFO ['-', '-', 'X', 'O', 'X', 'O', '-']
2018-03-15 00:16:03,207 INFO ['-', '-', 'O', 'X', 'X', 'X', '-']
2018-03-15 00:16:03,207 INFO ---------------
2018-03-15 00:16:14,175 INFO action: 16
2018-03-15 00:16:14,178 INFO ['-', '-', '-', '-', '-', '-', '-']
2018-03-15 00:16:14,179 INFO ['-', '-', '-', 'O', '-', '-', '-']
2018-03-15 00:16:14,179 INFO ['-', '-', 'X', 'O', 'O', 'O', '-']
2018-03-15 00:16:14,179 INFO ['-', '-', 'O', 'X', 'X', 'X', '-']
2018-03-15 00:16:14,179 INFO ['-', '-', 'X', 'O', 'X', 'O', '-']
2018-03-15 00:16:14,180 INFO ['-', '-', 'O', 'X', 'X', 'X', '-']
2018-03-15 00:16:14,180 INFO ---------------
2018-03-15 00:16:14,180 INFO player2 WINS!
2018-03-15 00:16:14,180 INFO ====================
2018-03-15 00:16:14,180 INFO EPISODE 2 OF 5
```

That's how you manually test the strength of a particular version of the model. Understanding the preceding board representation will also help you understand the iOS and Android code later. If you beat a model too easily, there are several things you can do to try to improve the model:

- Run the model in the run.ipynb (the second code block) Python notebook for a few days. In our tests, the version 19 of the model, after running for about a day on an older iMac, beats the version 1 or 4 10:0 (recall that version 1 and version 4 are at about the same level)

- To improve the strength of MCTS score formula: MCTS uses the Upper Confidence Tree (UCT) score during simulation to select which move to make, and the formula in the repo is like this (see the blog, `http://tim.hibal.org/blog/alpha-zero-how-and-why-it-works`, and the official AlphaZero paper for more detail):

    ```
    edge.stats['P'] * np.sqrt(Nb) / (1 + edge.stats['N'])
    ```

    If we change it to be more like what DeepMind used:

    ```
    edge.stats['P'] * np.sqrt(np.log(1+Nb) / (1 + edge.stats['N']))
    ```

    Then version 19 beats version 1 completely with 10:0, even with MCTS_SIMS set to only 10.

- Fine tune the deep neural network model to replicate AlphaZero as closely as possible

Going into the model detail is outside the scope of this book, but let's still take a look at how the model is built in Keras to appreciate it more when we later run it on iOS and Android (you can take a look at the rest of the main code in `agent.py`, `MCTS.py`, and `game.py` to have a better understanding of how the gameplay works).

Looking into the model-building code

In `model.py`, the Keras imports are as follows:

```
from keras.models import Sequential, load_model, Model
from keras.layers import Input, Dense, Conv2D, Flatten, BatchNormalization, Activation, LeakyReLU, add
from keras.optimizers import SGD
from keras import regularizers
```

The four key model-building methods are:

```
def residual_layer(self, input_block, filters, kernel_size)
def conv_layer(self, x, filters, kernel_size)
def value_head(self, x)
def policy_head(self, x)
```

They all have one or more `Conv2d` layers followed by `BatchNormalization` and `LeakyReLU` activation, as shown in Figure 10.1, but `value_head` and `policy_head` also have fully-connected layers, as shown in Figure 10.2, after the convolutional layers to generate the predicted value and policy probabilities for an input state we talked about before. In the `_build_model` method, the model input and output are defined:

```
main_input = Input(shape = self.input_dim, name = 'main_input')

vh = self.value_head(x)
ph = self.policy_head(x)

model = Model(inputs=[main_input], outputs=[vh, ph])
```

The deep neural network as wells as the model loss and optimizer are also defined in the `_build_model` method:

```
if len(self.hidden_layers) > 1:
    for h in self.hidden_layers[1:]:
        x = self.residual_layer(x, h['filters'], h['kernel_size'])

model.compile(loss={'value_head': 'mean_squared_error', 'policy_head': softmax_cross_entropy_with_logits}, optimizer=SGD(lr=self.learning_rate, momentum = config.MOMENTUM), loss_weights={'value_head': 0.5, 'policy_head': 0.5})
```

To find out the exact output node names (the input node name is specified as `'main_input'`), we can add `print(vh)` and `print(ph)` in model.py; now running python play.py will output the following two lines:

```
Tensor("value_head/Tanh:0", shape=(?, 1), dtype=float32)
Tensor("policy_head/MatMul:0", shape=(?, 42), dtype=float32)
```

We'll need them when freezing the TensorFlow checkpoint files and loading the model in mobile apps.

Freezing the model

First we need to create the TensorFlow checkpoint files – just uncomment the two lines for both player1 and player2 in funcs.py and run python play.py again:

```
if player1version > 0:
    player1_network = player1_NN.read(env.name, run_version, player1version)
    player1_NN.model.set_weights(player1_network.get_weights())
```

```
    # saver = tf.train.Saver()
    # saver.save(K.get_session(), '/tmp/alphazero19.ckpt')

if player2version > 0:
    player2_network = player2_NN.read(env.name, run_version,
player2version)
    player2_NN.model.set_weights(player2_network.get_weights())
    # saver = tf.train.Saver()
    # saver.save(K.get_session(), '/tmp/alphazero_4.ckpt')
```

This may look familiar to you as we did something similar in Chapter 8, *Predicting Stock Price with RNN*. Be sure to match the version number, such as 19 or 4, in alphazero19.ckpt and alphazero_4.ckpt with what's defined in play.py, such as playMatchesBetweenVersions(env, 1, 19, 4, 10, lg.logger_tourney, 0), and also with what's in run_archive/connect4/run0001/models directory—in this case, both version0019.h5 and version0004.h5 need to be there.

After running play.py, the alphazero19 checkpoint files will be generated in the /tmp directory:

```
-rw-r--r--  1 jeffmbair  wheel       99 Mar 13 18:17 checkpoint
-rw-r--r--  1 jeffmbair  wheel  1345545 Mar 13 18:17 alphazero19.ckpt.meta
-rw-r--r--  1 jeffmbair  wheel  7296096 Mar 13 18:17 alphazero19.ckpt.data-00000-of-00001
-rw-r--r--  1 jeffmbair  wheel     8362 Mar 13 18:17 alphazero19.ckpt.index
```

You can now go to the TensorFlow root source directory and run the freeze_graph script:

```
python tensorflow/python/tools/freeze_graph.py \
--input_meta_graph=/tmp/alphazero19.ckpt.meta \
--input_checkpoint=/tmp/alphazero19.ckpt \
--output_graph=/tmp/alphazero19.pb \
--output_node_names="value_head/Tanh,policy_head/MatMul" \
--input_binary=true
```

For simplicity, and because it's a small-sized model, we won't go through the graph transformation and memmapped conversion as we did in Chapter 6, *Describing Images in Natural Language* and Chapter 9, *Generating and Enhancing Images with GAN*. We're now ready to use the model on mobile and write code to play Connect 4 on iOS and Android devices.

Using the model in iOS to play Connect 4

For a newly frozen, and optionally transformed and memmapped, model, you can always try it with the TensorFlow pod to see whether you're lucky to be able to use it in the simple way. In our case, the `alphazero19.pb` model we generated would cause the following error when using the TensorFlow pod to load it:

```
Couldn't load model: Invalid argument: No OpKernel was registered to
support Op 'Switch' with these attrs. Registered devices: [CPU], Registered
kernels:
  device='GPU'; T in [DT_FLOAT]
  device='GPU'; T in [DT_INT32]
  device='GPU'; T in [DT_BOOL]
  device='GPU'; T in [DT_STRING]
  device='CPU'; T in [DT_INT32]
  device='CPU'; T in [DT_FLOAT]

     [[Node: batch_normalization_13/cond/Switch = Switch[T=DT_BOOL,
_output_shapes=[[], []]](batch_normalization_1/keras_learning_phase,
batch_normalization_1/keras_learning_phase)]]
```

You should know how to fix this type of error by now as it's been discussed in the previous chapters. To recap, simply make sure the kernel file for the Switch op is included in the `tensorflow/contrib/makefile/tf_op_files.txt` file. You can find out which kernel file is for Switch by running `grep 'REGISTER.*"Switch"' tensorflow/core/kernels/*.cc`, which should show `tensorflow/core/kernels/control_flow_ops.cc`. By default, since TensorFlow 1.4, the `control_flow_ops.cc` file is included in `tf_op_files.txt`, so all you need to do is build the TensorFlow iOS custom library by running `tensorflow/contrib/makefile/build_all_ios.sh`. If you have successfully run the iOS app in the last chapter, the library is already good and you don't need or want to run the time-consuming command again.

Now just create a new Xcode iOS project named AlphaZero, and drag and drop the `tensorflow_utils.mm` and `tensorflow_utils.h` files from the iOS project in the last chapter, as well as the `alphazero19.pb` model file generated in the last section, to the project. Rename `ViewController.m` to `ViewController.mm` and add some constants and variables. Your project should look like Figure 10.3:

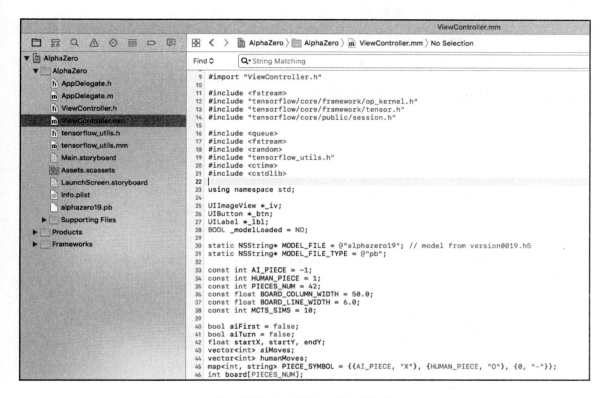

Figure 10.3 Showing the AlphaZero iOS app in Xcode

We only need to use three UI components:

- A `UIImageView` that shows the board and the played pieces.
- A `UILabel` that shows the game result and prompts the user for action.
- A `UIButton` to play or replay the game. As before, we create and position them programmatically in the `viewDidLoad` method.

When the **Play** or **Replay** button is tapped, randomly decide who goes first, reset the board represented as an integer array, clear the two vectors that store our moves and the AI's moves, and redraw the original board grid:

```
int n = rand() % 2;
aiFirst = (n==0);
if (aiFirst) aiTurn = true;
else aiTurn = false;

for (int i=0; i<PIECES_NUM; i++)
```

Building an AlphaZero-like Mobile Game App

```
        board[i] = 0;
    aiMoves.clear();
    humanMoves.clear();
    _iv.image = [self createBoardImageInRect:_iv.frame];
```

Then start the gameplay on a worker thread:

```
        dispatch_async(dispatch_get_global_queue(0, 0), ^{
            std::string result = playGame(withMCTS);
            dispatch_async(dispatch_get_main_queue(), ^{
                NSString *rslt = [NSString stringWithCString:result.c_str()
  encoding:[NSString defaultCStringEncoding]];
                [_lbl setText:rslt];
                _iv.image = [self createBoardImageInRect:_iv.frame];
            });
        });
```

In the `playGame` method, first check to see whether our model has been loaded and load it if not:

```
string playGame(bool withMCTS) {
    if (!_modelLoaded) {
        tensorflow::Status load_status;
        load_status = LoadModel(MODEL_FILE, MODEL_FILE_TYPE, &tf_session);
        if (!load_status.ok()) {
            LOG(FATAL) << "Couldn't load model: " << load_status;
            return "";
        }
        _modelLoaded = YES;
    }
```

If it's our turn, return and tell us so. Otherwise, convert the board state to a binary formatted input as expected by the model:

```
        if (!aiTurn) return "Tap the column for your move";
        int binary[PIECES_NUM*2];
        for (int i=0; i<PIECES_NUM; i++)
            if (board[i] == 1) binary[i] = 1;
            else binary[i] = 0;
        for (int i=0; i<PIECES_NUM; i++)
            if (board[i] == -1) binary[42+i] = 1;
            else binary[PIECES_NUM+i] = 0;
```

For example, if the board array is [0 1 1 -1 1 -1 0 0 1 -1 -1 -1 1 0 0 1 -1 1 -1 1 0 0 -1 -1 -1 1 -1 0 1 1 1 -1 -1 -1 -1 1 1 1 -1 1 1 -1], which represents the following board state ('X' for 1 and 'O' for -1, '-' for 0):

```
['-', 'X', 'X', 'O', 'X', 'O', '-']
['-', 'X', 'O', 'O', 'O', 'X', '-']
['-', 'X', 'O', 'X', 'O', 'X', '-']
['-', 'O', 'O', 'O', 'X', 'O', '-']
['X', 'X', 'X', 'O', 'O', 'O', 'O']
['X', 'X', 'X', 'O', 'X', 'X', 'O']
```

Then the binary array built with the preceding code snippet will be [0 1 1 0 1 0 0 0 1 0 0 0 1 0 0 1 0 1 0 0 0 0 1 0 0 1 1 1 0 0 0 0 1 1 1 0 1 1 0 0 0 0 1 0 1 0 0 0 1 1 1 0 0 0 0 1 0 1 0 0 0 1 1 1 0 1 0 0 0 0 1 1 1 1 0 0 0 1 0 0 1], which encodes both players' pieces on the board.

Still in the `playGame` method, call the `getProbs` method, which runs the frozen model with the `binary` input and returns the probability policy in `probs`, and finds the maximum probability value among the policy:

```
float *probs = new float[PIECES_NUM];
for (int i=0; i<PIECES_NUM; i++)
    probs[i] = -100.0;
if (getProbs(binary, probs)) {
    int action = -1;

    float max = 0.0;
    for (int i=0; i<PIECES_NUM; i++) {
        if (probs[i] > max) {
            max = probs[i];
            action = i;
        }
    }
}
```

The reason we initialize all `probs` array elements to -100.0 is that inside the `getProbs` method, which we'll show soon, the `probs` array will be changed, only for the allowed actions, to the values (all small ones around -1.0 to 1.0) returned in the policy, so the `probs` values for all the illegal actions will remain -100.0 and after the `softmax` function, which makes the probabilities for the illegal moves basically zero, we can just use the probabilities for the legal moves.

We only use the max probability value to guide the AI's move, without using MCTS, which would be necessary if we want AI to be really strong in a complicated game such as Chess or GO. If the policy returned from the trained model is perfect, as we mentioned before, we don't need to use MCTS. We'll leave the MCTS implementation in the book's source code repo for your reference instead of showing all the implementation detail of MCTS.

The rest of the code in the `playGame` method updates the board with the chosen action based on the maximum probability among all the legal moves returned by the model, calls the `printBoard` helper method to print the board on your Xcode output panel for better debugging, adds the action to the `aiMoves` vector so the board can be redrawn correctly, and returns the right status info if the game ends. By setting `aiTurn` to `false`, the touch event handler, which you'll see soon, will accept a human touch gesture for the move the human intends to make; if `aiTurn` is `true`, the touch handler will ignore all touch gestures:

```
        board[action] = AI_PIECE;
        printBoard(board);
        aiMoves.push_back(action);

        delete []probs;
        if (aiWon(board)) return "AI Won!";
        else if (aiLost(board)) return "You Won!";
        else if (aiDraw(board)) return "Draw";
    } else {
        delete []probs;
    }
    aiTurn = false;
    return "Tap the column for your move";
}
```

The `printBoard` helper method is as follows:

```
void printBoard(int bd[]) {
    for (int i = 0; i<6; i++) {
        for (int j=0; j<7; j++) {
            cout << PIECE_SYMBOL[bd[i*7+j]] << " ";
        }
        cout << endl;
    }
    cout << endl << endl;
}
```

So in the Xcode output panel, it'll print out something like this:

```
- - - - - - -
- - - - - - -
- - O - - - -
X - O - - - O
O O O X X - X
X X O O X - X
```

In the `getProbs` key method, first define the input and output node names, then prepare the input tensor using the values in `binary`:

```
bool getProbs(int *binary, float *probs) {
    std::string input_name = "main_input";
    std::string output_name1 = "value_head/Tanh";
    std::string output_name2 = "policy_head/MatMul";
    tensorflow::Tensor input_tensor(tensorflow::DT_FLOAT,
tensorflow::TensorShape({1,2,6,7}));
    auto input_mapped = input_tensor.tensor<float, 4>();
    for (int i = 0; i < 2; i++) {
        for (int j = 0; j<6; j++) {
            for (int k=0; k<7; k++) {
                input_mapped(0,i,j,k) = binary[i*42+j*7+k];
            }
        }
    }
```

Now run the model with the input and get the output:

```
    std::vector<tensorflow::Tensor> outputs;
    tensorflow::Status run_status = tf_session->Run({{input_name,
input_tensor}}, {output_name1, output_name2}, {}, &outputs);
    if (!run_status.ok()) {
        LOG(ERROR) << "Getting model failed:" << run_status;
        return false;
    }
    tensorflow::Tensor* value_tensor = &outputs[0];
    tensorflow::Tensor* policy_tensor = &outputs[1];
    const Eigen::TensorMap<Eigen::Tensor<float, 1, Eigen::RowMajor>,
Eigen::Aligned>& value = value_tensor->flat<float>();

    const Eigen::TensorMap<Eigen::Tensor<float, 1, Eigen::RowMajor>,
Eigen::Aligned>& policy = policy_tensor->flat<float>();
```

Set the probability values only for the allowed actions and then call `softmax` to make the sum of the `probs` values for the allowed actions be 1:

```
    vector<int> actions;
    getAllowedActions(board, actions);
    for (int action : actions) {
        probs[action] = policy(action);
    }
    softmax(probs, PIECES_NUM);
    return true;
}
```

The `getAllowedActions` function is defined as follows:

```cpp
void getAllowedActions(int bd[], vector<int> &actions) {
    for (int i=0; i<PIECES_NUM; i++) {
        if (i>=PIECES_NUM-7) {
            if (bd[i] == 0)
                actions.push_back(i);
        }
        else {
            if (bd[i] == 0 && bd[i+7] != 0)
                actions.push_back(i);
        }
    }
}
```

And the following is the `softmax` function, all pretty straightforward:

```cpp
void softmax(float vals[], int count) {
    float max = -FLT_MAX;
    for (int i=0; i<count; i++) {
        max = fmax(max, vals[i]);
    }
    float sum = 0.0;
    for (int i=0; i<count; i++) {
        vals[i] = exp(vals[i] - max);
        sum += vals[i];
    }
    for (int i=0; i<count; i++) {
        vals[i] /= sum;
    }
}
```

Several other helper functions are defined to test the game-end status:

```cpp
bool aiWon(int bd[]) {
    for (int i=0; i<69; i++) {
        int sum = 0;
        for (int j=0; j<4; j++)
            sum += bd[winners[i][j]];
        if (sum == 4*AI_PIECE ) return true;
    }
    return false;
}

bool aiLost(int bd[]) {
    for (int i=0; i<69; i++) {
        int sum = 0;
        for (int j=0; j<4; j++)
```

```
                sum += bd[winners[i][j]];
            if (sum == 4*HUMAN_PIECE ) return true;
        }
        return false;
}

bool aiDraw(int bd[]) {
    bool hasZero = false;
    for (int i=0; i<PIECES_NUM; i++) {
        if (bd[i] == 0) {
            hasZero = true;
            break;
        }
    }
    if (!hasZero) return true;
    return false;
}

bool gameEnded(int bd[]) {
    if (aiWon(bd) || aiLost(bd) || aiDraw(bd)) return true;
    return false;
}
```

Both the `aiWon` and `aiLost` functions use a constant array that defines all the 69 possible winning positions:

```
int winners[69][4] = {
    {0,1,2,3},
    {1,2,3,4},
    {2,3,4,5},
    {3,4,5,6},
    {7,8,9,10},
    {8,9,10,11},
    {9,10,11,12},
    {10,11,12,13},

    ......

    {3,11,19,27},
    {2,10,18,26},
    {10,18,26,34},
    {1,9,17,25},
    {9,17,25,33},
    {17,25,33,41},
    {0,8,16,24},
    {8,16,24,32},
    {16,24,32,40},
```

```
           {7,15,23,31},
           {15,23,31,39},
           {14,22,30,38}};
```

In the touch event handler, first make sure it's the human's turn. Then check to see whether the touch point values are within the board region, get the column tapped based on the touch position, and update the `board` array and `humanMoves` vector:

```
- (void) touchesEnded:(NSSet *)touches withEvent:(UIEvent *)event {
    if (aiTurn) return;
    UITouch *touch = [touches anyObject];
    CGPoint point = [touch locationInView:self.view];
    if (point.y < startY || point.y > endY) return;
    int column = (point.x-startX)/BOARD_COLUMN_WIDTH;
    for (int i=0; i<6; i++)
        if (board[35+column-7*i] == 0) {
            board[35+column-7*i] = HUMAN_PIECE;
            humanMoves.push_back(35+column-7*i);
            break;
        }
```

The rest of the touch handler redraws the ImageView by calling `createBoardImageInRect`, which uses `BezierPath` to draw or redraw the board and all the played pieces, check the game state and return the result if the game ends, or continue to play the game if not:

```
    _iv.image = [self createBoardImageInRect:_iv.frame];
    aiTurn = true;
    if (gameEnded(board)) {
        if (aiWon(board)) _lbl.text = @"AI Won!";
        else if (aiLost(board)) _lbl.text = @"You Won!";
        else if (aiDraw(board)) _lbl.text = @"Draw";
        return;
    }
    dispatch_async(dispatch_get_global_queue(0, 0), ^{
        std::string result = playGame(withMCTS));
        dispatch_async(dispatch_get_main_queue(), ^{
            NSString *rslt = [NSString stringWithCString:result.c_str()
encoding:[NSString defaultCStringEncoding]];
            [_lbl setText:rslt];
            _iv.image = [self createBoardImageInRect:_iv.frame];
        });
    });
}
```

The rest of the iOS code is all in the `createBoardImageInRect` method, which uses the `moveToPoint` and `addLineToPoint` methods from `UIBezierPath` to draw the board:

```
- (UIImage *)createBoardImageInRect:(CGRect)rect
{
    int margin_y = 170;

    UIGraphicsBeginImageContextWithOptions(CGSizeMake(rect.size.width, rect.size.height), NO, 0.0);
    UIBezierPath *path = [UIBezierPath bezierPath];

    startX = (rect.size.width - 7*BOARD_COLUMN_WIDTH)/2.0;
    startY = rect.origin.y+margin_y+30;
    endY = rect.origin.y - margin_y + rect.size.height;
    for (int i=0; i<8; i++) {
        CGPoint point = CGPointMake(startX + i * BOARD_COLUMN_WIDTH, startY);
        [path moveToPoint:point];
        point = CGPointMake(startX + i * BOARD_COLUMN_WIDTH, endY);
        [path addLineToPoint:point];
    }
    CGPoint point = CGPointMake(startX, endY);
    [path moveToPoint:point];
    point = CGPointMake(rect.size.width - startX, endY);
    [path addLineToPoint:point];

    path.lineWidth = BOARD_LINE_WIDTH;
    [[UIColor blueColor] setStroke];
    [path stroke];
```

The `bezierPathWithOvalInRect` method draws all the pieces moved by the AI and human – depending on who makes the first move, it starts drawing the pieces alternatively, but in a different order:

```
        int columnPieces[] = {0,0,0,0,0,0,0};
        if (aiFirst) {
            for (int i=0; i<aiMoves.size(); i++) {
                int action = aiMoves[i];
                int column = action % 7;
                CGRect r = CGRectMake(startX + column * BOARD_COLUMN_WIDTH, endY - BOARD_COLUMN_WIDTH - BOARD_COLUMN_WIDTH * columnPieces[column], BOARD_COLUMN_WIDTH, BOARD_COLUMN_WIDTH);
                UIBezierPath *path = [UIBezierPath bezierPathWithOvalInRect:r];
                UIColor *color = [UIColor redColor];
                [color setFill];
                [path fill];
                columnPieces[column]++;
```

```objc
                if (i<humanMoves.size()) {
                    int action = humanMoves[i];
                    int column = action % 7;
                    CGRect r = CGRectMake(startX + column * BOARD_COLUMN_WIDTH,
endY - BOARD_COLUMN_WIDTH - BOARD_COLUMN_WIDTH * columnPieces[column],
BOARD_COLUMN_WIDTH, BOARD_COLUMN_WIDTH);
                    UIBezierPath *path = [UIBezierPath
bezierPathWithOvalInRect:r];
                    UIColor *color = [UIColor yellowColor];
                    [color setFill];
                    [path fill];
                    columnPieces[column]++;
                }
            }
        }
        else {
            for (int i=0; i<humanMoves.size(); i++) {
                int action = humanMoves[i];
                int column = action % 7;
                CGRect r = CGRectMake(startX + column * BOARD_COLUMN_WIDTH,
endY - BOARD_COLUMN_WIDTH - BOARD_COLUMN_WIDTH * columnPieces[column],
BOARD_COLUMN_WIDTH, BOARD_COLUMN_WIDTH);
                UIBezierPath *path = [UIBezierPath bezierPathWithOvalInRect:r];
                UIColor *color = [UIColor yellowColor];
                [color setFill];
                [path fill];
                columnPieces[column]++;
                if (i<aiMoves.size()) {
                    int action = aiMoves[i];
                    int column = action % 7;
                    CGRect r = CGRectMake(startX + column * BOARD_COLUMN_WIDTH,
endY - BOARD_COLUMN_WIDTH - BOARD_COLUMN_WIDTH * columnPieces[column],
BOARD_COLUMN_WIDTH, BOARD_COLUMN_WIDTH);
                    UIBezierPath *path = [UIBezierPath
bezierPathWithOvalInRect:r];
                    UIColor *color = [UIColor redColor];
                    [color setFill];
                    [path fill];
                    columnPieces[column]++;
                }
            }
        }
    UIImage *image = UIGraphicsGetImageFromCurrentImageContext();
    UIGraphicsEndImageContext();
    return image;
}
```

Now run the app and you'll see screens similar to Figure 10.4:

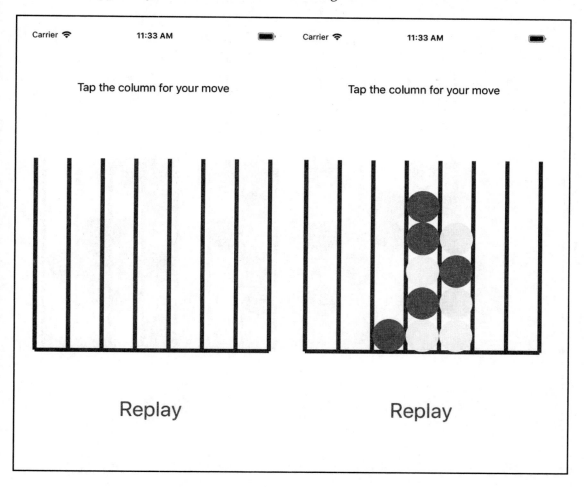

Figure 10.4 Playing with Connect 4 on iOS

Play a few games with the AI and Figure 10.5 shows some possible end games:

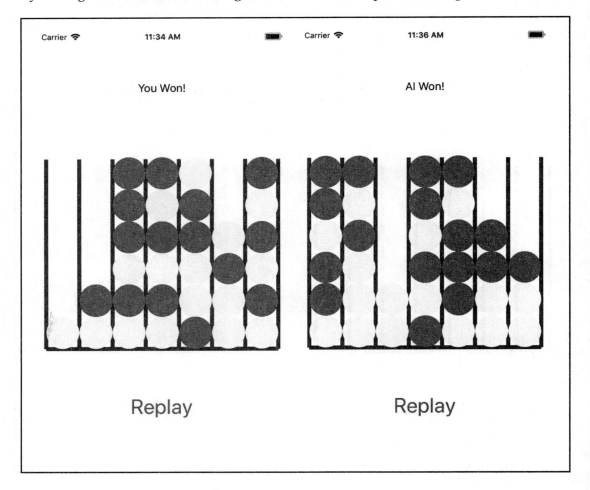

Figure 10.5 Some game results of Connect 4 on iOS

Before we pause, let's quickly take a look at the Android code that uses the model and plays the game.

Using the model in Android to play Connect 4

Not surprisingly, we don't need to use the custom Android library as we did in Chapter 7, *Recognizing Drawing with CNN and LSTM*, to load the model. Simply create a new Android Studio app with the name AlphaZero, copy the alphazero19.pb model file to the newly created assets folder, and add the compile 'org.tensorflow:tensorflow-android:+' line to the app's build.gradle file as we did before.

We'll first create a new class, BoardView, that extends View and is responsible for drawing the game board and the pieces AI and the user make:

```
public class BoardView extends View {
    private Path mPathBoard, mPathAIPieces, mPathHumanPieces;
    private Paint mPaint, mCanvasPaint;
    private Canvas mCanvas;
    private Bitmap mBitmap;
    private MainActivity mActivity;

    private static final float MARGINX = 20.0f;
    private static final float MARGINY = 210.0f;
    private float endY;
    private float columnWidth;

    public BoardView(Context context, AttributeSet attrs) {
        super(context, attrs);
        mActivity = (MainActivity) context;

        setPathPaint();
    }
```

We used three Path instances, mPathBoard, mPathAIPieces, and mPathHumanPieces, to draw the board, the moves the AI makes, and the moves the human makes, respectively, with different colors. The drawing feature of BoardView is implemented in the onDraw method using the moveTo and lineTo methods of Path and the drawPath method of Canvas:

```
protected void onDraw(Canvas canvas) {
    canvas.drawBitmap(mBitmap, 0, 0, mCanvasPaint);
    columnWidth = (canvas.getWidth() - 2*MARGINX) / 7.0f;

    for (int i=0; i<8; i++) {
        float x = MARGINX + i * columnWidth;
        mPathBoard.moveTo(x, MARGINY);
```

```
            mPathBoard.lineTo(x, canvas.getHeight()-MARGINY);
    }

    mPathBoard.moveTo(MARGINX, canvas.getHeight()-MARGINY);
    mPathBoard.lineTo(MARGINX + 7*columnWidth, canvas.getHeight()-
                                                            MARGINY);
    mPaint.setColor(0xFF0000FF);
    canvas.drawPath(mPathBoard, mPaint);
```

If the AI moves first, we start drawing the first AI move, then the first human move, if any, and alternate the drawing of the AI's moves and the human's moves:

```
    endY = canvas.getHeight()-MARGINY;
    int columnPieces[] = {0,0,0,0,0,0,0};

    for (int i=0; i<mActivity.getAIMoves().size(); i++) {
        int action = mActivity.getAIMoves().get(i);
        int column = action % 7;
        float x = MARGINX + column * columnWidth + columnWidth /
                                                            2.0f;
        float y = canvas.getHeight()-MARGINY-
                columnWidth*columnPieces[column]-columnWidth/2.0f;
        mPathAIPieces.addCircle(x,y, columnWidth/2,
                                            Path.Direction.CW);
        mPaint.setColor(0xFFFF0000);
        canvas.drawPath(mPathAIPieces, mPaint);
        columnPieces[column]++;

        if (i<mActivity.getHumanMoves().size()) {
            action = mActivity.getHumanMoves().get(i);
            column = action % 7;
            x = MARGINX + column * columnWidth + columnWidth /
                                                            2.0f;
            y = canvas.getHeight()-MARGINY-
                columnWidth*columnPieces[column]-columnWidth/2.0f;
            mPathHumanPieces.addCircle(x,y, columnWidth/2,
                        Path.Direction.CW);
            mPaint.setColor(0xFFFFFF00);
            canvas.drawPath(mPathHumanPieces, mPaint);
            columnPieces[column]++;
        }
    }
}
```

Chapter 10

If the human moves first, a similar drawing code is applied, as in the iOS code. Inside `public boolean onTouchEvent(MotionEvent event)` of `BoardView`, which returns if it's the AI's turn, we check which column has been tapped, and if the column has not been filled up with all six possible pieces, add the new human move to the `humanMoves` Vector of `MainActivity` and redraw the view:

```
public boolean onTouchEvent(MotionEvent event) {
    if (mActivity.getAITurn()) return true;

    float x = event.getX();
    float y = event.getY();

    switch (event.getAction()) {
        case MotionEvent.ACTION_DOWN:
            break;
        case MotionEvent.ACTION_MOVE:
            break;
        case MotionEvent.ACTION_UP:
            if (y < MARGINY || y > endY) return true;

            int column = (int)((x-MARGINX)/columnWidth);
            for (int i=0; i<6; i++)
                if (mActivity.getBoard()[35+column-7*i] == 0) {
                    mActivity.getBoard()[35+column-7*i] =
                                    MainActivity.HUMAN_PIECE;
                    mActivity.getHumanMoves().add(35+column-7*i);
                    break;
                }

            invalidate();
```

After that, set the turn to be the AI's and return if the game ends; otherwise, launch a new thread to continue to play the game by letting the AI make the next move based on the model's policy return before the human can touch and select their next move:

```
            mActivity.setAiTurn();
            if (mActivity.gameEnded(mActivity.getBoard())) {
                if (mActivity.aiWon(mActivity.getBoard()))
                    mActivity.getTextView().setText("AI Won!");
                else if (mActivity.aiLost(mActivity.getBoard()))
                    mActivity.getTextView().setText("You Won!");
                else if (mActivity.aiDraw(mActivity.getBoard()))
                    mActivity.getTextView().setText("Draw");
                return true;
            }
            Thread thread = new Thread(mActivity);
            thread.start();
```

[313]

```
            break;
        default:
            return false;
    }

    return true;
}
```

The main UI layout is defined in `activity_main.xml`, consisting of three UI elements, a `TextView`, a custom `BoardView`, and a `Button`:

```
<TextView
    android:id="@+id/textview"
    android:layout_width="wrap_content"
    android:layout_height="wrap_content"
    android:text=""
    android:textAlignment="center"
    android:textColor="@color/colorPrimary"
    android:textSize="24sp"
    android:textStyle="bold"
    app:layout_constraintBottom_toBottomOf="parent"
    app:layout_constraintHorizontal_bias="0.5"
    app:layout_constraintLeft_toLeftOf="parent"
    app:layout_constraintRight_toRightOf="parent"
    app:layout_constraintTop_toTopOf="parent"
    app:layout_constraintVertical_bias="0.06"/>

<com.ailabby.alphazero.BoardView
    android:id="@+id/boardview"
    android:layout_width="fill_parent"
    android:layout_height="fill_parent"
    app:layout_constraintBottom_toBottomOf="parent"
    app:layout_constraintLeft_toLeftOf="parent"
    app:layout_constraintRight_toRightOf="parent"
    app:layout_constraintTop_toTopOf="parent"/>

<Button
    android:id="@+id/button"
    android:layout_width="wrap_content"
    android:layout_height="wrap_content"
    android:text="Play"
    app:layout_constraintBottom_toBottomOf="parent"
    app:layout_constraintHorizontal_bias="0.5"
    app:layout_constraintLeft_toLeftOf="parent"
    app:layout_constraintRight_toRightOf="parent"
    app:layout_constraintTop_toTopOf="parent"
    app:layout_constraintVertical_bias="0.94" />
```

In `MainActivity.java`, first define some constants and fields:

```
public class MainActivity extends AppCompatActivity implements Runnable {

    private static final String MODEL_FILE =
    "file:///android_asset/alphazero19.pb";

    private static final String INPUT_NODE = "main_input";
    private static final String OUTPUT_NODE1 = "value_head/Tanh";
    private static final String OUTPUT_NODE2 = "policy_head/MatMul";

    private Button mButton;
    private BoardView mBoardView;
    private TextView mTextView;

    public static final int AI_PIECE = -1;
    public static final int HUMAN_PIECE = 1;
    private static final int PIECES_NUM = 42;

    private Boolean aiFirst = false;
    private Boolean aiTurn = false;

    private Vector<Integer> aiMoves = new Vector<>();
    private Vector<Integer> humanMoves = new Vector<>();

    private int board[] = new int[PIECES_NUM];
    private static final HashMap<Integer, String> PIECE_SYMBOL;
    static
    {
        PIECE_SYMBOL = new HashMap<Integer, String>();
        PIECE_SYMBOL.put(AI_PIECE, "X");
        PIECE_SYMBOL.put(HUMAN_PIECE, "O");
        PIECE_SYMBOL.put(0, "-");
    }

    private TensorFlowInferenceInterface mInferenceInterface;
```

Then define all the winning positions as we did in the iOS version of the app:

```
    private final int winners[][] = {
        {0,1,2,3},
        {1,2,3,4},
        {2,3,4,5},
        {3,4,5,6},

        {7,8,9,10},
        {8,9,10,11},
        {9,10,11,12},
```

```
        {10,11,12,13},
        ...
        {0,8,16,24},
        {8,16,24,32},
        {16,24,32,40},
        {7,15,23,31},
        {15,23,31,39},
        {14,22,30,38}};
```

A few getters and setters for the `BoardView` class to use:

```
    public boolean getAITurn() {
        return aiTurn;
    }
    public boolean getAIFirst() {
        return aiFirst;
    }
    public Vector<Integer> getAIMoves() {
        return aiMoves;
    }
    public Vector<Integer> getHumanMoves() {
        return humanMoves;
    }
    public int[] getBoard() {
        return board;
    }
    public void setAiTurn() {
        aiTurn = true;
    }
```

And some helpers, which are direct ports of the iOS code, to check the game states:

```
    public boolean aiWon(int bd[]) {
        for (int i=0; i<69; i++) {
            int sum = 0;
            for (int j=0; j<4; j++)
                sum += bd[winners[i][j]];
            if (sum == 4*AI_PIECE ) return true;
        }
        return false;
    }

    public boolean aiLost(int bd[]) {
        for (int i=0; i<69; i++) {
            int sum = 0;
            for (int j=0; j<4; j++)
                sum += bd[winners[i][j]];
            if (sum == 4*HUMAN_PIECE ) return true;
```

```
        }
        return false;
    }

    public boolean aiDraw(int bd[]) {
        boolean hasZero = false;
        for (int i=0; i<PIECES_NUM; i++) {
            if (bd[i] == 0) {
                hasZero = true;
                break;
            }
        }
        if (!hasZero) return true;
        return false;
    }


    public boolean gameEnded(int[] bd) {
        if (aiWon(bd) || aiLost(bd) || aiDraw(bd)) return true;

        return false;
    }
```

The `getAllowedActions` method, also a direct port of the iOS code, sets all the allowed actions for a given board position to the `actions` vector:

```
void getAllowedActions(int bd[], Vector<Integer> actions) {
    for (int i=0; i<PIECES_NUM; i++) {
        if (i>=PIECES_NUM-7) {
            if (bd[i] == 0)
                actions.add(i);
        }
        else {
            if (bd[i] == 0 && bd[i+7] != 0)
                actions.add(i);
        }
    }
}
```

In the `onCreate` method, instantiate the three UI elements, and set the button click listener so it randomly decides who makes the first move. The button is also tapped when the user wants to replay the game, so we need to reset the `aiMoves` and `humanMoves` vectors before drawing the board and starting a thread to play the game:

```
protected void onCreate(Bundle savedInstanceState) {
    super.onCreate(savedInstanceState);
    setContentView(R.layout.activity_main);
```

```java
        mButton = findViewById(R.id.button);
        mTextView = findViewById(R.id.textview);
        mBoardView = findViewById(R.id.boardview);

        mButton.setOnClickListener(new View.OnClickListener() {
            @Override
            public void onClick(View v) {
                mButton.setText("Replay");
                mTextView.setText("");

                Random rand = new Random();
                int n = rand.nextInt(2);
                aiFirst = (n==0);
                if (aiFirst) aiTurn = true;
                else aiTurn = false;

                if (aiTurn)
                    mTextView.setText("Waiting for AI's move");
                else
                    mTextView.setText("Tap the column for your move");

                for (int i=0; i<PIECES_NUM; i++)
                    board[i] = 0;
                aiMoves.clear();
                humanMoves.clear();
                mBoardView.drawBoard();

                Thread thread = new Thread(MainActivity.this);
                thread.start();
            }
        });
    }
```

The thread starts the `run` method, which further calls the `playGame` method to first convert the board position to a `binary` integer array to be used as the input of the model:

```java
public void run() {
    final String result = playGame();
    runOnUiThread(
            new Runnable() {
                @Override
                public void run() {
                    mBoardView.invalidate();
                    mTextView.setText(result);
                }
            });
}
```

```
String playGame() {
    if (!aiTurn) return "Tap the column for your move";
    int binary[] = new int[PIECES_NUM*2];
    for (int i=0; i<PIECES_NUM; i++)
        if (board[i] == 1) binary[i] = 1;
        else binary[i] = 0;

    for (int i=0; i<PIECES_NUM; i++)
        if (board[i] == -1) binary[42+i] = 1;
        else binary[PIECES_NUM+i] = 0;
```

The rest of the `playGame` method is also pretty much a straight port of the iOS code, which calls the `getProbs` method to get the max probability value among all allowed actions, using the probability values returned for all actions, 42 in total including both legal and illegal ones, in the model's policy output:

```
    float probs[] = new float[PIECES_NUM];
    for (int i=0; i<PIECES_NUM; i++)
        probs[i] = -100.0f;
    getProbs(binary, probs);
    int action = -1;
    float max = 0.0f;
    for (int i=0; i<PIECES_NUM; i++) {
        if (probs[i] > max) {
            max = probs[i];
            action = i;
        }
    }

    board[action] = AI_PIECE;
    printBoard(board);
    aiMoves.add(action);

    if (aiWon(board)) return "AI Won!";
    else if (aiLost(board)) return "You Won!";
    else if (aiDraw(board)) return "Draw";

    aiTurn = false;
    return "Tap the column for your move";

}
```

The `getProbs` method loads the model if it hasn't been loaded, runs the model with the current board state as input, and gets the output policy before calling `softmax` to get the true probability values, which sum to 1, for the allowed actions:

```
void getProbs(int binary[], float probs[]) {
    if (mInferenceInterface == null) {
        AssetManager assetManager = getAssets();
        mInferenceInterface = new
        TensorFlowInferenceInterface(assetManager, MODEL_FILE);
    }

    float[] floatValues = new float[2*6*7];
    for (int i=0; i<2*6*7; i++) {
        floatValues[i] = binary[i];
    }

    float[] value = new float[1];
    float[] policy = new float[42];

    mInferenceInterface.feed(INPUT_NODE, floatValues, 1, 2, 6, 7);
    mInferenceInterface.run(new String[] {OUTPUT_NODE1, OUTPUT_NODE2},
                                                            false);
    mInferenceInterface.fetch(OUTPUT_NODE1, value);
    mInferenceInterface.fetch(OUTPUT_NODE2, policy);

    Vector<Integer> actions = new Vector<>();
    getAllowedActions(board, actions);
    for (int action : actions) {
        probs[action] = policy[action];
    }

    softmax(probs, PIECES_NUM);
}
```

The `softmax` method is defined almost the same as in the iOS version:

```
void softmax(float vals[], int count) {
    float maxval = -Float.MAX_VALUE;
    for (int i=0; i<count; i++) {
        maxval = max(maxval, vals[i]);
    }
    float sum = 0.0f;
    for (int i=0; i<count; i++) {
        vals[i] = (float)exp(vals[i] - maxval);
        sum += vals[i];
    }
    for (int i=0; i<count; i++) {
```

Chapter 10

```
        vals[i] /= sum;
    }
}
```

Now run the app on an Android virtual or real device and play the game with the app, you'll see the initial screen and some game results:

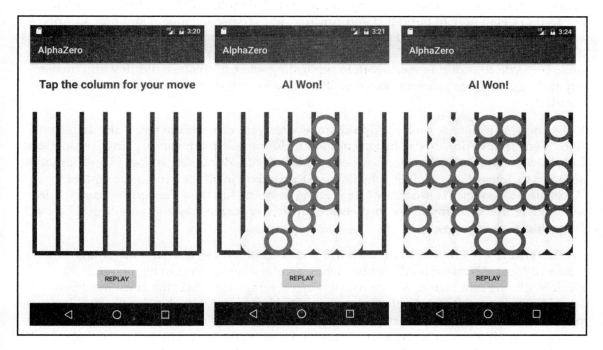

Figure 10.6 Showing the game board and some results on Android

As you play the game on iOS and Android with the preceding code, you'll soon find the policy returned by the model is not that strong—the main reason is that MCTS, not shown in detail here due to the scope limit, is not used along with the deep neural network model. You're strongly recommended to research and implement MCTS yourself, or use our implementation in the source code repo as a reference. You should also apply the network model and MCTS to other games of your interest—after all, AlphaZero used general-purpose MCTS and self-play reinforcement learning with no domain knowledge to make the superhuman learning easily ported to other problem domains. By combining MCTS with a deep neural network model, you can achieve what AlphaZero did.

Summary

In this chapter, we introduced the amazing world of AlphaZero, the latest and greatest achievement of DeepMind as of December 2017. We showed you how to train an AlphaZero-like model for Connect 4, using the powerful Keras API with the TensorFlow backend, and how to test and possibly improve such a model. Then we froze the model and covered in detail how to build iOS and Android apps to use the model and play Connect 4 with the model-powered AI. It's not the exact AlphaZero model that can beat human Chess or GO champions yet, but we hope this chapter provides you with a solid foundation and motivates you to continue your work in replicating what AlphaZero did first and further expanding it to other problem domains. It'll take a lot of hard work, but will be totally worth it.

If the latest AI progress, such as AlphaZero, excites you, chances are you'll also find the latest TensorFlow-powered solutions or toolkits for mobile platforms exciting. TensorFlow Lite, as we mentioned in Chapter 1, *Getting Started with Mobile TensorFlow*, is an alternative solution to TensorFlow Mobile, which we've focused on in all the previous chapters. According to Google, TensorFlow Lite is going to be the future of TensorFlow on mobile, although at this time and in the foreseeable future, TensorFlow Mobile should still be used for production cases.

While TensorFlow Lite works for both iOS and Android, it can also take advantage of the Android Neural Networks API for hardware acceleration when running on Android devices. On the other hand, iOS developers can leverage Core ML, the latest machine learning framework from Apple targeted on iOS 11 or above, which supports running many powerful pre-trained deep learning models, as well as models built with classical machine learning algorithms and Keras, in an optimized way on devices, with minimized app binary size. In the next chapter, we'll cover how to use TensorFlow Lite and Core ML in your iOS and Android apps.

11
Using TensorFlow Lite and Core ML on Mobile

In the previous nine chapters, we've used TensorFlow Mobile to run all kinds of powerful deep learning models, built with TensorFlow and Keras, on mobile. As we mentioned in Chapter 1, *Getting Started with Mobile TensorFlow*, Google also offers TensorFlow Lite, an alternative to TensorFlow Mobile, to run models on mobile. Although it's still in the developer preview as of Google I/O 2018, it's intended by Google to "greatly simplify the developer experience of targeting a model for small devices." So it's worth taking a look at TensorFlow Lite in detail and being ready for the future.

If you're an iOS developer or work with both iOS and Android, Apple's annual **Worldwide Developers Conference** (**WWDC**) is an event you don't want to miss. In WWDC 2017, Apple announced the new Core ML framework to support the running of both deep learning models and standard machine learning models on iOS (and all the other Apple OS platforms: macOS, tvOS, and watchOS). Core ML is available since iOS 11, which already takes more than 80% marker share as of May 2018. To get at least a basic understanding of what you can do with Core ML in your iOS apps is definitely meaningful.

So we'll cover both TensorFlow Lite and Core ML in this chapter, showing the strengths and limits of both, with the following topics:

- TensorFlow Lite - an overview
- Using TensorFlow Lite in iOS
- Using TensorFlow Lite in Android
- CoreML for iOS - an overview
- Using CoreML with Scikit-Learn Machine Learning
- Using CoreML with Keras and TensorFlow

TensorFlow Lite – an overview

TensorFlow Lite (`https://www.tensorflow.org/mobile/tflite`) is a lightweight solution that enables running deep learning models on mobile and embedded devices. If a model built in TensorFlow or Keras can be successfully converted to the TensorFlow Lite format, a new model format based on FlatBuffers (`https://google.github.io/flatbuffers`), which is similar but faster and a lot smaller in size than ProtoBuffers, which we talked about in `Chapter 3`, *Detecting Objects and Their Locations*, then you can expect the model to run with low latency and a smaller binary size. The basic workflow of using TensorFlow Lite in your mobile apps is as follows:

1. Build and train (or retrain) a TensorFlow model with TensorFlow or Keras with TensorFlow as the backend, such as the models we trained in the previous chapters.

    You can also pick a prebuilt TensorFlow Lite model, such as the MobileNet models available at `https://github.com/tensorflow/models/blob/master/research/slim/nets/mobilenet_v1.md`, which we used for retraining in `Chapter 2`, *Classifying Images with Transfer Learning*. Each of the MobileNet model `tgz` files that you can download there contains a converted TensorFlow Lite model. For example, the `MobileNet_v1_1.0_224.tgz` file contains a `mobilenet_v1_1.0_224.tflite` file that you can use directly on mobile. If you use such a prebuilt TensorFlow Lite model, you can skip steps 2 and 3.

2. Build the TensorFlow Lite Converter tool. If you download the TensorFlow 1.5 or 1.6 release from `https://github.com/tensorflow/tensorflow/releases`, you can run `bazel build tensorflow/contrib/lite/toco:toco` on a Terminal from your TensorFlow source root directory. If you use later releases or get the latest TensorFlow repo, you should be able to do so with this `build` command, but check the documentation of that new release if not.

3. Use the TensorFlow Lite converter tool to convert your TensorFlow model to the TensorFlow Lite model. You'll see a detailed example in the next section.

4. Deploy the TensorFlow Lite model on iOS or Android—for iOS, use the C++ API to load and run the model; for Android, use the Java API, a wrapper around the C++ API, to load and run the model. Both the C++ and Java APIs use the TensorFlow-lite-specific `Interpreter` class, unlike the `Session` class we used in TensorFlow Mobile projects before, to make inference on the model. We'll show you both the iOS C++ code and Android Java code to use the `Interpreter` in the next two sections.

If you run a TensorFlow Lite model on Android, and if the Android device is Android 8.1 (API level 27) or above and supports hardware acceleration with a dedicated neural network hardware, a GPU, or some other digital signal processors, then the `Interpreter` will use the Android Neural Networks API (https://developer.android.com/ndk/guides/neuralnetworks/index.html) to speed up the model running. For example, Google's Pixel 2 phone has a custom chip optimized for image processing, which can be turned on with Android 8.1, and support hardware acceleration.

Let's now see how to use TensorFlow Lite in iOS.

Using TensorFlow Lite in iOS

Before we show you how to create a new iOS app and add the TensorFlow Lite support to it, let's first take a look at a couple of sample TensorFlow iOS apps using TensorFlow Lite.

Running the example TensorFlow Lite iOS apps

There are two TensorFlow Lite example apps for iOS, named simple and camera, similar to the TensorFlow Mobile iOS apps simple and camera, but implemented in the TensorFlow Lite API, in the official releases of TensorFlow 1.5 - 1.8 at https://github.com/tensorflow/tensorflow/releases, and likely also in the latest TensorFlow repo. You can run the following commands to prepare and run the two apps, similarly documented under "iOS Demo App" at https://github.com/tensorflow/tensorflow/tree/master/tensorflow/contrib/lite:

```
cd tensorflow/contrib/lite/examples/ios
./download_models.sh
sudo gem install cocoapods
cd camera
```

```
pod install
open tflite_camera_example.xcworkspace
cd ../simple
pod install
open simple.xcworkspace
```

Now you'll have two Xcode iOS projects, simple and camera (named `tflite_simple_example` and `tflite_camera_example`, respectively, in Xcode), launched and you can install and run them on your iOS device (the simple app can also run on your iOS simulator).

`download_models.sh` will download a zip file that contains the `mobilenet_quant_v1_224.tflite` model file and `labels.txt` label file, then copy them to the simple/data and camera/data directories. Notice that somehow this script is not included in the official TensorFlow 1.5.0 and 1.6.0 releases. You'll need to do `git clone https://github.com/tensorflow/tensorflow` and clone the latest source (as of March 2018) to get it.

You can take a look at the source code in the Xcode `tflite_camera_example` project's `CameraExampleViewController.mm` file and the `tflite_simple_example` `RunModelViewController.mm` file to get an idea of how to use the TensorFlow Lite API to load and run a TensorFlow Lite model. Before we walk you through a step-by-step tutorial of how to create a new iOS app and add the TensorFlow Lite support to it to run a prebuilt TensorFlow Lite model, we'll quickly show you in concrete numbers one of the benefits of using TensorFlow Lite, as we mentioned earlier—the app binary size:

The `tensorflow_inception.graph.pb` model file used in the `tf_camera_example` TensorFlow Mobile example app, located in the `tensorflow/examples/ios/camera` folder, is 95.7 MB, while the `mobilenet_quant_v1_224.tflite` TensorFlow Lite model file used in the `tflite_camera_example` TensorFlow Lite example app, located in the `tensorflow/contrib/lite/examples/ios/camera` folder, is only 4.3 MB. The quantized version of the TensorFlow Mobile retrained Inception 3 model file, as we've seen in the HelloTensorFlow app in Chapter 2, *Classifying Images with Transfer Learning*, is about 22.4 MB, and the retrained MobileNet TensorFlow Mobile model file is 17.6 MB. In summary, the following lists the sizes of four different types of models:

- TensorFlow Mobile Inception 3 model: 95.7 MB
- Quantized and retrained TensorFlow Mobile Inception 3 model: 22.4 MB
- Retrained TensorFlow Mobile MobileNet 1.0 224 model: 17.6 MB
- TensorFlow Lite MobileNet 1.0 224 model: 4.3 MB

If you install and run the two apps on your iPhone, you'll see from your iPhone's settings that the app size for `tflite_camera_example` is about 18.7 MB, and the size for `tf_camera_example` is about 44.2 MB.

It's true that the accuracy with the Inception 3 model is a bit higher than with a MobileNet model, but in many use cases, the small accuracy difference can be ignored. Also, admittedly, mobile apps these days can easily take dozens of MB and the difference of 20 or 30 MB in app size may not sound like a big deal in some use cases, but size would be more sensitive in smaller embedded devices, and if we can achieve about the same accuracy with a faster speed and smaller size, without going through too much trouble, it'd always be a nice thing for users.

Using a prebuilt TensorFlow Lite model in iOS

Perform the following steps to create a new iOS app and add TensorFlow Lite support to it, using a prebuilt TensorFlow Lite model for image classification:

1. Create a new Xcode iOS project with Single View called HelloTFLite, set Objective-C as the language, then add the `ios_image_load.mm` and `ios_image_load.h` files from the `tensorflow/contrib/lite/examples/ios` folder to the project.

> If you prefer Swift as the programming language, you can refer to Chapter 2, *Classifying Images with Transfer Learning*, or Chapter 5, *Understanding Simple Speech Commands*, after following the steps here, to see how to convert the Objective-C app to a Swift app. But be aware that the TensorFlow Lite inference code still needs to be in C++ so you'll end up with a mix of Swift, Objective-C, and C++ code, with your Swift code mainly responsible for the UI and pre- and post-processing of the TensorFlow Lite inference.

2. Add the model file and label file generated with the preceding download_models.sh script from the `tensorflow/contrib/lite/examples/ios/simple/data` folder, as well as a test image, such as `lab1.jpg` from Chapter 2's source code folder, to the project.

3. Close the project and create a new file named `Podfile` with the following content:

   ```
   platform :ios, '8.0'

   target 'HelloTFLite'
           pod 'TensorFlowLite'
   ```

   Run `pod install`. Then open `HelloTFLite.xcworkspace` in Xcode, rename `ViewController.m` to `ViewController.mm`, and add the necessary C++ headers and TensorFlow Lite headers. Your Xcode project should look like the following screenshot:

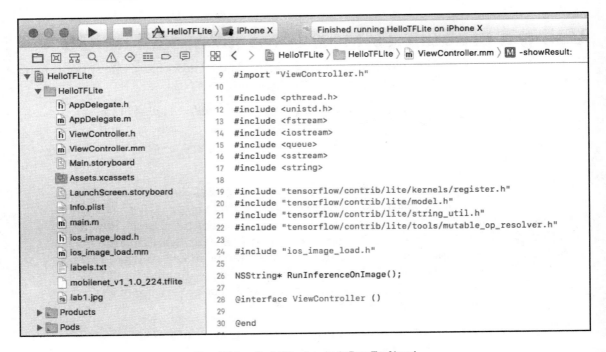

Figure 11.1 A new Xcode iOS project using the TensorFlow Lite pod

We're only showing you how to use the TensorFlow Lite pod in your iOS apps. There's another way to add TensorFlow Lite to iOS, similar to building the custom TensorFlow Mobile iOS library that we've done many times in the previous chapters. For more information on how to build your own custom TensorFlow Lite iOS library, see the documentation at https://github.com/tensorflow/tensorflow/blob/master/tensorflow/contrib/lite/g3doc/ios.md.

4. Copy the similar UI code from the `HelloTensorFlow` iOS app in Chapter 2, *Classifying Images with Transfer Learning,* to `ViewController.mm`, which uses `UITapGestureRecognizer` to capture the user's gesture on the screen, and then call the `RunInferenceOnImage` method, which loads the TensorFlow Lite model file:

```
NSString* RunInferenceOnImage() {
    NSString* graph = @"mobilenet_v1_1.0_224";
    std::string input_layer_type = "float";
    std::vector<int> sizes = {1, 224, 224, 3};
    const NSString* graph_path = FilePathForResourceName(graph, @"tflite");
    std::unique_ptr<tflite::FlatBufferModel>
model(tflite::FlatBufferModel::BuildFromFile([graph_path UTF8String]));
    if (!model) {
        NSLog(@"Failed to mmap model %@.", graph);
        exit(-1);
    }
```

5. Create an instance of the `Interpreter` class and set its input:

```
tflite::ops::builtin::BuiltinOpResolver resolver;
std::unique_ptr<tflite::Interpreter> interpreter;
tflite::InterpreterBuilder(*model, resolver)(&interpreter);
if (!interpreter) {
    NSLog(@"Failed to construct interpreter.");
    exit(-1);
}
interpreter->SetNumThreads(1);

int input = interpreter->inputs()[0];
interpreter->ResizeInputTensor(input, sizes);

if (interpreter->AllocateTensors() != kTfLiteOk) {
    NSLog(@"Failed to allocate tensors.");
    exit(-1);
}
```

Unlike with TensorFlow Mobile, TensorFlow Lite uses `interpreter->inputs()[0]` instead of specific input node names when feeding a TensorFlow Lite model for inference.

6. After loading the `labels.txt` file in the same way as we did in the `HelloTensorFlow` app, the image to be classified is loaded also in the same way, but prepared using the `typed_tensor` method of TensorFlow Lite's `Interpreter` instead of the TensorFlow Mobile's `Tensor` class and its `tensor` method. Figure 11.2 compares the TensorFlow Mobile and Lite code for loading and processing the image file data:

```
NSString* image_path = FilePathForResourceName(@"lab1", @"jpg");
int image_width;
int image_height;
int image_channels;
std::vector<tensorflow::uint8> image_data = LoadImageFromFile([image_path UTF8String],
    &image_width, &image_height, &image_channels);
const int wanted_channels = 3;
const float input_mean = 128.0f;
const float input_std = 128.0f;

assert(image_channels >= wanted_channels);
tensorflow::Tensor image_tensor(tensorflow::DT_FLOAT,tensorflow::TensorShape({1,
    wanted_height, wanted_width, wanted_channels}));
auto image_tensor_mapped = image_tensor.tensor<float, 4>();
tensorflow::uint8* in = image_data.data();
float* out = image_tensor_mapped.data();
for (int y = 0; y < wanted_height; ++y) {
    const int in_y = (y * image_height) / wanted_height;
    tensorflow::uint8* in_row = in + (in_y * image_width * image_channels);
    float* out_row = out + (y * wanted_width * wanted_channels);
    for (int x = 0; x < wanted_width; ++x) {
        const int in_x = (x * image_width) / wanted_width;
        tensorflow::uint8* in_pixel = in_row + (in_x * image_channels);
        float* out_pixel = out_row + (x * wanted_channels);
        for (int c = 0; c < wanted_channels; ++c) {
            out_pixel[c] = (in_pixel[c] - input_mean) / input_std;
        }
    }
}
```

```
NSString* image_path = FilePathForResourceName(@"lab1", @"jpg");
int image_width;
int image_height;
int image_channels;
std::vector<uint8_t> image_data = LoadImageFromFile([image_path UTF8String],
    &image_width, &image_height, &image_channels);
const int wanted_width = 224;
const int wanted_height = 224;
const int wanted_channels = 3;
const float input_mean = 127.5f;
const float input_std = 127.5f;
assert(image_channels >= wanted_channels);
uint8_t* in = image_data.data();
float* out = interpreter->typed_tensor<float>(input);
for (int y = 0; y < wanted_height; ++y) {
    const int in_y = (y * image_height) / wanted_height;
    uint8_t* in_row = in + (in_y * image_width * image_channels);
    float* out_row = out + (y * wanted_width * wanted_channels);
    for (int x = 0; x < wanted_width; ++x) {
        const int in_x = (x * image_width) / wanted_width;
        uint8_t* in_pixel = in_row + (in_x * image_channels);
        float* out_pixel = out_row + (x * wanted_channels);
        for (int c = 0; c < wanted_channels; ++c) {
            out_pixel[c] = (in_pixel[c] - input_mean) / input_std;
        }
    }
}
```

Figure 11.2 The TensorFlow Mobile (left) and Lite code of loading and processing the image input

7. Call the `Invoke` method on `Interpreter` to run the model and the `typed_out_tensor` method to get the model's output, before calling the `GetTopN` helper method to get the top N classification results. The code difference between TensorFlow Mobile and Lite for this is shown in Figure 11.3:

```
std::vector<tensorflow::Tensor> outputs;
tensorflow::Status run_status = session->Run({{input_layer, image_tensor}},
                                    {output_layer}, {}, &outputs);
if (!run_status.ok()) {
    result = @"Error running model";
    return result;
}

tensorflow::Tensor* output = &outputs[0];
const int kNumResults = 5;
const float kThreshold = 0.01f;
std::vector<std::pair<float, int> > top_results;
GetTopN(output->flat<float>(), kNumResults, kThreshold, &top_results);
```

```
if (interpreter->Invoke() != kTfLiteOk) {
    NSLog(@"Failed to invoke!");
    exit(-1);
}

float* output = interpreter->typed_output_tensor<float>(0);
const int output_size = 1000;
const int kNumResults = 5;
const float kThreshold = 0.1f;
std::vector<std::pair<float, int> > top_results;
GetTopN(output, output_size, kNumResults, kThreshold, &top_results);
```

Figure 11.3 The TensorFlow Mobile (left) and Lite code of running the model and getting the output

8. Implement the `GetTopN` method in a similar way to the method in HelloTensorFlow, with the `const float* prediction` type for TensorFlow Lite instead of `const Eigen::TensorMap<Eigen::Tensor<float, 1, Eigen::RowMajor>, Eigen::Aligned>& prediction` for TensorFlow Mobile. The comparison of the `GetTopN` method in TensorFlow Mobile and Lite is shown in Figure 11.4:

```
static void GetTopN(
                const Eigen::TensorMap<Eigen::Tensor<float, 1,
                Eigen::RowMajor>,
                Eigen::Aligned>& prediction,
                const int num_results, const float threshold,
                std::vector<std::pair<float, int> >* top_results) {
    // Will contain top N results in ascending order.
    std::priority_queue<std::pair<float, int>,
    std::vector<std::pair<float, int> >,
    std::greater<std::pair<float, int> > > top_result_pq;

    const long count = prediction.size();
    for (int i = 0; i < count; ++i) {
        const float value = prediction(i);

        // Only add it if it beats the threshold and has a chance at being in
        // the top N.
        if (value < threshold) {
            continue;
        }

        top_result_pq.push(std::pair<float, int>(value, i));
```

```
static void GetTopN(const float* prediction, const int prediction_size, const int
    num_results,
                const float threshold, std::vector<std::pair<float, int> >*
                top_results) {
    // Will contain top N results in ascending order.
    std::priority_queue<std::pair<float, int>, std::vector<std::pair<float, int> >,
    std::greater<std::pair<float, int> > >
    top_result_pq;

    const long count = prediction_size;
    for (int i = 0; i < count; ++i) {
        const float value = prediction[i];

        // Only add it if it beats the threshold and has a chance at being in
        // the top N.
        if (value < threshold) {
            continue;
        }

        top_result_pq.push(std::pair<float, int>(value, i));

        // If at capacity, kick the smallest value out.
```

Figure 11.4 The TensorFlow Mobile (left) and Lite code of processing the model output to return the top results

9. Use a simple `UIAlertController` to show the top results with the confidence values, returned by the TensorFlow Lite model, if the values are larger than the threshold (set as 0.1f):

```
-(void) showResult:(NSString *)result {
    UIAlertController* alert = [UIAlertController
alertControllerWithTitle:@"TFLite Model Result" message:result
preferredStyle:UIAlertControllerStyleAlert];
    UIAlertAction* action = [UIAlertAction actionWithTitle:@"OK"
style:UIAlertActionStyleDefault handler:nil];
    [alert addAction:action];
    [self presentViewController:alert animated:YES completion:nil];
}
-(void)tapped:(UITapGestureRecognizer *)tapGestureRecognizer {
    NSString *result = RunInferenceOnImage();
    [self showResult:result];
}
```

Run the iOS app now and tap the screen to run the model. For the `lab1.jpg` test image, you'll see the model's result in Figure 11.5:

Figure 11.5 The test image and model inference result

That's how you can use a prebuilt MobileNet TensorFlow Lite model in a new iOS app. Now let's see how we can use a retrained TensorFlow model.

Using a retrained TensorFlow model for TensorFlow Lite in iOS

In `Chapter 2`, *Classifying Images with Transfer Learning*, we retrained a MobileNet TensorFlow model for the task of dog breed recognition, and to use such model with TensorFlow Lite, we first need to convert it to the TensorFlow Lite format using the TensorFlow Lite converter tool:

```
bazel build tensorflow/contrib/lite/toco:toco

bazel-bin/tensorflow/contrib/lite/toco/toco \
  --input_file=/tmp/dog_retrained_mobilenet10_224_not_quantized.pb \
  --input_format=TENSORFLOW_GRAPHDEF --output_format=TFLITE \
  --output_file=/tmp/dog_retrained_mobilenet10_224_not_quantized.tflite --
```

```
    inference_type=FLOAT \
      --input_type=FLOAT --input_array=input \
      --output_array=final_result --input_shape=1,224,224,3
```

We have to use `--input_array` and `--output_array` to specify the input node name and output node name. For detailed command-line parameters of the converter tool, refer to https://github.com/tensorflow/tensorflow/blob/master/tensorflow/contrib/lite/toco/g3doc/cmdline_examples.md.

After adding the converted `dog_retrained_mobilenet10_224_not_quantized.tflite` TensorFlow Lite model file, as well as the same `dog_retrained_labels.txt` labels file from `HelloTensorFlow`, to the Xcode project, simply change the line in step 4 from `NSString* graph = @"mobilenet_v1_1.0_224";` to `NSString* graph = @"dog_retrained_mobilenet10_224_not_quantized";` and `const int output_size = 1000;` to `const int output_size = 121;` (recall that the MobileNet model classifies 1,000 objects and our retrained dog model classifies 121 dog breeds), and run the app again using the retrained model in the TensorFlow Lite format. The result will be about the same.

So it's pretty straightforward to use a retrained MobileNet TensorFlow model, after we successfully convert it to the TensorFlow Lite model. What about all those custom models covered in the book and elsewhere?

Using a custom TensorFlow Lite model in iOS

We have trained many custom TensorFlow models and frozen them for mobile use in the previous chapters. Unfortunately, if you try to use the `bazel-bin/tensorflow/contrib/lite/toco/toco` TensorFlow Lite converter tool built in the last section to convert the models from the TensorFlow format to the TensorFlow Lite format, they all will fail, except for the retrained models in Chapter 2, *Classifying Images with Transfer Learning*, which we covered in the previous subsection; most of the errors are of the "Converting unsupported operation" type. For example, the following command tries to convert the TensorFlow Object Detection model in Chapter 3, *Detecting Objects and Their Locations*, to the TensorFlow Lite format:

```
bazel-bin/tensorflow/contrib/lite/toco/toco \
  --input_file=/tmp/ssd_mobilenet_v1_frozen_inference_graph.pb \
  --input_format=TENSORFLOW_GRAPHDEF  --output_format=TFLITE \
  --output_file=/tmp/ssd_mobilenet_v1_frozen_inference_graph.tflite --
inference_type=FLOAT \
  --input_type=FLOAT --input_arrays=image_tensor \
```

Using TensorFlow Lite and Core ML on Mobile

```
    --
    output_arrays=detection_boxes,detection_scores,detection_classes,num_detect
ions \
    --input_shapes=1,224,224,3
```

But you'll get many errors in TensorFlow 1.6, including:

```
Converting unsupported operation: TensorArrayV3
Converting unsupported operation: Enter
Converting unsupported operation: Equal
Converting unsupported operation: NonMaxSuppressionV2
Converting unsupported operation: ZerosLike
```

The following command tries to convert the neural style transfer model in `Chapter 4`, *Transforming Pictures with Amazing Art Styles*, to the TensorFlow Lite format:

```
bazel-bin/tensorflow/contrib/lite/toco/toco \
   --input_file=/tmp/stylize_quantized.pb \
   --input_format=TENSORFLOW_GRAPHDEF --output_format=TFLITE \
   --output_file=/tmp/stylize_quantized.tflite --inference_type=FLOAT \
   --inference_type=QUANTIZED_UINT8 \
   --input_arrays=input,style_num \
   --output_array=transformer/expand/conv3/conv/Sigmoid \
   --input_shapes=1,224,224,3:26
```

The following command tries to convert the model in `Chapter 10`, *Building an AlphaZero-like Mobile Game App*:

```
bazel-bin/tensorflow/contrib/lite/toco/toco \
   --input_file=/tmp/alphazero19.pb \
   --input_format=TENSORFLOW_GRAPHDEF --output_format=TFLITE \
   --output_file=/tmp/alphazero19.tflite --inference_type=FLOAT \
   --input_type=FLOAT --input_arrays=main_input \
   --output_arrays=value_head/Tanh,policy_head/MatMul \
   --input_shapes=1,2,6,7
```

But you'll also get many "Converting unsupported operation" errors.

TensorFlow Lite is still in the developer preview, as of March 2018, and in TensorFlow 1.6, but more operations will be supported in future releases, so if you want to try TensorFlow Lite in TensorFlow 1.6, you should pretty much limit yourself to the pre-trained and retrained Inception and MobileNet models, while keeping an eye on future TensorFlow Lite releases. It's possible that more TensorFlow models covered in previous chapters in the book and elsewhere will be successfully converted to the TensorFlow Lite format in TensorFlow 1.7 or by the time you read this book.

But at least for now, for a custom and complicated model built with TensorFlow or Keras, most likely you won't be able to get a successful TensorFlow Lite conversion, so you should stay with TensorFlow Mobile as we've covered in the previous chapters for now, unless you're committed to making them work with TensorFlow Lite and don't mind helping to add more operations to be supported by TensorFlow Lite—after all, TensorFlow is an open source project.

Before we finish our coverage of TensorFlow Lite, we'll take a look at how to use TensorFlow Lite in Android.

Using TensorFlow Lite in Android

For simplicity, we'll just show how to add TensorFlow Lite with a prebuilt TensorFlow Lite MobileNet model in a new Android app, uncovering some helpful tips along the way. There's an example Android app using TensorFlow Lite that you may want to run with Android Studio first (https://www.tensorflow.org/mobile/tflite/demo_android), on an Android device with an API Level of at least 15 (version at least 4.0.3), before going through the following steps to use TensorFlow Lite in a new Android app. If you successfully build and run the demo app, you should be able to see the recognized objects by the device camera and the TensorFlow Lite MobileNet model, when you move around your Android device.

Now perform the following steps to create a new Android app and add the TensorFlow Lite support to classify an image, as we did in the HelloTensorFlow Android app in Chapter 2, *Classifying Images with Transfer Learning*:

1. Create a new Android Studio project and name the app `HelloTFLite`. Set the minimum SDK as API 15: Android 4.0.3, and accept all the other defaults.
2. Create a new `assets` folder, drag and drop the `mobilenet_quant_v1_224.tflite` TensorFlow Lite file and the `labels.txt` file from the demo app, located in the `tensorflow/contrib/lite/java/demo/app/src/main/assets` folder, as well as a test image, to the HelloTFLite app's `assets` folder.
3. Drag and drop the `ImageClassifier.java` file from the `tensorflow/contrib/lite/java/demo/app/src/main/java/com/example/android/tflitecamerademo` folder to the HelloTFLite app in Android Studio. `ImageClassifier.java` contains all the code using the TensorFlow Lite Java API to load and run the TensorFlow Lite model and we'll look at it in detail soon.

Using TensorFlow Lite and Core ML on Mobile

4. Open the app's `build.gradle` file, add `compile 'org.tensorflow:tensorflow-lite:0.1'` to the end of the `dependencies` section, and the following three lines after the `buildTypes` section inside `android`:

```
aaptOptions {
    noCompress "tflite"
}
```

This is required to avoid the following error you'd get when running the app:

```
10185-10185/com.ailabby.hellotflite W/System.err:
java.io.FileNotFoundException: This file can not be opened as a file
descriptor; it is probably compressed
03-20 00:32:28.805 10185-10185/com.ailabby.hellotflite W/System.err: at
android.content.res.AssetManager.openAssetFd(Native Method)
03-20 00:32:28.806 10185-10185/com.ailabby.hellotflite W/System.err: at
android.content.res.AssetManager.openFd(AssetManager.java:390)
03-20 00:32:28.806 10185-10185/com.ailabby.hellotflite W/System.err: at
com.ailabby.hellotflite.ImageClassifier.loadModelFile(ImageClassifier.java:
173)
```

Now your HelloTFLite app in Android Studio should look like Figure 11.6:

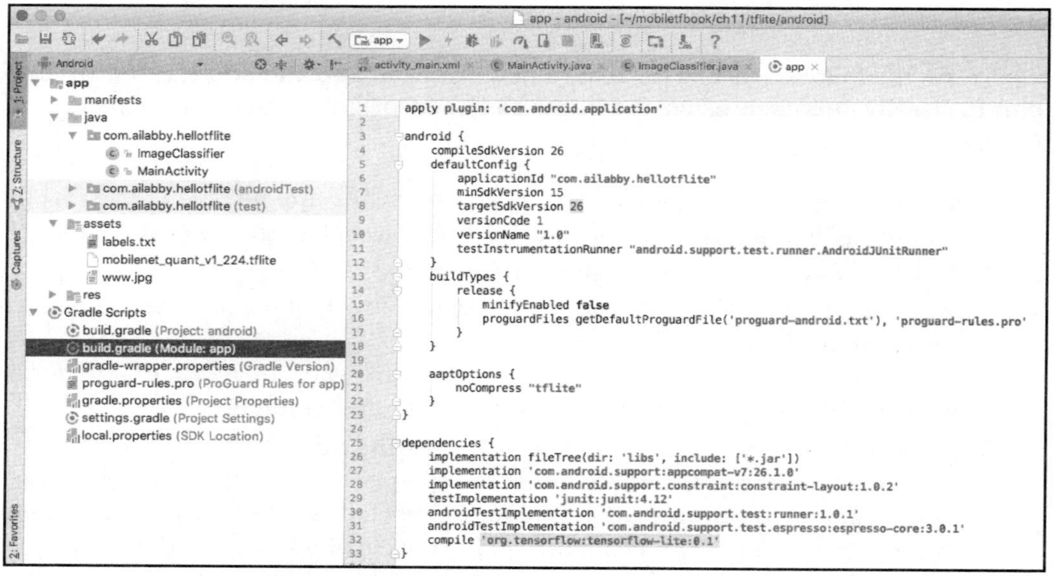

Figure 11.6 New Android app using TensorFlow Lite and the prebuilt MobileNet image classification model

[336]

5. Add `ImageView` and a `Button` in the `activity_main.xml`, as we did several times before, then in the `onCreate` method of the `MainActivity.java`, set the `ImageView` to the test image's content, and set the click listener of `Button` to launch a new thread, and instantiate an `ImageClassifier` instance called `classifier`:

```
private ImageClassifier classifier;

@Override
protected void onCreate(Bundle savedInstanceState) {
...
    try {
        classifier = new ImageClassifier(this);
    } catch (IOException e) {
        Log.e(TAG, "Failed to initialize an image classifier.");
    }
```

6. The thread's `run` method reads the test image data into `Bitmap`, calls the `classifyFrame` method of `ImageClassifier`, and shows the result as a `Toast`:

```
Bitmap bitmap =
BitmapFactory.decodeStream(getAssets().open(IMG_FILE));
Bitmap croppedBitmap = Bitmap.createScaledBitmap(bitmap,
INPUT_SIZE, INPUT_SIZE, true);
if (classifier == null ) {
    Log.e(TAG, "Uninitialized Classifier or invalid context.");
    return;
}
final String result = classifier.classifyFrame(croppedBitmap);
runOnUiThread(
        new Runnable() {
            @Override
            public void run() {
                mButton.setText("TF Lite Classify");
                Toast.makeText(getApplicationContext(), result,
Toast.LENGTH_LONG).show();
            }
        });
```

Using TensorFlow Lite and Core ML on Mobile

If you run the app now, you'll see the test image and a button with the title "TF Lite Classify". Tap on it and you'll see the classification result, such as "Labrador retriever: 0.86 pug: 0.05 dalmatian: 0.04".

The TensorFlow Lite-related code in `ImageClassifier` that uses the core `org.tensorflow.lite.Interpreter` class and its `run` method to run the model is as follows:

```
import org.tensorflow.lite.Interpreter;

public class ImageClassifier {

    private Interpreter tflite;
    private byte[][] labelProbArray = null;

    ImageClassifier(Activity activity) throws IOException {
        tflite = new Interpreter(loadModelFile(activity));
        ...
    }

    String classifyFrame(Bitmap bitmap) {
        if (tflite == null) {
            Log.e(TAG, "Image classifier has not been initialized;
                                                    Skipped.");
            return "Uninitialized Classifier.";
        }
        convertBitmapToByteBuffer(bitmap);
        tflite.run(imgData, labelProbArray);
        ...
    }
```

And the `loadModelFile` method is defined:

```
private MappedByteBuffer loadModelFile(Activity activity) throws IOException {
    AssetFileDescriptor fileDescriptor = activity.getAssets().openFd(MODEL_PATH);
    FileInputStream inputStream = new FileInputStream(fileDescriptor.getFileDescriptor());
    FileChannel fileChannel = inputStream.getChannel();
    long startOffset = fileDescriptor.getStartOffset();
    long declaredLength = fileDescriptor.getDeclaredLength();
    return fileChannel.map(FileChannel.MapMode.READ_ONLY, startOffset, declaredLength);
}
```

Recall that in step 4, we have to add `noCompress "tflite"` in the build.gradle file, otherwise the `openFd` method will cause an error. The method returns a memmapped version of the model, and we talked about using the `convert_graphdef_memmapped_format` tool to convert TensorFlow Mobile models to the memmapped format in `Chapter 6`, *Describing Images in Natural Language*, and `Chapter 9`, *Generating and Enhancing Images with GAN*.

That's all it takes to load and run a prebuilt TensorFlow Lite model in a new Android app. If you're interested in using the retrained and converted TensorFlow Lite model, as we did in the iOS app, in the Android app, or a custom TensorFlow Lite model if you successfully get a converted one, you can try them on top of the HelloTFLite app. We'll leave the cutting-edge TensorFlow Lite as is for now, and move on to another cool and WWDC-heavyweight topic for iOS developers.

Core ML for iOS – an overview

Apple's Core ML framework (https://developer.apple.com/documentation/coreml) lets iOS developers easily use trained machine learning models in their iOS apps, running iOS 11 or above, and built with Xcode 9 or above. You can download and use pre-trained models, already in the Core ML format, offered by Apple at https://developer.apple.com/machine-learning, or use a Python tool called coremltools, the Core ML community tools, at https://apple.github.io/coremltools, to convert other machine learning and deep learning models to the Core ML format.

Pre-trained models in the Core ML format include the popular MobileNet and Inception V3 models, as well as the more recent ResNet50 model (we talked about residual networks briefly in `Chapter 10`, *Building an AlphaZero-like Mobile Game App*). Models that can be converted to the Core ML format include deep learning models built with Caffe or Keras, as well as traditional machine learning models such as Linear Regression, Support Vector Machine, and Decision Trees built with Scikit Learn (http://scikit-learn.org), a very popular machine learning library for Python.

So if you want to use traditional machine learning models in iOS, then Scikit Learn and Core ML is definitely the way to go. Although this is a book on mobile TensorFlow, building intelligent apps sometimes doesn't require deep learning; classical machine learning makes total sense in some use cases. Besides, the support for Scikit Learn models in Core ML is so smooth that we can't refuse to take a quick look so you'd know when to put your mobile TensorFlow skills on a brief break if necessary.

If you want to use Apple's pre-trained MobileNet Core ML models, check out Apple's nice sample code project *Classifying Images with Vision and Core ML* at `https://developer.apple.com/documentation/vision/classifying_images_with_vision_and_core_ml` and also watch the WWDC 2017 videos on Core ML listed at `https://developer.apple.com/machine-learning`.

In the next two sections, we'll show you two tutorials on how to convert and use Scikit Learn models and the stock prediction RNN model in Keras, with TensorFlow as backend, which we built in `Chapter 8`, *Predicting Stock Price with RNN*. You'll see complete iOS apps built from scratch with source code, in both Objective-C and Swift, to use the converted Core ML models. If the phrase "from scratch" excites you and reminds you of AlphaZero, you probably enjoyed the previous chapter, `Chapter 10`, *Building an AlphaZero-like Mobile Game App*.

Using Core ML with Scikit-Learn machine learning

Linear Regression and Support Vector Machine are two of the most common classical machine learning algorithms, supported by Scikit Learn of course. We'll take a look at how to build models for house price prediction using the two algorithms.

Building and converting the Scikit Learn models

First, let's get a dataset of house prices, available for download at `https://wiki.csc.calpoly.edu/datasets/wiki/Houses`. The downloaded `RealEstate.csv` file looks like this:

```
MLS,Location,Price,Bedrooms,Bathrooms,Size,Price/SQ.Ft,Status
132842,Arroyo Grande,795000.00,3,3,2371,335.30,Short Sale
134364,Paso Robles,399000.00,4,3,2818,141.59,Short Sale
135141,Paso Robles,545000.00,4,3,3032,179.75,Short Sale
...
```

We'll use Pandas (https://pandas.pydata.org), a popular open source Python data analysis library, to parse the csv file. To install Scikit Learn and Pandas, simply run the following commands, preferably from the TensorFlow and Keras virtual environment you created before:

```
pip install scikit-learn
pip install pandas
```

Now enter the following code to read and parse the `RealEstate.csv` file, use all the rows under columns four to six (Bedrooms, Bathrooms, and Size) as the input data, and all the rows with column three (Price) as the target output:

```
from sklearn.linear_model import LinearRegression
from sklearn.svm import LinearSVR
import pandas as pd
import sklearn.model_selection as ms

data = pd.read_csv('RealEstate.csv')
X, y = data.iloc[:, 3:6], data.iloc[:, 2]
```

Split the dataset into a training set and test set, and train the dataset with Scikit Learn's Linear Regression model using the standard `fit` method:

```
X_train, X_test, y_train, y_test = ms.train_test_split(X, y,
test_size=0.25)

lr = LinearRegression()
lr.fit(X_train, y_train)
```

Test with three new inputs (3 bedrooms, 2 bathrooms, 1,560 square feet, and so on) with the trained model, using the `predict` method:

```
X_new = [[ 3, 2, 1560.0],
         [3, 2, 1680],
         [5, 3, 2120]]

print(lr.predict(X_new))
```

This will output three values as the predicted house price: `[319289.9552276 352603.45104977 343770.57498118]`.

To train a Support Vector Machine model and test it with the `X_new` input, similarly add the following code:

```
svm = LinearSVR(random_state=42)
svm.fit(X_train, y_train)

print(svm.predict(X_new))
```

This will output the predicted house prices using the Support Vector Machine model as [298014.41462535 320991.94354092 404822.78465954]. We won't discuss which model is better, how to make a Linear Regression or Support Vector Machine model work better, or how to choose a better model among all the algorithms supported by Scikit Learn—there are many good books and online resources covering these topics.

To convert the two Scikit Learn models, `lr` and `svm`, to the Core ML format that can be used in your iOS apps, you need to first install the Core ML tools (https://github.com/apple/coremltools). We recommend you install these using `pip install -U coremltools` in the TensorFlow and Keras virtual environment we created and used in Chapter 8, *Predicting Stock Price with RNN*, and Chapter 10, *Building an AlphaZero-powered Mobile Game App*, because we'll also use it to convert our Keras model in the next section.

Now simply run the following code to convert the two Scikit Learn models to the Core ML format:

```
import coremltools
coreml_model = coremltools.converters.sklearn.convert(lr, ["Bedrooms", "Bathrooms", "Size"], "Price")
coreml_model.save("HouseLR.mlmodel")

coreml_model = coremltools.converters.sklearn.convert(svm, ["Bedrooms", "Bathrooms", "Size"], "Price")
coreml_model.save("HouseSVM.mlmodel")
```

For more details on the converter tool, see its online documentation at https://apple.github.io/coremltools/coremltools.converters.html. We can now add the two models to an Objective-C or Swift iOS app, but we'll just show the Swift example here; you'll see both the Objective-C and Swift examples using the stock prediction Core ML model converted from the Keras and TensorFlow model in the next section.

Using the converted Core ML models in iOS

After adding the two Core ML model files, `HouseLR.mlmodel` and `HouseSVM.mlmodel`, to a new Swift-based Xcode iOS project, `HouseLR.mlmodel` looks like Figure 11.7:

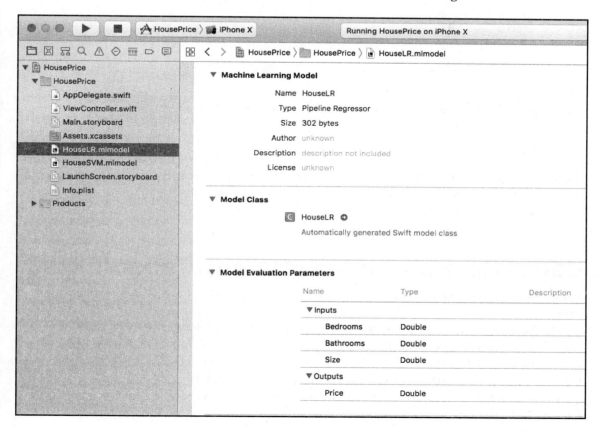

Figure 11.7 Showing the Swift iOS project and the Linear Regression Core ML model

The other `HouseSVM.mlmodel` model looks exactly the same, except the Machine Learning Model Name and the Model Class are changed from `HouseLR` to `HouseSVM`.

Add the following code to `class ViewController` in `ViewController.swift`:

```
private let lr = HouseLR()
private let svm = HouseSVM()

override func viewDidLoad() {
    super.viewDidLoad()
```

```
        let lr_input = HouseLRInput(Bedrooms: 3, Bathrooms: 2, Size: 1560)
        let svm_input = HouseSVMInput(Bedrooms: 3, Bathrooms: 2, Size: 1560)
        guard let lr_output = try? lr.prediction(input: lr_input) else {
            return
        }
        print(lr_output.Price)
        guard let svm_output = try? svm.prediction(input: svm_input) else {
            return
        }
        print(svm_output.Price)
}
```

It should all be pretty straightforward. Running the app will print:

```
319289.955227601
298014.414625352
```

They're the same as the first two numbers in the two arrays output by the Python script in the last subsection, as we use the first input in the X_new value of the Python code for the HouseLR and HouseSVM's prediction input.

Using Core ML with Keras and TensorFlow

The coremltools tool also officially supports converting models built with Keras (see the keras.convert link at https://apple.github.io/coremltools/coremltools.converters.html). The latest version of coremltools, 0.8, as of March 2018, works with TensorFlow 1.4 and Keras 2.1.5, which we used to build the Keras stock prediction model in Chapter 8, *Predicting Stock Price with RNN*. There are two ways you can use coremltools to generate the Core ML format of the model. The first is to call coremltools' convert and save methods directly in the Python Keras code after the model has been trained. For example, add the last three lines of code below to the ch8/python/keras/train.py file after model.fit:

```
model.fit(
    X_train,
    y_train,
    batch_size=512,
    epochs=epochs,
    validation_split=0.05)

import coremltools
coreml_model = coremltools.converters.keras.convert(model)
coreml_model.save("Stock.mlmodel")
```

For our model conversion, you can ignore the following warning when running the new script:

```
WARNING:root:Keras version 2.1.5 detected. Last version known to be
fully compatible of Keras is 2.1.3.
```

When you drag and drop the generated `Stock.mlmodel` file to your Xcode 9.2 iOS project, it'll use the default input name, `input1`, and default output name, `output1`, as shown in Figure 11.8, which is fine with both Objective-C- and Swift-based iOS apps:

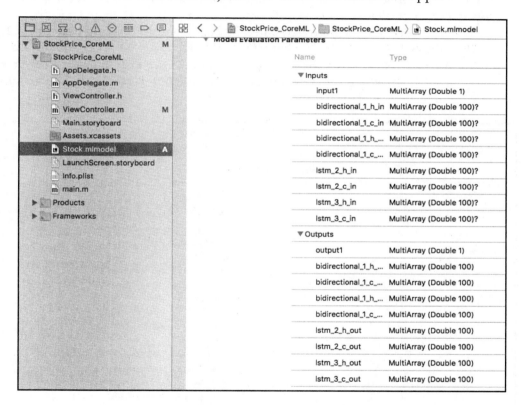

Figure 11.8 Showing the stock prediction Core ML model converted from Keras and TensorFlow in an Objective-C app

The other way to use coremltools to generate the Core ML format of the model is to first save the model built with Keras with the Keras HDF5 model format, the format we used in Chapter 10, *Building an AlphaZero-like Mobile Game App*, before converting to AlphaZero TensorFlow checkpoint files. To do that, simply run `model.save('stock.h5')`.

Using TensorFlow Lite and Core ML on Mobile

Then you can use the following code snippet to convert the Keras `.h5` model to a Core ML model:

```
import coremltools
coreml_model = coremltools.converters.keras.convert('stock.h5',
 input_names = ['bidirectional_1_input'],
 output_names = ['activation_1/Identity'])
coreml_model.save('Stock.mlmodel')
```

Note that we use the same input and output names here as we did when freezing the TensorFlow checkpoint files. If you drag and drop `Stock.mlmodel` to an Objective-C project, there'll be an error in the auto-generated Stock.h, because of a bug in Xcode 9.2 unable to correctly deal with the the "/" character in the `activation_1/Identity` output name. If it's a Swift iOS object, then the auto-generated `Stock.swift` file correctly changes the "/" character to "_", avoiding the compiler error, as shown in Figure 11.9 with a Swift app:

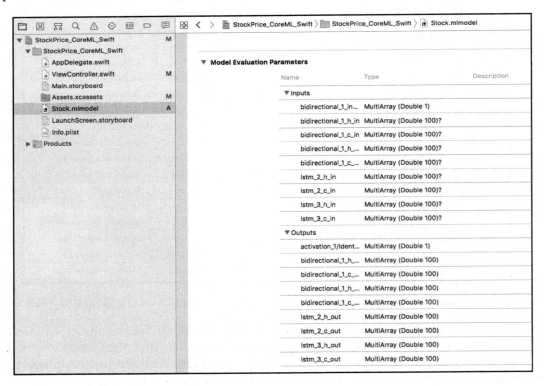

Figure 11.9 Showing the stock prediction Core ML model converted from Keras and TensorFlow in a Swift app

To use the model in Objective-C, create a `Stock` object and an `MLMultiArray` object with the data type and shape specified, then populate the array object with some input data, and call the `predictionFromFeatures` method using a `StockInput` instance initialized with the `MLMultiArray` data:

```objc
#import "ViewController.h"
#import "Stock.h"

@interface ViewController ()
@end

@implementation ViewController

- (void)viewDidLoad {
    [super viewDidLoad];

    Stock *stock = [[Stock alloc] init];
    double input[] = {
        0.40294855,
        0.39574954,
        0.39789235,
        0.39879138,
        0.40368535,
        0.41156033,
        0.41556879,
        0.41904324,
        0.42543786,
        0.42040193,
        0.42384258,
        0.42249741,
        0.4153998 ,
        0.41925279,
        0.41295281,
        0.40598363,
        0.40289448,
        0.44182321,
        0.45822208,
        0.44975226};

    NSError *error = nil;
    NSArray *shape = @[@20, @1, @1];
    MLMultiArray *mlMultiArray = [[MLMultiArray alloc]
 initWithShape:(NSArray*)shape dataType:MLMultiArrayDataTypeDouble error:&error] ;

    for (int i = 0; i < 20; i++) {
        [mlMultiArray setObject:[NSNumber numberWithDouble:input[i]]
```

Using TensorFlow Lite and Core ML on Mobile

```
atIndexedSubscript:(NSInteger)i];
    }
    StockOutput *output = [stock predictionFromFeatures:[[StockInput alloc]
initWithInput1:mlMultiArray] error:&error];
    NSLog(@"output = %@", output.output1 );
}
```

We used hardcoded normalized input and NSLog here just for a demonstration of how to use the Core ML model. If you run the app now, you'll see the output value of 0.4486984312534332, which after denormalization, shows the predicted next-day stock price.

The Swift version of the preceding code is as follows:

```
import UIKit
import CoreML

class ViewController: UIViewController {
    private let stock = Stock()
    override func viewDidLoad() {
        super.viewDidLoad()
        let input = [
            0.40294855,
            0.39574954,
            ...
            0.45822208,
            0.44975226]
        guard let mlMultiArray = try? MLMultiArray(shape:[20,1,1],
dataType:MLMultiArrayDataType.double) else {
            fatalError("Unexpected runtime error. MLMultiArray")
        }
        for (index, element) in input.enumerated() {
            mlMultiArray[index] = NSNumber(floatLiteral: element)
        }
        guard let output = try? stock.prediction(input:
StockInput(bidirectional_1_input:mlMultiArray)) else {
            return
        }
        print(output.activation_1_Identity)
    }
}
```

Notice we use `bidirectional_1_input` and `activation_1_Identity`, as we did with the TensorFlow Mobile iOS app, to set the input and get the output.

If you try to convert the AlphaZero model we built and trained in Keras in `Chapter 10`, *Building an AlphaZero-like Mobile Game App*, you'll get an error, `ValueError: Unknown loss function:softmax_cross_entropy_with_logits`. If you try to convert other TensorFlow models we've built in the book, the best unofficial tool you can use is the TensorFlow to Core ML Converter at `https://github.com/tf-coreml/tf-coreml`. Unfortunately, similar to TensorFlow Lite, it only supports a limited set of TensorFlow ops, some of the reasons being the limits of Core ML, others being the limits of the tf-coreml converter. We won't go into the details of converting TensorFlow models to Core ML models. But at least you've seen how to convert and use traditional machine learning models built with Scikit Learn, as well as a Keras-based RNN model, which hopefully provides you with a good foundation to build and use Core ML models. Of course, if you like Core ML, you should keep an eye on its future improved versions, as well as future releases of coremltools and the tf-coreml converter. There's still a lot we haven't covered on Core ML—to understand its exact features, see its complete API documentation at `https://developer.apple.com/documentation/coreml/core_ml_api`.

Summary

In this chapter, we covered two cutting-edge tools of using machine learning and deep learning models on mobile and embedded devices: TensorFlow Lite and Core ML. While TensorFlow Lite is still in developer preview, with limited support for TensorFlow operations, its future releases will support more and more TensorFlow features, while keeping the latency low and app size small. We offered step-by-step tutorials on how to develop TensorFlow Lite iOS and Android apps to classify an image from scratch. Core ML is Apple's framework for mobile developers to integrate machine learning in iOS apps, and it has great support for converting and using classical machine learning models built with Scikit Learn, as well as good support for Keras-based models. We also showed how to convert Scikit Learn and Keras models to Core ML models and use them in Objective-C and Swift apps. Both TensorFlow Lite and Core ML have some serious limits now, causing that they are unable to convert the complicated TensorFlow and Keras models we've built in the book. But they already have their use cases today, and their futures will simply be better. The best we can do is to be aware of their uses, their limits, and their potential, so we can choose the most appropriate tools, now or in the future, for different tasks. After all, we have more than just a hammer, so not everything has to look like a nail.

In the next and last chapter of the book, we'll pick some models we've built before and add the power of reinforcement learning – a key technology behind the success of AlphaGo and AlphaZero and one of the 10 breakthrough technologies in 2017 according to MIT Review – to the cool Raspberry Pi platform, a tiny, affordable yet powerful computer - who doesn't like the combinations of the three? We'll see how much intelligence—seeing, listening, walking, balancing, and of course, learning – we can add to a small Raspberry-Pi-powered robot in a single chapter. If a self-driving car is one of the hottest AI technologies these days, a self-walking robot could be one of our coolest toys at home.

12
Developing TensorFlow Apps on Raspberry Pi

According to Wikipedia, "The Raspberry Pi is a series of small single-board computers developed in the United Kingdom by the Raspberry Pi Foundation to promote the teaching of basic computer science in schools and in developing countries." The official site of Raspberry Pi (https://www.raspberrypi.org) describes it as "a small and affordable computer that you can use to learn programming." If you have never heard of or used Raspberry Pi before, just go its website and chances are you'll quickly fall in love with the cool little thing. Little yet powerful—in fact, developers of TensorFlow made TensorFlow available on Raspberry Pi from early versions around mid-2016, so we can run complicated TensorFlow models on the tiny computer that you can buy for about $35. This is probably beyond "the teaching of basic computer science" or "to learn programming," but on the other hand, if we think about all the rapid advances in mobile devices in the past few years, we shouldn't be surprised to see how greater and greater features have been implemented in smaller and smaller devices.

In this chapter, we'll enter the fun and exciting world of Raspberry Pi, the smallest device officially supported by TensorFlow. We'll first cover how to get and set up a new Raspberry Pi 3 B board, including all the necessary accessories used in this chapter to make it see, listen, and speak. Then we'll cover how to use the GoPiGo Robot Base Kit (https://www.dexterindustries.com/shop/gopigo3-robot-base-kit) to turn a Raspberry Pi board into a robot that can move. After that, we'll provide the simplest working steps to set up TensorFlow 1.6 on Raspberry Pi and build its example Raspberry Pi apps. We'll also discuss how to integrate image classification, the model we used in Chapter 2, *Classifying Images with Transfer Learning*, with text to speech to make the robot tell us what it recognizes, and how to integrate audio recognition, the model we used in Chapter 5, *Understanding Simple Speech Commands*, with the GoPiGo API to let you use voice commands to control the movement of your robot.

Finally, we'll show you how to use TensorFlow and OpenAI Gym, a Python toolkit for developing and comparing reinforcement learning algorithms, to implement a powerful reinforcement learning algorithm in a simulated environment to get our robot ready to move and balance in a real physical environment.

In Google I/O 2016, there's a session called *How to build a smart RasPi Bot with Cloud Vision and Speech API* (you can watch the video on YouTube). It uses Google's Cloud APIs to perform image classification and speech recognition and synthesis. In this chapter, we'll see how we can implement the tasks in the demo, as well as reinforcement learning, offline on device, showing the power of TensorFlow on Raspberry Pi.

In summary, we'll cover the following topics in this chapter to build a robot that moves, sees, listens, speaks, and learns:

- Setting up Raspberry Pi and making it move
- Setting up TensorFlow on Raspberry Pi
- Image recognition and text to speech
- Audio recognition and robot movement
- Reinforcement learning on Raspberry Pi

Setting up Raspberry Pi and making it move

The series of small, single-board Raspberry Pi computers include Raspberry Pi 3 B+, 3 B, 2B, 1 B+, 1 A+, Zero, and Zero W (for detailed info see `https://www.raspberrypi.org/products/#buy-now-modal`). We'll use the Pi 3 B motherboard here, which you can purchase for about $35 from the preceding link or on Amazon (`https://www.amazon.com/gp/product/B01CD5VC92`). Accessories that we have used and tested with the board with their prices are as follows:

- CanaKit 5V 2.5A Raspberry Pi Power Supply for about $10 (`https://www.amazon.com/gp/product/B00MARDJZ4`) to use during development.
- Kinobo - USB 2.0 Mini Microphone for about $4 (`https://www.amazon.com/gp/product/B00IR8R7WQ`) to record your voice commands.
- USHONK USB Mini Speaker for about $12 (`https://www.amazon.com/gp/product/B075M7FHM1`) to play a synthesized voice.
- Arducam 5 Megapixels 1080p Sensor OV5647 Mini Camera for about $14 (`https://www.amazon.com/gp/product/B012V1HEP4`) to support image classification.

- 16 GB MicroSD and Adapter for about $10 (https://www.amazon.com/gp/product/B00TDBLTWK) to store the installation files for Raspbian, the official operating system for Raspberry Pi, and act as the hard drive after installtion.
- A USB disk, such as SanDisk 32GB USB Drive for $9 (https://www.amazon.com/gp/product/B008AF380Q), to be used as swap partition (see the next section for details) so we can manually build the TensorFlow library, required to build and run the TensorFlow C++ code.
- The GoPiGo Robot Base Kit for $110 (https://www.amazon.com/gp/product/B00NYB3J0A or the official site at https://www.dexterindustries.com/shop) to turn the Raspberry Pi board into a robot that can move.

You'll also need an HDMI cable to connect your Raspberry Pi board to your computer monitor, a USB keyboard, and a USB mouse. That's a total of about $200, including the $110 GoPiGo, to build a Raspberry Pi robot that moves, sees, listens, and speaks. Although the GoPiGo kit seems a little expensive compared to the powerful Raspberry Pi computer, without it, a motionless Raspberry Pi would probably lose much of its appeal.

There's an older blog, *How to build a robot that "sees" with $100 and TensorFlow* (https://www.oreilly.com/learning/how-to-build-a-robot-that-sees-with-100-and-tensorflow), written by *Lukas Biewald* in September 2016, that covers how to use TensorFlow and Raspberry Pi 3 with some alternative parts to build a robot that sees and speaks. It's a fun read. What we cover here offers more detailed steps to set up Raspberry Pi 3 with GoPiGo, the user-friendly and Google-recommended toolkit to turn Pi into a robot, and the newer version of TensorFlow 1.6, in addition to adding voice command recognition and reinforcement learning.

Now let's first see how to set up Raspbian, the operating system for the Raspberry Pi board.

Setting up Raspberry Pi

The easiest way is to follow the Raspbian software installation guide at https://www.raspberrypi.org/learning/software-guide/quickstart, which is, in summary, a simple three-step process:

1. Download and install the SD Formatter (https://www.sdcard.org/downloads/formatter_4/index.html) for your Windows or Mac.
2. Use the SD Formatter to format a MicroSD card.

3. Download the offline ZIP version of New Out Of Box Software (NOOBS), the official easy installer for Raspbian, at `https://www.raspberrypi.org/downloads/noobs`, unzip it, and then drag and drop all the files in the extracted `NOOBS` folder to the formatted MicroSD card.

Now eject the MicroSD card and insert it into the Raspberry Pi board. Connect your HDMI cable from your monitor, as well as your USB keyboard and mouse to the board. Power the board with the power supply, then follow the steps on the screen to complete the installation of Raspbian, including setting up the Wifi network. The whole installation takes less than an hour to complete. After it's done, you can open a Terminal and enter `ifconfig` to find out the IP address of the board, then use `ssh pi@<board_ip_address>` from your compute to access it, which, as we'll see later, is really convenient and required to test control the Raspberry Pi robot on the move—you don't want to or can't take the keyboard, mouse, and the monitor with the board when it's moving around.

But SSH by default is not enabled, so you'll get an "SSH connection refused" error when you try to ssh to your Pi board for the first time. The quickest way to enable it is to run the following two commands:

```
sudo systemctl enable ssh
sudo systemctl start ssh
```

After that, you can ssh using the default password for the `pi` login, which is `raspberry`. Of course, you can change the default password to a new one using the `passwd` command.

Now that we have Raspbian installed, let's insert the USB mini microphone, USB mini speaker, and the mini camera to the Pi board. Both the USB microphone and speaker are plug and play. After they're inserted, you can use the `aplay -l` command to find out the supported audio playback devices:

```
aplay -l
**** List of PLAYBACK Hardware Devices ****
card 0: Device_1 [USB2.0 Device], device 0: USB Audio [USB Audio]
  Subdevices: 1/1
  Subdevice #0: subdevice #0
card 2: ALSA [bcm2835 ALSA], device 0: bcm2835 ALSA [bcm2835 ALSA]
  Subdevices: 8/8
  Subdevice #0: subdevice #0
  Subdevice #1: subdevice #1
  Subdevice #2: subdevice #2
  Subdevice #3: subdevice #3
  Subdevice #4: subdevice #4
  Subdevice #5: subdevice #5
  Subdevice #6: subdevice #6
```

```
    Subdevice #7: subdevice #7
card 2: ALSA [bcm2835 ALSA], device 1: bcm2835 ALSA [bcm2835 IEC958/HDMI]
    Subdevices: 1/1
    Subdevice #0: subdevice #0
```

The Pi board also has an audio jack, which you can use to get audio output during development. But the USB speaker is certainly more convenient.

To find out the supported recording device, use the `arecord -l` command:

```
arecord -l
**** List of CAPTURE Hardware Devices ****
card 1: Device [USB PnP Sound Device], device 0: USB Audio [USB Audio]
    Subdevices: 1/1
    Subdevice #0: subdevice #0
```

Now you can test the audio recording using the following command:

```
arecord -D plughw:1,0 -d 3 test.wav
```

`-D` specifies the audio input device, and here it means it's a plug and play device with card 1, device 0, as shown in the output of the `arecord -l` command. `-d` specifies the duration of the recording in seconds.

To play the recorded audio on the USB speaker, you first need to create a file named `.asoundrc` in your home directory with the following content:

```
pcm.!default {
        type plug
        slave {
                pcm "hw:0,0"
        }
}

ctl.!default {
        type hw
        card 0
}
```

Note that `"hw:0,0"` matches the card 0, device 0 information returned by `aplay -l` for the USB speaker device. Now you can test the recorded audio playback on the speaker with the `aplay test.wav` command.

Occasionally, after the Pi board reboots, the card number for the USB speaker gets changed by the system automatically and you won't hear the audio when running `aplay test.wav`. In that case, you can run `aplay -l` again to find the new card number set for the USB speaker and update the `~/.asoundrc` file accordingly.

If you want to adjust the speaker volume, use the `amixer set PCM -- 100%` command, where 100% sets the volume to the maximum.

To load the driver for the camera, run the `sudo modprobe bcm2835-v4l2` command. After that, to verify that the camera has been detected, run the `vcgencmd get_camera` command, which should return `supported=1 detected=1`. To load the camera driver every time the board boots, which is what we need, run `sudo vi /etc/modules` and add a line, `bcm2835-v4l2`, to the end of `/etc/modules` (or you can just run `sudo bash -c "echo 'bcm2835-v4l2' >> /etc/modules"` to do this). We'll test the camera in a later section when we run the TensorFlow image classification example.

That's all there is to get Raspberry Pi set up for our tasks. Now let's see how to make it move.

Making Raspberry Pi move

GoPiGo is a popular toolkit to turn your Raspberry Pi board into a moving robot. After you purchase and receive the GoPiGo Robot Base Kit as we mentioned earlier, follow the detailed steps at https://www.dexterindustries.com/GoPiGo/get-started-with-the-gopigo3-raspberry-pi-robot/1-assemble-gopigo3 to assemble it together with your Pi board. This should take you about one or two hours, depending on whether you're also watching March Madness or the NBA playoff games at the same time.

After you're done, your Raspberry Pi robot, along with all the accessories we listed earlier, should look as follows:

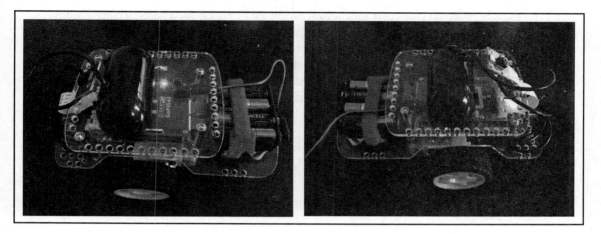

Figure 12.1 Raspberry Pi robot with GoPiGo Kit and camera, USB speaker, and USB microphone

Now use the Raspberry Pi power supply to turn on the Pi robot, and after it boots, connect to it using `ssh pi@<your_pi_board_ip>`. To install the GoPiGo Python library so we can use the GoPiGo's Python API (http://gopigo3.readthedocs.io/en/master/api-basic.html) to control the robot, run the following command, which will execute a shell script that creates a new `/home/pi/Dexter` directory and install all the library and firmware files there:

```
sudo sh -c "curl -kL dexterindustries.com/update_gopigo3 | bash"
```

You should also go to the `~/Dexter` directory and run the following command to update the GoPiGo board's firmware:

```
bash GoPiGo3/Firmware/gopigo3_flash_firmware.sh
```

Now run `sudo reboot` to reboot the board to make the changes take effect. After the Pi board reboots, you can test the GoPiGo and Raspberry Pi movement from iPython, which you can install with `sudo pip install ipython`.

To test the basic GoPiGo Python API, run iPython first, then enter the following code line by line:

 Be sure to put your GoPiGo Raspberry Pi robot on a safe surface as it'll start moving around. During the final test, you should use the GoPiGo battery pack to power the robot so it can move freely. But in the development and initial tests, you definitely should use the power adapter to save the battery unless you use a rechargeable battery. Always be careful if you put your robot on your desk as it may fall over if you issue a command causing it to make a bad move.

```
import easygopigo3 as easy
gpg3_obj = easy.EasyGoPiGo3()

gpg3_obj.drive_cm(5)
gpg3_obj.drive_cm(-5)
gpg3_obj.turn_degrees(30)
gpg3_obj.turn_degrees(-30)
gpg3_obj.stop()
```

`drive_cm` moves the robot forward or backward, depending on whether its parameter value is positive or negative. `turn_degrees` turns the robot clockwise or counterclockwise, depending on whether its parameter value is positive or negative. So the preceding example code moves the robot forward 5 cm, then backward 5 cm, turns clockwise 30 degrees, then counterclockwise 30 degrees. These calls are by default blocking calls, so they won't return until the robot finishes moving. To make a non-blocking call, add the `False` parameter, like this:

```
gpg3_obj.drive_cm(5, False)
gpg3_obj.turn_degrees(30, False)
```

You can also use `forward`, `backward`, and many other API calls, as documented at http://gopigo3.readthedocs.io/en/master/api-basic.html, to control the robot's movement, but in this chapter we'll only use `drive_cm` and `turn_degrees`.

We're now ready to use TensorFlow to add more intelligence to the robot.

Setting up TensorFlow on Raspberry Pi

To use TensorFlow in Python, as we'll do in the *Audio recognition* and *Reinforcement learning* sections later, we can install the TensorFlow 1.6 nightly build for Pi at the TensorFlow Jenkins continuous integrate site (http://ci.tensorflow.org/view/Nightly/job/nightly-pi/223/artifact/output-artifacts):

```
sudo pip install
http://ci.tensorflow.org/view/Nightly/job/nightly-pi/lastSuccessfulBuild/ar
tifact/output-artifacts/tensorflow-1.6.0-cp27-none-any.whl
```

This method is more common and described in detail in a nice blog entry, *Cross-compiling TensorFlow for the Raspberry Pi* (https://petewarden.com/2017/08/20/cross-compiling-tensorflow-for-the-raspberry-pi), by *Pete Warden*.

A more complicated method is to use the `makefile`, required when you need to build and use the TensorFlow library. The Raspberry Pi section of the official TensorFlow makefile documentation (https://github.com/tensorflow/tensorflow/tree/master/tensorflow/contrib/makefile) has detailed steps to build the TensorFlow library, but it may not work with every release of TensorFlow. The steps there work perfectly with an earlier version of TensorFlow (0.10), but would cause many "undefined reference to `google::protobuf`" errors with the TensorFlow 1.6.

The following steps have been tested with the TensorFlow 1.6 release, downloadable at https://github.com/tensorflow/tensorflow/releases/tag/v1.6.0; you can certainly try a newer version in the TensorFlow releases page, or clone the latest TensorFlow source by `git clone https://github.com/tensorflow/tensorflow`, and fix any possible hiccups.

After `cd` to your TensorFlow source root, and run the following commands:

```
tensorflow/contrib/makefile/download_dependencies.sh
sudo apt-get install -y autoconf automake libtool gcc-4.8 g++-4.8
cd tensorflow/contrib/makefile/downloads/protobuf/
./autogen.sh
./configure
make CXX=g++-4.8
sudo make install
sudo ldconfig  # refresh shared library cache
cd ../../../../..
export HOST_NSYNC_LIB=`tensorflow/contrib/makefile/compile_nsync.sh`
export TARGET_NSYNC_LIB="$HOST_NSYNC_LIB"
```

Make sure you run `make CXX=g++-4.8`, instead of just `make`, as documented in the official TensorFlow Makefile documentation, because Protobuf must be compiled with the same `gcc` version as that used for building the following TensorFlow library, in order to fix those "undefined reference to `google::protobuf`" errors. Now try to build the TensorFlow library using the following command:

```
make -f tensorflow/contrib/makefile/Makefile HOST_OS=PI TARGET=PI \
    OPTFLAGS="-Os -mfpu=neon-vfpv4 -funsafe-math-optimizations -ftree-vectorize" CXX=g++-4.8
```

After a few hours of building, you'll likely get an error such as "virtual memory exhausted: Cannot allocate memory" or the Pi board will just freeze due to running out of memory. To fix this, we need to set up a swap, because without the swap, when an application runs out of the memory, the application will get killed due to a kernel panic. There are two ways to set up a swap: swap file and swap partition. Raspbian uses a default swap file of 100 MB on the SD card, as shown here using the `free` command:

```
pi@raspberrypi:~/tensorflow-1.6.0 $ free -h
total used free shared buff/cache available
Mem:   927M 45M 843M 660K 38M 838M
Swap:  99M 74M 25M
```

To improve the swap file size to 1 GB, modify the `/etc/dphys-swapfile` file via `sudo vi /etc/dphys-swapfile`, changing `CONF_SWAPSIZE=100` to `CONF_SWAPSIZE=1024`, then restart the swap file service:

```
sudo /etc/init.d/dphys-swapfile stop
sudo /etc/init.d/dphys-swapfile start
```

After this, `free -h` will show the Swap total to be 1.0 GB.

A swap partition is created on a separate USB disk and is preferred because a swap partition can't get fragmented but a swap file on the SD card can get fragmented easily, causing slower access. To set up a swap partition, plug a USB stick with no data you need on it to the Pi board, then run `sudo blkid`, and you'll see something like this:

```
/dev/sda1: LABEL="EFI" UUID="67E3-17ED" TYPE="vfat" PARTLABEL="EFI System Partition" PARTUUID="622fddad-da3c-4a09-b6b3-11233a2ca1f6"
/dev/sda2: UUID="E67F-6EAB" TYPE="vfat" PARTLABEL="NO NAME" PARTUUID="a045107a-9e7f-47c7-9a4b-7400d8d40f8c"
```

`/dev/sda2` is the partition we'll use as the swap partition. Now unmount and format it to be a swap partition:

```
sudo umount /dev/sda2
sudo mkswap /dev/sda2
mkswap: /dev/sda2: warning: wiping old swap signature.
Setting up swapspace version 1, size = 29.5 GiB (31671701504 bytes)
no label, UUID=23443cde-9483-4ed7-b151-0e6899eba9de
```

You'll see a UUID output in the `mkswap` command; run `sudo vi /etc/fstab`, add a line as follows to the `fstab` file with the UUID value:

```
UUID=<UUID value> none swap sw,pri=5 0 0
```

Save and exit the fstab file and then run `sudo swapon -a`. Now if you run `free -h` again, you'll see the Swap total to be close to the USB storage size. We definitely don't need all that size for swap—in fact, the recommended maximum swap size for the Raspberry Pi 3 board with 1 GB memory is 2 GB, but we'll leave it as is because we just want to successfully build the TensorFlow library.

With either of the swap setting changes, we can rerun the `make` command:

```
make -f tensorflow/contrib/makefile/Makefile HOST_OS=PI TARGET=PI \
  OPTFLAGS="-Os -mfpu=neon-vfpv4 -funsafe-math-optimizations -ftree-
  vectorize" CXX=g++-4.8
```

After this completes, the TensorFlow library will be generated as `tensorflow/contrib/makefile/gen/lib/libtensorflow-core.a`, which should look familiar to you if you have gone through the previous chapters where we manually built the TensorFlow library. Now we can build the image classification example using the library.

Image recognition and text to speech

There are two TensorFlow Raspberry Pi example apps (https://github.com/tensorflow/tensorflow/tree/master/tensorflow/contrib/pi_examples) located in `tensorflow/contrib/pi_examples`: `label_image` and `camera`. We'll modify the camera example app to integrate text to speech so the app can speak out its recognized images when moving around. Before we build and test the two apps, we need to install some libraries and download the prebuilt TensorFlow Inception model file:

```
sudo apt-get install -y libjpeg-dev
sudo apt-get install libv4l-dev
```

```
curl
https://storage.googleapis.com/download.tensorflow.org/models/inception_dec
_2015_stripped.zip -o /tmp/inception_dec_2015_stripped.zip

cd ~/tensorflow-1.6.0
unzip /tmp/inception_dec_2015_stripped.zip -d
tensorflow/contrib/pi_examples/label_image/data/
```

To build the `label_image` and camera apps, run:

```
make -f tensorflow/contrib/pi_examples/label_image/Makefile
make -f tensorflow/contrib/pi_examples/camera/Makefile
```

You may encounter the following error when building the apps:

```
./tensorflow/core/platform/default/mutex.h:25:22: fatal error: nsync_cv.h:
No such file or directory
 #include "nsync_cv.h"
                    ^
compilation terminated.
```

To fix this, run `sudo cp tensorflow/contrib/makefile/downloads/nsync/public/nsync*.h /usr/include`.

Then edit the `tensorflow/contrib/pi_examples/label_image/Makefile` or `tensorflow/contrib/pi_examples/camera/Makefile` file, add the following library, and include paths before running the `make` command again:

```
-L$(DOWNLOADSDIR)/nsync/builds/default.linux.c++11 \

-lnsync \
```

To test run the two apps, run the apps directly:

```
tensorflow/contrib/pi_examples/label_image/gen/bin/label_image
tensorflow/contrib/pi_examples/camera/gen/bin/camera
```

Take a look at the C++ source code, `tensorflow/contrib/pi_examples/label_image/label_image.cc` and `tensorflow/contrib/pi_examples/camera/camera.cc`, and you'll see they use the similar C++ code as in our iOS apps in the previous chapters to load the model graph file, prepare input tensor, run the model, and get the output tensor.

By default, the camera example also uses the prebuilt Inception model unzipped in the `label_image/data` folder. But for your own specific image classification task, you can provide your own model retrained via transfer learning like the one we did in Chapter 2, *Classifying Images with Transfer Learning*, using the `--graph` parameter when running the two example apps.

In general, voice is a Raspberry Pi robot's main UI to interact with us. Ideally, we should run a TensorFlow-powered natural-sounding **Text-to-Speech** (**TTS**) model such as WaveNet (https://deepmind.com/blog/wavenet-generative-model-raw-audio) or Tacotron (https://github.com/keithito/tacotron), but it'd be beyond the scope of this chapter to run and deploy such a model. It turns out that we can use a much simpler TTS library called **Flite** by CMU (http://www.festvox.org/flite), which offers pretty decent TTS, and it takes just one simple command to install it: `sudo apt-get install flite`. If you want to install the latest version of Flite to hopefully get a better TTS quality, just download the latest Flite source from the link and build it.

To test Flite with our USB speaker, run flite with the `-t` parameter followed by a double quoted text string such as `flite -t "i recommend the ATM machine"`. If you don't like the default voice, you can find other supported voices by running `flite -lv`, which should return `Voices available: kal awb_time kal16 awb rms slt`. Then you can specify a voice used for TTS: `flite -voice rms -t "i recommend the ATM machine"`.

To let the camera app speak out the recognized objects, which should be the desired behavior when the Raspberry Pi robot moves around, you can use this simple `pipe` command:

```
tensorflow/contrib/pi_examples/camera/gen/bin/camera | xargs -n 1 flite -t
```

You'll likely hear too much voice. To fine tune the TTS result of image classification, you can also modify the camera.cc file and add the following code to the `PrintTopLabels` function before rebuilding the example using `make -f tensorflow/contrib/pi_examples/camera/Makefile`:

```
std::string cmd = "flite -voice rms -t \"";
cmd.append(labels[label_index]);
cmd.append("\"");
system(cmd.c_str());
```

Now that we have completed the image classification and speech synthesis tasks, without using any Cloud APIs, of that *How to build a smart RasPi Bot with Cloud Vision and Speech API demo*, let's see how we can do audio recognition on Raspberry Pi, using the same model we used in Chapter 5, *Understanding Simple Speech Commands*.

Audio recognition and robot movement

To use the pre-trained audio recognition model in the TensorFlow tutorial (https://www.tensorflow.org/tutorials/audio_recognition) or its retrained model we described before, we'll reuse a `listen.py` Python script from https://gist.github.com/aallan, and add the GoPiGo API calls to control the robot movement after it recognizes four basic audio commands: "left," "right," "go," and "stop." The other six commands supported by the pre-trained model—"yes," "no," "up," "down," "on," and "off"—don't apply well in our example, and if you want, you can use a retrained model as shown in Chapter 5, *Understanding Simple Speech Commands*, to support other voice commands for your specific task.

To run the script, first download the pre-trained audio recognition model from http://download.tensorflow.org/models/speech_commands_v0.01.zip and unzip it to `/tmp` for example, or `scp` the model we used in Chapter 5, *Understanding Simple Speech Commands*, to the Pi board's `/tmp` directory, then run:

```
python listen.py --graph /tmp/conv_actions_frozen.pb --labels /tmp/conv_actions_labels.txt -I plughw:1,0
```

Or you can run:

```
python listen.py --graph /tmp/speech_commands_graph.pb --labels /tmp/conv_actions_labels.txt -I plughw:1,0
```

Note that `plughw` value `1,0` should match the card number and device number of your USB microphone, which can be found using the `arecord -l` command we showed before.

The `listen.py` script also supports many other parameters. For example, we can use `--detection_threshold 0.5` instead of the default detection threshold 0.8.

Let's now take a quick look at how `listen.py` works before we add the GoPiGo API calls to make the robot move. `listen.py` uses Python's `subprocess` module and its `Popen` class to spawn a new process of running the `arecord` command with appropriate parameters. The `Popen` class has an `stdout` attribute that specifies the `arecord` executed command's standard output file handle, which can be used to read the recorded audio bytes.

The Python code to load the trained model graph is as follows:

```
with tf.gfile.FastGFile(filename, 'rb') as f:
    graph_def = tf.GraphDef()
    graph_def.ParseFromString(f.read())
    tf.import_graph_def(graph_def, name='')
```

A TensorFlow session is created using `tf.Session()` and after the graph is loaded and session created, the recorded audio buffer gets sent, along with the sample rate, as the input data to the TensorFlow session's `run` method, which returns the prediction of the recognition:

```
run(softmax_tensor, {
  self.input_samples_name_: input_data,
  self.input_rate_name_: self.sample_rate_
})
```

Here, `softmax_tensor` is defined as the TensorFlow graph's `get_tensor_by_name(self.output_name_)`, and `output_name_`, `input_samples_name_`, and `input_rate_name_` are defined as `labels_softmax, decoded_sample_data:0, decoded_sample_data:1`, respectively, the same names as what we used in the iOS and Android apps in Chapter 5, *Understanding Simple Speech Commands*.

In the previous chapters, we primarily used Python to train and test the TensorFlow models before running the models in iOS using C++ or Android using the Java interface code to the native TensorFlow C++ library. On Raspberry Pi, you can choose to run the TensorFlow models on Pi using the TensorFlow Python API directly, or C++ API (as in the label_image and camera examples), although normally you'd still train the models on a more powerful computer. For the complete TensorFlow Python API documentation, see https://www.tensorflow.org/api_docs/python.

To use the GoPiGo Python API to make the robot move based on your voice command, first add the following two lines to `listen.py`:

```
import easygopigo3 as gpg

gpg3_obj = gpg.EasyGoPiGo3()
```

Then add the following code to the end of the `def add_data` method:

```
if current_top_score > self.detection_threshold_ and time_since_last_top > self.suppression_ms_:
  self.previous_top_label_ = current_top_label
  self.previous_top_label_time_ = current_time_ms
  is_new_command = True
  logger.info(current_top_label)

  if current_top_label=="go":
    gpg3_obj.drive_cm(10, False)
  elif current_top_label=="left":
    gpg3_obj.turn_degrees(-30, False)
```

```
elif current_top_label=="right":
  gpg3_obj.turn_degrees(30, False)
elif current_top_label=="stop":
  gpg3_obj.stop()
```

Now put your Raspberry Pi robot on the ground, connect to it with `ssh` from your computer, and run the following script:

```
python listen.py --graph /tmp/conv_actions_frozen.pb --labels /tmp/conv_actions_labels.txt -I plughw:1,0 --detection_threshold 0.5
```

You'll see output like this:

```
INFO:audio:started recording
INFO:audio:_silence_
INFO:audio:_silence_
```

Then you can say left, right, stop, go, and stop to see the commands get recognized and the robot moves accordingly:

```
INFO:audio:left
INFO:audio:_silence_
INFO:audio:_silence_
INFO:audio:right
INFO:audio:_silence_
INFO:audio:stop
INFO:audio:_silence_
INFO:audio:go
INFO:audio:stop
```

You can run the camera app in a separate Terminal, so while the robot moves around based on your voice commands, it'll recognize new images it sees and speak out the results. That's all it takes to build a basic Raspberry Pi robot that listens, moves, sees, and speaks—what the Google I/O 2016 demo does but without using any Cloud APIs. It's far from a fancy robot that can understand natural human speech, engage in interesting conversations, or perform useful and non-trivial tasks. But powered with pre-trained, retrained, or other powerful TensorFlow models, and using all kinds of sensors, you can certainly add more and more intelligence and physical power to the Pi robot we have built.

Having seen how to run pre-trained and retrained TensorFlow models on Pi, in the next section, we'll show you how to add a powerful reinforcement learning model to the robot, built and trained with TensorFlow. After all, the trial-and-error way of reinforcement learning and its very nature of interacting with an environment for maximum rewards make reinforcement learning a very appropriate machine learning method for robots.

Reinforcement learning on Raspberry Pi

OpenAI Gym (https://gym.openai.com) is an open source Python toolkit that offers many simulated environments to help you develop, compare, and train reinforcement learning algorithms, so you don't have to buy all the sensors and train your robot in the real environment, which can be costly in both time and money. In this section, we'll show you how to develop, compare, and train different reinforcement learning models on Raspberry Pi using TensorFlow in an OpenAI Gym's simulated environment called CartPole (https://gym.openai.com/envs/CartPole-v0).

To install OpenAI Gym, run the following commands:

```
git clone https://github.com/openai/gym.git
cd gym
sudo pip install -e .
```

You can verify that you have TensorFlow 1.6 and gym installed by running `pip list` (the last part of the *Setting up TensorFlow on Raspberry Pi* section covered how to install TensorFlow 1.6):

```
pi@raspberrypi:~ $ pip list
gym (0.10.4, /home/pi/gym)
tensorflow (1.6.0)
```

Or you can start iPython then import TensorFlow and gym:

```
pi@raspberrypi:~ $ ipython
Python 2.7.9 (default, Sep 17 2016, 20:26:04)
IPython 5.5.0 -- An enhanced Interactive Python.

In [1]: import tensorflow as tf

In [2]: import gym

In [3]: tf.__version__
Out[3]: '1.6.0'

In [4]: gym.__version__
Out[4]: '0.10.4'
```

We're now all set to use TensorFlow and gym to build some interesting reinforcement learning model running on Raspberry Pi.

Understanding the CartPole simulated environment

CartPole is an environment that can be used to train a robot to stay in balance—if it carries something and wants to keep it put while moving around. Due to the scope of this chapter, we'll only build models that work in the simulated CartPole environment, but the model and how it's built and trained can certainly be applied to a real physical environment similar to CartPole.

In the CartPole environment, a pole is attached to a cart, which moves horizontally along a track. You can take an action of 1 (accelerating right) or 0 (accelerating left) to the cart. The pole starts upright, and the goal is to prevent it from falling over. A reward of 1 is provided for every time step that the pole remains upright. An episode ends when the pole is more than 15 degrees from vertical, or the cart moves more than 2.4 units from the center.

Let's play with the CartPole environment now. First, create a new environment and find out the possible actions an agent can take in the environment:

```
env = gym.make("CartPole-v0")
env.action_space
# Discrete(2)
env.action_space.sample()
# 0 or 1
```

Every observation (state) consists of four values about the cart: its horizontal position, its velocity, its pole's angle, and its angular velocity:

```
obs=env.reset()
obs
# array([ 0.04052535, 0.00829587, -0.03525301, -0.00400378])
```

Each step (action) in the environment will result in a new observation, a reward of the action, whether the episode is done (if it is then you can't take any further steps), and some additional information:

```
obs, reward, done, info = env.step(1)

obs
# array([ 0.04069127, 0.2039052 , -0.03533309, -0.30759772])
```

Remember action (or step) 1 means moving right, and 0 left. To see how long an episode can last when you keep moving the cart right, run:

```
while not done:
    obs, reward, done, info = env.step(1)
    print(obs)

#[ 0.08048328  0.98696604 -0.09655727 -1.54009127]
#[ 0.1002226   1.18310769 -0.12735909 -1.86127705]
#[ 0.12388476  1.37937549 -0.16458463 -2.19063676]
#[ 0.15147227  1.5756628  -0.20839737 -2.52925864]
#[ 0.18298552  1.77178219 -0.25898254 -2.87789912]
```

Let's now manually go through a series of actions from start to end and print out the observation's first value (the horizontal position) and third value (the pole's angle in degrees from vertical) as they're the two values that determine whether an episode is done.

First, reset the environment and accelerate the cart right a few times:

```
import numpy as np

obs=env.reset()
obs[0], obs[2]*360/np.pi
# (0.008710582898326602, 1.4858315848689436)

obs, reward, done, info = env.step(1)
obs[0], obs[2]*360/np.pi
# (0.009525842685697472, 1.5936049816642313)

obs, reward, done, info = env.step(1)
obs[0], obs[2]*360/np.pi
# (0.014239775393474322, 1.040038643681757)

obs, reward, done, info = env.step(1)
obs[0], obs[2]*360/np.pi
# (0.0228521194217381, -0.17418034908781568)
```

You can see that the cart's position value gets bigger and bigger as it's moved right, the pole's vertical degree gets smaller and smaller, and the last step shows a negative degree, meaning the pole is going to the left side of the center. All this makes sense, with just a little vivid picture in your mind of your favorite dog pushing a cart with a pole. Now change the action to accelerate the cart left (0) a few times:

```
obs, reward, done, info = env.step(0)
obs[0], obs[2]*360/np.pi
# (0.03536432554326476, -2.0525933052704954)
```

```
obs, reward, done, info = env.step(0)
obs[0], obs[2]*360/np.pi
# (0.04397450935915654, -3.261322987287562)

obs, reward, done, info = env.step(0)
obs[0], obs[2]*360/np.pi
# (0.04868738508385764, -3.812330822419413)

obs, reward, done, info = env.step(0)
obs[0], obs[2]*360/np.pi
# (0.04950617929263011, -3.7134404042580687)

obs, reward, done, info = env.step(0)
obs[0], obs[2]*360/np.pi
# (0.04643238384389254, -2.968245724428785)

obs, reward, done, info = env.step(0)
obs[0], obs[2]*360/np.pi
# (0.039465670006712444, -1.5760901885345346)
```

You may be surprised at first to see the 0 action causes the positions (obs[0]) to continue to get bigger for several times, but remember that the cart is moving at a velocity and one or several actions of moving the cart to the other direction won't decrease the position value immediately. But if you keep moving the cart to the left, you'll see, as shown in the last two preceding steps, that the cart's position starts becoming smaller (toward the left). Now continue the 0 action and you'll see the position gets smaller and smaller, with a negative value meaning the cart enters the left side of the center, while the pole's angle gets bigger and bigger:

```
obs, reward, done, info = env.step(0)
obs[0], obs[2]*360/np.pi
# (0.028603948219811447, 0.46789197320636305)

obs, reward, done, info = env.step(0)
obs[0], obs[2]*360/np.pi
# (0.013843572459953138, 3.1726728882727504)

obs, reward, done, info = env.step(0)
obs[0], obs[2]*360/np.pi
# (-0.00482029774222077, 6.551160678086707)

obs, reward, done, info = env.step(0)
obs[0], obs[2]*360/np.pi
# (-0.02739315127299434, 10.619948631208114)
```

As we stated earlier, the CartPole environment is defined in a way that an episode "ends when the pole is more than 15 degrees from vertical," so let's make a few more moves and print out the `done` value too this time:

```
obs, reward, done, info = env.step(0)
obs[0], obs[2]*360/np.pi, done
# (-0.053880356973985064, 15.39896478042983, False)

obs, reward, done, info = env.step(0)
obs[0], obs[2]*360/np.pi, done
# (-0.08428612474261402, 20.9109976051126, False)

obs, reward, done, info = env.step(0)
obs[0], obs[2]*360/np.pi, done
# (-0.11861214326416822, 27.181070460526062, True)

obs, reward, done, info = env.step(0)
# WARN: You are calling 'step()' even though this environment has already
returned done = True. You should always call 'reset()' once you receive
'done = True' -- any further steps are undefined behavior.
```

There's some delay on when the environment decides it should return `True` for `done` —although the first two steps already return a degree larger than 15 (and an episode ends when the pole is more than 15 degrees from vertical), you can still take a 0 action on the environment. The third step returns `done` as `True`, and a further step (the last one) on the environment would result in a warning as the environment has already completed the episode.

For the CartPole environment, the `reward` value returned in each `step` call is always 1, and the info is always {}. So that's all there's to know about the CartPole simulated environment. Now that we understand how CartPole works, let's see what kinds of policies we can come up with so at each state (observation), we can let the policy tell us which action (step) to take so we can keep the pole upright for as long as possible, in other words, so we can maximize our rewards. Remember that a policy in reinforcement learning is just a function that takes the state an agent is in as input and outputs the action the agent should take next, for the goal of maximizing values, or long-term rewards.

Starting with basic intuitive policy

Obviously, taking the same kind of action (all 0s or 1s) every time won't keep the pole upright for too long. For a baseline comparison, run the following code to see the average rewards over 1,000 episodes when the same action is applied during each episode:

```
# single_minded_policy.py

import gym
import numpy as np

env = gym.make("CartPole-v0")
total_rewards = []
for _ in range(1000):
  rewards = 0
  obs = env.reset()
  action = env.action_space.sample()
  while True:
    obs, reward, done, info = env.step(action)
    rewards += reward
    if done:
      break
  total_rewards.append(rewards)

print(np.mean(total_rewards))
# 9.36
```

So the average reward for all 1,000 episodes is about 10. Note that `env.action_space.sample()` samples a 0 or 1 action, which is the same as a random output of 0 or 1. You can verify this by evaluating `np.sum([env.action_space.sample() for _ in range(10000)])`, which should be close to 5,000.

To see how a different policy can work better, let's use a simple and intuitively-sound policy that takes the 1 action (move the cart right) when the pole degree is positive (on the right of vertical), and 0 (move the cart left) when the pole degree is negative (on the left of vertical). This policy makes sense as it's likely what we'd do to keep a cart pole balancing for as long as possible:

```
# simple_policy.py

import gym
import numpy as np

env = gym.make("CartPole-v0")
```

```
total_rewards = []
for _ in range(1000):
  rewards = 0
  obs = env.reset()
  while True:
    action = 1 if obs[2] > 0 else 0
    obs, reward, done, info = env.step(action)
    rewards += reward
    if done:
      break
  total_rewards.append(rewards)

print(np.mean(total_rewards))
# 42.19
```

The average reward for 1,000 episodes is now 42, a good improvement over 9.36.

Let's now see whether we can develop an even better, more sophisticated policy. Recall that a policy is just a map or function from a state to action. One thing we've learned in the neural network resurgence over the past few years is that if it's unclear how to define a complicated function, such as a policy in reinforcement learning, think of a neural network, which is after all a universal function approximator (see the section *A visual proof that neural nets can compute any function* by *Michael Nelson* at `http://neuralnetworksanddeeplearning.com/chap4.html` for a detailed explanation).

We covered AlphaGo and AlphaZero in the last chapter, and Jim Fleming wrote an interesting blog entry titled Before AlphaGo there was TD-Gammon (`https://medium.com/jim-fleming/before-alphago-there-was-td-gammon-13deff866197`), which was the first reinforcement learning application that trains itself using a neural network as an evaluation function to beat human Backgammon champions. Both the blog entry and the book, *Reinforcement Learning: An Introduction* by *Sutton* and *Barto*, have in-depth descriptions of TD-Gammon; you can also Google "Temporal Difference Learning and TD-Gammon" for the original paper if you want to know more about using neural networks as a powerful universal function.

Using neural networks to build a better policy

Let's first see how to build a random policy using a simple fully connected (dense) neural network, which takes 4 values in an observation as input, uses a hidden layer of 4 neurons, and outputs the probability of the 0 action, based on which, the agent can sample the next action between 0 and 1:

```
# nn_random_policy.py

import tensorflow as tf
import numpy as np
import gym

env = gym.make("CartPole-v0")

num_inputs = env.observation_space.shape[0]
inputs = tf.placeholder(tf.float32, shape=[None, num_inputs])
hidden = tf.layers.dense(inputs, 4, activation=tf.nn.relu)
outputs = tf.layers.dense(hidden, 1, activation=tf.nn.sigmoid)
action = tf.multinomial(tf.log(tf.concat([outputs, 1-outputs], 1)), 1)

with tf.Session() as sess:
  sess.run(tf.global_variables_initializer())

  total_rewards = []
  for _ in range(1000):
    rewards = 0
    obs = env.reset()
    while True:
      a = sess.run(action, feed_dict={inputs: obs.reshape(1, num_inputs)})
      obs, reward, done, info = env.step(a[0][0])
      rewards += reward
      if done:
        break
    total_rewards.append(rewards)

  print(np.mean(total_rewards))
```

Note that we use the `tf.multinomial` function to sample an action based on the probability distribution of action 0 and 1, defined as `outputs` and `1-outputs`, respectively (the sum of the two probabilities is 1). The mean of the total rewards will be around 20-something, better than the single-minded policy but worse than the simple intuitive policy in the previous subsection. This is a neural network this is generating a random policy, with no training at all.

To train the network, we use `tf.nn.sigmoid_cross_entropy_with_logits` to define the loss function between the network output and the desired `y_target` action, defined using the basic simple policy in the previous subsection, so we expect this neural network policy to achieve about the same rewards as the basic non-neural-network policy:

```python
# nn_simple_policy.py

import tensorflow as tf
import numpy as np
import gym

env = gym.make("CartPole-v0")

num_inputs = env.observation_space.shape[0]
inputs = tf.placeholder(tf.float32, shape=[None, num_inputs])
y = tf.placeholder(tf.float32, shape=[None, 1])
hidden = tf.layers.dense(inputs, 4, activation=tf.nn.relu)
logits = tf.layers.dense(hidden, 1)
outputs = tf.nn.sigmoid(logits)
action = tf.multinomial(tf.log(tf.concat([outputs, 1-outputs], 1)), 1)

cross_entropy = tf.nn.sigmoid_cross_entropy_with_logits(labels=y, logits=logits)
optimizer = tf.train.AdamOptimizer(0.01)
training_op = optimizer.minimize(cross_entropy)

with tf.Session() as sess:
  sess.run(tf.global_variables_initializer())

  for _ in range(1000):
    obs = env.reset()

    while True:
      y_target = np.array([[1. if obs[2] < 0 else 0.]])
      a, _ = sess.run([action, training_op], feed_dict={inputs: obs.reshape(1, num_inputs), y: y_target})
      obs, reward, done, info = env.step(a[0][0])
      if done:
        break
  print("training done")
```

We define `outputs` as a `sigmoid` function of the `logits` net output, that is, the probability of action 0, and then use the `tf.multinomial` to sample an action. Note that we use the standard `tf.train.AdamOptimizer` and its `minimize` method to train the network. To test and see how good the policy is, run the following code:

```
total_rewards = []
for _ in range(1000):
  rewards = 0
  obs = env.reset()

  while True:
    y_target = np.array([1. if obs[2] < 0 else 0.])
    a = sess.run(action, feed_dict={inputs: obs.reshape(1, num_inputs)})
    obs, reward, done, info = env.step(a[0][0])
    rewards += reward
    if done:
      break
  total_rewards.append(rewards)

print(np.mean(total_rewards))
```

The mean of the total rewards will be around 40-something, about the same as that of using the simple policy with no neural network, which is exactly what we expected as we specifically used the simple policy, with `y: y_target` in the training phase, to train the network.

We're now all set to explore how we can implement a policy gradient method on top of this to make our neural network perform much better, getting rewards several times larger.

The basic idea of a policy gradient is that in order to train a neural work to generate a better policy, when all an agent knows from the environment is the rewards it can get when taking an action from any given state (meaning we can't use supervised learning for training), we can adopt two new mechanisms:

- Discounted rewards: Each action's value needs to consider its future action rewards. For example, an action that gets an immediate reward, 1, but ends the episode two actions (steps) later should have fewer long-term rewards than an action that gets an immediate reward, 1, but ends the episode 10 steps later. The typical formula of discounted rewards for an action is the sum of its immediate reward plus the multiple of each of its future rewards and a discount rate powered by the steps into the future. So if an action sequence has 1, 1, 1, 1, 1 rewards before the end of an episode, the discounted rewards for the first action is *1+(1*discount_rate)+(1*discount_rate**2)+(1*discount_rate**3)+(1*discount_rate**4)*.

- Test run the current policy and see which actions lead to higher discounted rewards, then update the current policy's gradients (of the loss for weights) with the discounted rewards, in a way that an action with higher discounted rewards will, after the network update, have a higher probability of being chosen next time. Repeat such test runs and update the process many times to train a neural network for a better policy.

For a more detailed discussion and a walkthrough of policy gradients, see *Andrej Karpathy's* blog entry, *Deep Reinforcement Learning: Pong from Pixels* (http://karpathy.github.io/2016/05/31/rl). Let's now see how to implement a policy gradient for our CartPole problem in TensorFlow.

First, import tensorflow, numpy, and gym, and define a helper method that calculates the normalized and discounted rewards:

```
import tensorflow as tf
import numpy as np
import gym

def normalized_discounted_rewards(rewards):
    dr = np.zeros(len(rewards))
    dr[-1] = rewards[-1]
    for n in range(2, len(rewards)+1):
        dr[-n] = rewards[-n] + dr[-n+1] * discount_rate
    return (dr - dr.mean()) / dr.std()
```

For example, if discount_rate is 0.95, then the discounted reward for the first action in a reward list [1,1,1] is 1+1*0.95+1*0.95**2=2.8525, and the discounted rewards for the second and the last elements are 1.95 and 1; the discounted reward for the first action in a reward list [1,1,1,1,1] is 1+1*0.95+1*0.95**2 + 1*0.95**3 + 1*0.95**4=4.5244, for the rest of actions are 3.7099, 2.8525, 1.95, and 1. The normalized discounted rewards for [1,1,1] and [1,1,1,1,1] are [1.2141, 0.0209, -1.2350] and [1.3777, 0.7242, 0.0362, -0.6879, -1.4502]. Each normalized discounted list is in the decreasing order, meaning the longer an action lasts (before the end of an episode), the larger its reward is.

Next, create the CartPole gym environment, define the `learning_rate` and `discount_rate` hyper-parameters, and build the network with four input neurons, four hidden neurons and one output neuron as before:

```
env = gym.make("CartPole-v0")

learning_rate = 0.05
discount_rate = 0.95
```

```
num_inputs = env.observation_space.shape[0]
inputs = tf.placeholder(tf.float32, shape=[None, num_inputs])
hidden = tf.layers.dense(inputs, 4, activation=tf.nn.relu)
logits = tf.layers.dense(hidden, 1)
outputs = tf.nn.sigmoid(logits)
action = tf.multinomial(tf.log(tf.concat([outputs, 1-outputs], 1)), 1)

prob_action_0 = tf.to_float(1-action)
cross_entropy = tf.nn.sigmoid_cross_entropy_with_logits(logits=logits,
labels=prob_action_0)
optimizer = tf.train.AdamOptimizer(learning_rate)
```

Note that we don't use the `minimize` function, as we did in the previous simple neural network policy example, anymore here because we need to manually fine-tune the gradients to take into consideration the discounted rewards for each action. This requires us to first use the `compute_gradients` method, then update the gradients the way we want, and finally call the `apply_gradients` method (the `minimize` method that we should use most of the time actually calls `compute_gradients` and `apply_gradients` behind the scenes—see https://github.com/tensorflow/tensorflow/blob/master/tensorflow/python/training/optimizer.py for more information).

So let's now compute the gradients of the cross-entropy loss for the network parameters (weights and biases), and set up gradient placeholders, which are to be fed later with the values that consider both the computed gradients and the discounted rewards of the actions taken using the current policy during test run:

```
gvs = optimizer.compute_gradients(cross_entropy)
gvs = [(g, v) for g, v in gvs if g != None]
gs = [g for g, _ in gvs]

gps = []
gvs_feed = []
for g, v in gvs:
    gp = tf.placeholder(tf.float32, shape=g.get_shape())
    gps.append(gp)
    gvs_feed.append((gp, v))
training_op = optimizer.apply_gradients(gvs_feed)
```

The `gvs` returned from `optimizer.compute_gradients(cross_entropy)` is a list of tuples, and each tuple consists of the gradient (of the `cross_entropy` for a trainable variable) and the trainable variable. For example, if you take a look at `gvs` after the whole program runs, you'll see something like this:

```
[(<tf.Tensor 'gradients/dense/MatMul_grad/tuple/control_dependency_1:0'
shape=(4, 4) dtype=float32>,
```

```
   <tf.Variable 'dense/kernel:0' shape=(4, 4) dtype=float32_ref>),
  (<tf.Tensor 'gradients/dense/BiasAdd_grad/tuple/control_dependency_1:0'
shape=(4,) dtype=float32>,
   <tf.Variable 'dense/bias:0' shape=(4,) dtype=float32_ref>),
  (<tf.Tensor 'gradients/dense_2/MatMul_grad/tuple/control_dependency_1:0'
shape=(4, 1) dtype=float32>,
   <tf.Variable 'dense_1/kernel:0' shape=(4, 1) dtype=float32_ref>),
  (<tf.Tensor 'gradients/dense_2/BiasAdd_grad/tuple/control_dependency_1:0'
shape=(1,) dtype=float32>,
   <tf.Variable 'dense_1/bias:0' shape=(1,) dtype=float32_ref>)]
```

Note that kernel is just another name for weight, and (4, 4), (4,), (4, 1), and (1,) are the shapes of the weights and biases for the first (input to hidden) and second (hidden to output) layers. If you run the script multiple times from iPython, the default graph of the `tf` object will contain trainable variables from previous runs, so unless you call `tf.reset_default_graph()`, you need to use `gvs = [(g, v) for g, v in gvs if g != None]` to remove those obsolete training variables, which would return None gradients (for more information about `computer_gradients`, see https://www.tensorflow.org/api_docs/python/tf/train/AdamOptimizer#compute_gradients).

Now, play some games and save the rewards and gradient values:

```
with tf.Session() as sess:
    sess.run(tf.global_variables_initializer())

    for _ in range(1000):
        rewards, grads = [], []
        obs = env.reset()
        # using current policy to test play a game
        while True:
            a, gs_val = sess.run([action, gs], feed_dict={inputs:
                                    obs.reshape(1, num_inputs)})
            obs, reward, done, info = env.step(a[0][0])
            rewards.append(reward)
            grads.append(gs_val)
            if done:
                break
```

After the test play of a game, update the gradients with discounted rewards and train the network (remember that `training_op` is defined as `optimizer.apply_gradients(gvs_feed)`):

```
        # update gradients and do the training
        nd_rewards = normalized_discounted_rewards(rewards)
        gp_val = {}
        for i, gp in enumerate(gps):
```

```
                gp_val[gp] = np.mean([grads[k][i] * reward for k, reward in
                                enumerate(nd_rewards)], axis=0)
        sess.run(training_op, feed_dict=gp_val)
```

Finally, after 1,000 iterations of test play and updates, we can test the trained model:

```
total_rewards = []

for _ in range(100):
    rewards = 0
    obs = env.reset()

    while True:
        a = sess.run(action, feed_dict={inputs: obs.reshape(1,
                                            num_inputs)})
        obs, reward, done, info = env.step(a[0][0])
        rewards += reward
        if done:
            break
    total_rewards.append(rewards)

print(np.mean(total_rewards))
```

Note that we now use the trained policy network and `sess.run` to get the next action with the current observation as input. The output mean of the total rewards will be about 200, a big improvement over our simple intuitive policy, using a neural network or not.

You can also save a trained model after the training using `tf.train.Saver`, as we did many times before in the previous chapters:

```
saver = tf.train.Saver()
saver.save(sess, "./nnpg.ckpt")
```

Then you can reload it in a separate test program with:

```
with tf.Session() as sess:
    saver.restore(sess, "./nnpg.ckpt")
```

All the preceding policy implementations run on Raspberry Pi, even the one that uses TensorFlow to train a reinforcement learning policy gradient model, which takes about 15 minutes to finish. Here are the total rewards returned after running on Pi each policy we've covered:

```
pi@raspberrypi:~/mobiletf/ch12 $ python single_minded_policy.py
9.362

pi@raspberrypi:~/mobiletf/ch12 $ python simple_policy.py
```

```
42.535

pi@raspberrypi:~/mobiletf/ch12 $ python nn_random_policy.py
21.182

pi@raspberrypi:~/mobiletf/ch12 $ python nn_simple_policy.py
41.852

pi@raspberrypi:~/mobiletf/ch12 $ python nn_pg.py
199.116
```

Now that you have a powerful neural-network-based policy model that can help your robot keep in balance, fully tested in a simulated environment, you can deploy it in a real physical environment, after replacing the simulated environment API returns with real environment data, of course—but the code to build and train the neural network reinforcement learning model can certainly be easily reused.

Summary

In this chapter, we first went through detailed steps to set up Raspberry Pi with all the necessary accessories and operating system, as well as the GoPiGo toolkit that turns a Raspberry Pi board into a moving robot. We then covered how to install TensorFlow on Raspberry Pi and build the TensorFlow library, and how to integrate TTS with image classification, and audio command recognition with GoPiGO APIs, resulting in a Raspberry Pi robot that moves, sees, listens, and speaks, all without using Cloud APIs. Finally, we introduced the OpenAI Gym toolkit for reinforcement learning and showed you how to build and train a powerful reinforcement learning neural network model using TensorFlow to keep your robot in balance in a simulated environment.

Final words

So it's about time to say Goodbye. In this book, we started with three pre-trained TensorFlow models of image classification, object detection, and neural-style transfer, and discussed in detail how we can retrain the models and use them in iOS and Android apps. Then we covered three interesting models from the TensorFlow tutorials built with Python —audio recognition, image captioning, and quick drawing—and showed how to retrain and run the models on mobile.

After that, we developed RNN models from scratch for stock price prediction in TensorFlow and Keras, two GAN models for digit recognition and pixel translation, and an AlphaZero-like model for Connect 4, along with complete iOS and Android apps using all those TensorFlow models. We then covered how to use TensorFlow Lite, as well as Apple's Core ML with standard machine learning models and converted TensorFlow models, showing their potential as well as their limits. Finally, we explored how to build a Raspberry Pi robot with TensorFlow that moves, sees, listens, speaks, and learns with a powerful reinforcement learning algorithm.

We also showed Objective-C and Swift iOS apps using both the TensorFlow pod and the manually built TensorFlow library, and Android apps using the ready-to-use TensorFlow library as well as the manually built library, to fix all kinds of errors you may encounter when deploying and running TensorFlow models on mobile.

We've covered so much, yet there's so much more to cover. New releases of TensorFlow have been coming out at a rapid speed. New TensorFlow models implementing the latest research papers have been built and made available quickly. The primary goal of this book is to show you enough working iOS and Android apps using all kinds of intelligent TensorFlow models, along with all the practical troubleshooting and debugging tips, so you can quickly deploy and run your favorite TensorFlow models on mobile for your next killer mobile AI apps.

If you want to build your own great models using TensorFlow or Keras, implementing the algorithms and networks that excite you most, you'll need to continue your learning after the book, as we didn't show you in detail how to do that, but hopefully we've motivated you enough to start the journey, with the assurance gained from the book that once you get the models built and trained, you know how to quickly deploy and run them on mobile, anytime, anywhere.

As for which journey to take and which AI tasks to tackle, Ian Goodfellow's advice in his interview with Andrew Ng probably says it best: *ask yourself what's the best thing for you to do next and which path is the best for you to take: reinforcement learning, unsupervised learning, or generative adversarial network*. No matter what, it'll be a great path full of excitement, backed by hard work of course, and the skills you've learned from the book will be like your smartphone, ready to be of service to you anytime, and ready for you to make your sweet and smart, little device even sweeter and smarter.

Other Books You May Enjoy

If you enjoyed this book, you may be interested in these other books by Packt:

TensorFlow 1.x Deep Learning Cookbook
Antonio Gulli, Amita Kapoor

ISBN: 978-1-78829-359-4

- Install TensorFlow and use it for CPU and GPU operations
- Implement DNNs and apply them to solve different AI-driven problems.
- Leverage different data sets such as MNIST, CIFAR-10, and Youtube8m with TensorFlow and learn how to access and use them in your code.
- Use TensorBoard to understand neural network architectures, optimize the learning process, and peek inside the neural network black box.
- Use different regression techniques for prediction and classification problems
- Build single and multilayer perceptrons in TensorFlow
- Implement CNN and RNN in TensorFlow, and use it to solve real-world use cases.
- Learn how restricted Boltzmann Machines can be used to recommend movies.
- Understand the implementation of Autoencoders and deep belief networks, and use them for emotion detection.
- Master the different reinforcement learning methods to implement game playing agents.
- GANs and their implementation using TensorFlow.

Other Books You May Enjoy

Mastering TensorFlow 1.x
Armando Fandango

ISBN: 978-1-78829-206-1

- Master advanced concepts of deep learning such as transfer learning, reinforcement learning, generative models and more, using TensorFlow and Keras
- Perform supervised (classification and regression) and unsupervised (clustering) learning to solve machine learning tasks
- Build end-to-end deep learning (CNN, RNN, and Autoencoders) models with TensorFlow
- Scale and deploy production models with distributed and high-performance computing on GPU and clusters
- Build TensorFlow models to work with multilayer perceptrons using Keras, TFLearn, and R
- Learn the functionalities of smart apps by building and deploying TensorFlow models on iOS and Android devices
- Supercharge TensorFlow with distributed training and deployment on Kubernetes and TensorFlow Clusters

Leave a review - let other readers know what you think

Please share your thoughts on this book with others by leaving a review on the site that you bought it from. If you purchased the book from Amazon, please leave us an honest review on this book's Amazon page. This is vital so that other potential readers can see and use your unbiased opinion to make purchasing decisions, we can understand what our customers think about our products, and our authors can see your feedback on the title that they have worked with Packt to create. It will only take a few minutes of your time, but is valuable to other potential customers, our authors, and Packt. Thank you!

Index

A

AlphaZero algorithm
  components 286
AlphaZero-like model
  building code 295
  freezing 296
  testing 292, 294
  testing, for Connect 4 288
  training 288
  training, for Connect 4 288
  using, in Android to play Connect 4 311
  using, in iOS to play Connect 4 298, 305
Android app
  retrained models, using 43
  TensorFlow, adding 59, 62
Android Studio
  reference 20
  setting up 21, 22
Android
  AlphaZero-like model, used for playing Connect 4 311, 313, 316, 317, 320, 321
  fast neural-style transfer model 108
  fast neural-style transfer model, using 113
  GAN models, using 277
  image captioning model, using 185, 186, 188, 191
  Keras RNN model, executing 254, 258
  reference 335
  speech recognition model, using 134
  TensorFlow Lite, using 335, 339
  TensorFlow Magenta multi-style model, using 122, 125
  TensorFlow RNN model, executing 254, 258
AphaZero-like model
  using, in Android to play Connect 4 313
Application Binary Interface (ABI) 217

audio recognition model
  about 366
  reference 364
automatic speech recognition (ASR) 127

C

CartPole
  reference 367
  simulated environment 368, 371
Cloud ML Engine
  reference 16
CMU Sphinx
  reference 129
COCO API
  reference 158
Codelab
  reference 122
Connect 4
  AlphaZero-like model, testing 288
  AlphaZero-like model, training 288
  playing, Android model used 311, 314, 317, 321
  playing, iOS model used 298, 301, 304, 306, 310
  reference 286
Connectionist Temporal Classification (CTC) 129
Convolutional Neural Networks (CNNs) 28
Core ML
  converted models, using in iOS 343
  for iOS 339
  reference 339
  tools, reference 342
  using, with Keras 344
  using, with Scikit-Learn machine learning 340
  using, with TensorFlow 344

D

darkflow library
　reference 93
Darknet
　reference 93
deep learning
　reference 15
Deep Speech
　reference 129
detection model zoo
　reference 70
Discrete Fourier Transform (DFT)
　about 130
　reference 131
drawing classification model, using in Android
　about 216
　Android app, developing 218, 222, 226
　custom TensorFlow library, building 216
drawing classification model, using in iOS
　about 206
　custom TensorFlow library, building 206
　iOS app, developing 207, 211, 214
drawing classification model
　about 194
　predicting 196, 197, 199
　preparing 196, 200, 201, 203, 205
　reference 196
　training 196
　using, in Android 215
　using, in iOS 206
　working 194, 195

F

Fast Fourier Transform (FFT) 130
fast neural-style transfer models
　adding 102, 105
　testing 102, 105
　training 99
　used, for viewing iOS code 106
　using, for iOS 102
　using, in Android 107, 113
faster RCNN model
　retraining 75, 78, 80
Flite
　reference 363

G

GAN models, using in Android
　advanced model, using 282, 284
　basic model, using 280
　using 278
GAN models, using in iOS
　advanced model, using 274, 277
　basic model, using 272
　using 270
GAN models
　building, with TensorFlow 262
　image resolution, enhancing 265, 268
　of generating handwritten digits 262, 264
　training, with TensorFlow 262
　using, in iOS 272
Gated Recurrent Unit (GRU) 231
Generative Adversarial Network (GAN)
　about 259, 260
　need for 260
　reference 261
GoPiGo Robot Base Kit
　reference 351
GoPiGo's Python API
　reference 357
GPU-powered Ubuntu
　TensorFlow, setting up 15, 18
graph transform tool
　reference 38
Graphical Processing Unit (GPU) 15

H

Hidden Markov Models (HMM) 128

I

im2txt model
　reference 158
image captioning model
　about 158
　caption generation, testing 160, 162
　caption generation, training 160, 162
　freezing 160, 163, 165, 166, 167, 169
　optimizing 169

training 160
transforming 169
using, in Android 185, 186, 188, 191
using, in iOS 175, 177, 179, 182, 184
working 158
image recognition model 361, 363
Inception v3 model
  used, for retraining 29, 32, 35, 37
Inception
  reference 23
iOS app
  libraries, using 82
  object detection feature, adding 86, 89, 90
  retrained models, using 40, 42
  TensorFlow, adding 44
iOS
  AlphaZero-like model, used for playing Connect 4 298, 301, 310
  converted Core ML models, using 343
  drawing classification model, using 206
  fast neural-style transfer models, using 102
  GAN models, using 269, 272
  image captioning model, using 175, 177, 179, 182, 184
  Keras RNN model, executing 246
  libraries, building manually 81
  libraries, using in app 85
  object detection models, using 81
  speech recognition model, using with Objective-C 143
  TensorFlow Lite, using 325
  TensorFlow Magenta multi-style model, using 114, 121
  TensorFlow RNN models, executing 246

K

Kaldi
  reference 129
Keras RNN LSTM API
  using, for stock price prediction 239
Keras RNN model
  executing, in Android 254, 258
  executing, on iOS 246, 250, 253
  testing 244, 246
  training 240, 243

Keras
  Core ML, using with 344, 345

L

Long Short Term Memory (LSTM)
  reference 159

M

MacOS
  TensorFlow, setting up 13
makefile documentation
  reference 359
Matplotlib
  reference 233
mobile deployment
  reference 150
MobileNet models
  reference 38
  used, for retraining 38, 40
Monte-Carlo Tree Search (MCTS) algorithm 286

N

Native Development Kit (NDK) 22
Neural Algorithm of Artistic Style
  reference 98
neural-style transfer models
  overview 98
  using, in iOS 102
New Out Of Box Software (NOOBS)
  reference 354
numpy
  reference 233

O

object detection models
  using, in iOS 81
object
  detecting 66
Objective-C iOS app
  TensorFlow, adding 44
OpenAI Gym
  reference 367
OpenEars
  reference 129

Out of Memory (OOM) 269

P

Pandas
 reference 341
PASCAL VOC
 reference 67
PocketSphinx
 reference 129
pretrained models
 reference 29
previous
 drawing classification model, using 215
Protocol Buffers
 release 3.4.0, download link 86
PyCharm
 reference 164

Q

Quick Draw
 reference 193

R

Raspberry Pi
 accessories 352
 example apps, reference 361
 installation link 353
 reference 351
 reinforcement learning 367
 setting up 352, 353, 355
 setting, in motion 356, 358
 TensorFlow, setting up 359, 361
 using 352
Raspbian
 reference 354
RCNN
 reference 66
recurrent neural network (RNN)
 about 128, 230
 reference 129
reinforcement learning
 basic intuitive policy, starting with 372, 373
 CartPole simulated environment 368, 371
 neural networks, using 374, 376, 378, 380

on Raspberry Pi 367
 reference 377
retrained models
 using, in Android app 43
 using, in iOS app 40, 42
robot movement
 controlling 364, 366

S

Scikit Learn models
 building 340, 342
 converting 340
 reference 339
Scikit-Learn machine learning
 Core ML, using with 340
SD Formatter
 reference 353
simple commands recognition model
 training 130, 134
Single Shot Multibox Detector 25
speech recognition model
 model-loading errors, fixing with tf_op_files.txt 150
 reference 128
 results, displaying 140, 142
 used, for building new app 135, 139, 143, 146, 148
 using, in Android 134
 using, in iOS with Objective-C 143
 using, in iOS with Swift 151, 155
SSD-MobileNet model
 retraining 75, 78, 80
stock price prediction
 about 230
 Keras RNN LSTM API, using 239
 TensorFlow RNN API, using 232
Swift iOS app
 TensorFlow, adding 52, 56, 58
Swift
 speech recognition model, using in iOS 151, 155
synchronization methods
 reference 83
System Integrity Protection (SIP) 15

T

Tacotron
  reference 363
TensorFlow 1.6 release
  reference 359
TensorFlow core library
  reference 170
TensorFlow for Poets 2
  reference 172
TensorFlow Lite, using in Android
  about 335
  using 339
TensorFlow Lite, using in iOS
  custom model, using 333, 334
  example apps, executing 325
  prebuilt model, using 327, 330, 332
  retrained TensorFlow model, using 332
TensorFlow Lite
  about 324
  overview 324
  using 324
  using, in iOS 325
TensorFlow Magenta multi-style model
  using, in Android 122, 125
  using, in iOS 114, 119, 121
TensorFlow Magenta project
  reference 99, 114
TensorFlow Mobile
  website link 173
TensorFlow Object Detection API
  interface, executing 68
  interface, installing 68
  pre-trained models, using 69, 74
  setting up 68
TensorFlow Playground
  reference 231
TensorFlow RNN API
  using, for stock price prediction 232
TensorFlow RNN model
  executing, in Android 254, 258
  executing, on iOS 246, 250, 253
  testing 237, 239
  training 233
TensorFlow
  adding, to Android app 58, 62
  adding, to iOS app 44
  Android apps, reference 25
  Core ML, using with 344
  installation link 13
  Mobile, versus Lite 23
  reference 11, 12, 36
  sample Android apps, executing 24
  sample iOS apps, executing 23
  setting up 12
  setting up, on GPU-powered Ubuntu 15, 18
  setting up, on MacOS 13
  setting up, Raspberry Pi 359, 361
  TensorFlow, adding 47, 51
  version 1.5 or 1.6, download link 324
Text-to-Speech (TTS) model 361, 363
transfer learning
  about 28
  reference 29
transform_graph tool
  reference 203
transformed image captioning model
  optimizing 173
  used, for fixing errors 170, 172, 173

V

VGG-19 model
  reference 98

W

WaveNet
  reference 363
Worldwide Developers Conference (WWDC) 323

X

Xcode
  download link 20
  setting up 20

Y

YOLO2
  reference 92
  using 92, 95